BASIC VIROLOGY

Edited by

Natassia Correa

Toronto Academic Press

BASIC VIROLOGY

Editor: **Natassia Correa**

Toronto Academic Press
4164 Lakeshore Road
Burlington ON L7L 1A4
Canada
www.tap-books.com

Email: orders@arclereducation.com

ISBN: 978-1-77956-327-9

Toronto Academic Press publishes wide variety of books and eBooks. For more information about Toronto Academic Press and its products, visit our website at www.tap-books.com.

ABOUT THE EDITOR

Natássia Corrêa is a biologist with a background in Molecular Biology, Cellular Biology, and Biotechnology. She is currently a public servant at the Federal University of Uberlândia, Minas Gerais, Brazil. Her academic journey includes earning her Master's and Doctorate degrees in Molecular Biology from the Federal University of Minas Gerais. In addition to working in the biotechnology industry, she completed her postdoctoral studies at FIOCRUZ and served as a faculty member at the Federal University of Paraná. Dr. Natássia Corrêa has always valued the diversity of her learning experiences, driven by a passion for new knowledge, as evidenced by her diverse and in-depth curriculum in Molecular and Cellular Biology.

Table of Contents

List of Figures

receptor binding) results in anchoring the fusion peptide in the cell membrane (B), forcing the viral membrane and cell membrane into a hemifusion state (C). Subsequently, the fusion protein is refolded into a highly stable postfusion conformation, which leads to a fusion pore formation (D). For simplicity, only dimers are represented, but the fusion proteins are always trimeric: HA of influenza virus and F (fusion) protein of paramyxovirus

Figure 5.8. Cytoplasmic trafficking. Two distinct viruses are used to explain how the entry is linked to cytoplasmic trafficking: (A) adenovirus (naked) and (B) herpes virus (enveloped). Incoming viruses can enter cells by endocytosis (A) or direct fusion (B). Following penetration into the cytoplasm, either endocytic vesicles or viral capsids exploit dynein motors to traffic toward the minus ends of microtubules. Either the endocytic vesicles (A) or the capsids (B) interact directly with the microtubules. The virus can also lyse the endocytic membrane, releasing the capsid into the cytosol (A)

Figure 5.9. Strategies for the nuclear entry of DNA viruses. (A) Polyomavirus capsids are small enough to enter the nucleus via the nuclear pore complex without disassembly. Uncoating of the polyomavirus genome takes place in the nucleus. (B) The adenovirus capsids are partially disrupted upon binding to the nuclear pore complex, allowing the transit of the DNA genome into the nucleus. (C) For herpesvirus, the nucleocapsids are minimally disassembled to allow transit of the DNA genome into the nucleus

Figure 5.10. (a) The surface of a "typical" animal cell. The lipid bilayer plasma membrane is penetrated by cell surface proteins of various functions. The proteins that extend from the surface (mainly glycoproteins) can be utilized by different viruses as "tether points" or "anchors" for bringing the virus close enough to the cell surface to initiate the entry process. This interaction between a cell surface protein serving as a virus receptor and the virus itself is highly specific between proteins. Integral membrane proteins, such as those mediating the transport of small molecules and ions across the plasma membrane, tend not to project as far into the extracellular matrix and are not as commonly utilized by viruses as receptors. Some viral receptors are listed in Table 5.2. (b) The interaction between a cellular surface protein (receptor) and a ligand or co-receptor can lead to chemical and structural changes that transmit signals between the exterior and interior of the cell. This is the process of signal transduction. Here, for example, ligand binding with two monomeric receptor proteins leads to dimerization, activating a protein kinase in the cytoplasm. This in turn leads to phosphorylation of a target protein, leading to further changes in the cell

Figure 5.11. Schematic of receptor-mediated endocytosis utilized by poliovirus for entry into the host cell. The endocytotic vesicle forms due to close association between the poliovirus-receptor complex and the plasma membrane

Figure 5.12. (a) The two basic modes of entry of an enveloped animal virus into the host cell. Membrane-associated viral glycoproteins can either interact with cellular receptors to initiate a fusion between the viral membrane and the cell plasma membrane or can induce endocytosis. The fate of the input virus membrane differs in the two processes. (b) The fusion of pseudorabies virus with the plasma membrane of an infected cultured cell is shown in this series of electron micrographs (the bars represent 150nm). Although each electron micrograph represents a single event "frozen in time," a logical progression from the initial association between viral envelope glycoproteins and the cellular receptor on the plasma membrane through the fusion event is shown. The final micrograph contains colloidal gold particles bound to antibodies against the viral envelope glycoproteins (dense dots). With them, the envelope can be seen clearly to remain at the infected cell surface

Figure 5.13. Entry of T4 bacteriophage DNA into an Escherichia coli cell. Initial attachment is between the fibers to the ompC lipopolysaccharide receptor on the bacterial cell wall (a). The binding of protein pins on the base plate to the cell wall leads to the contraction of the tail fibers and sheath proteins, leading to the insertion of the tail tube through the cell wall. As shown in the electron micrograph (b), phage pilot protein (arrow) allows the highly charged viral DNA genome to penetrate the bacterial plasma membrane and enter the cell. Phage DNA can be seen as shadowy lines emanating from the tail tube

Figure 5.14. Expression of a varicella-zoster virus protein following transfection of a cell with the viral gene under the control of an active promoter in the uninfected cell. (a) The basic process. The cell membrane is treated with agents that allow it to readily take up large aggregates of protein and nucleic acids by phagocytosis. The transfecting DNA is caused to form aggregates with the use of calcium phosphate (Ca3(PO4)2), and then mixed with cells that have been appropriately treated. While most of the DNA taken up by the cell is degraded, some gets to the nucleus by nonspecific cellular transport of macromolecules, and this DNA can be transcribed, and any genes within it expressed as proteins. (b) An actual experiment. Cells were made permeable and transfected with DNA containing the varicella-zoster virus glycoprotein L gene. The protein encoded in this gene was expressed following its transcription into mRNA. Cells were treated with a fluorescent antibody reactive with the glycoprotein at (clockwise from the top left) 0, 12, and 24 hours after infection. The expression of the glycoprotein in the cytoplasm is clearly evident from the green fluorescence

Figure 5.15. Relationship between capsid assembly and genome packaging. (A) Sequential mechanism. For picornavirus, the procapsid, a precursor of the capsids, is preassembled without an RNA genome. Subsequently, the RNA genome penetrates the procapsid via a pore. (B) Coupled mechanism. For adenovirus, the DNA genome is packaged into the capsid during capsid assembly

Figure 5.16. Relationship between capsid assembly and envelopment. (A) Sequential mechanism. The capsid assembly occurs prior to the envelopment. The assembled capsid is then targeted to the membrane for envelopment. Togavirus constitutes a family of positive-strand RNA viruses. (B) Coupled mechanism. Capsid proteins and the viral genome are recruited to the budding site on the membrane. Capsid assembly and the envelopment of the capsid proceed simultaneously. The envelopment process can be divided into three steps: bud formation, bud growth, and finally, membrane fusion

Figure 5.17. Late domain and its function in MVBs pathway. (A) Three kinds of L domain found in enveloped viruses such as retroviruses (i.e., HIV, HTLV-1, and RSV) and rhabdoviruses (i.e., vesicular stomatitis virus and rabies virus): PTAP motif, PPPY motif, and PPEY motif. A (alanine), E (glutamic acid), P (proline), T (threonine), and Y (tyrosine). (B) L domain and ESCRT complexes involved in MVBs pathway. The budding of retroviral Gag is facilitated by ESCRT complexes, which are normally involved in the MVB pathway. In HIV, Gag interacts with Tsg101 and Alix, an adapter protein, leading to recruitment of the additional components of the MVB pathway, that is, ESCRT-II (green), ESCRT-III (purple), to assemble into a functional complex. Vps4 (red) is involved in recycling the MVB machinery

expression cannot occur in the nonpermissive cell. Still, the large T antigen (T-Ag) can interfere with cellular growth control (tumor suppressor) proteins, leading to cell replication. Stable transformation requires a second step, the integration of the viral DNA. This is a random (stochastic) occurrence with SV40, and integration is random throughout the genome. A similar path is followed in the transformation of nonpermissive cells by other polyomaviruses. t-Ag = small t antigen

Figure 10.5. The human papillomavirus (HPV)-16 genome. The 7-kbp circular genome contains a number of translational reading frames that are expressed from spliced mRNAs. Unlike the related polyomaviruses, papillomaviruses encode all proteins on the same DNA strand. The actual details of mRNA expression also appear to differ among different papillomaviruses. Analysis of the bovine papillomavirus genome and expressed transcripts shows that both early and late transcripts are produced from a limited number of promoters as pre-mRNAs. Unlike polyomaviruses, papillomaviruses exhibit extensive splicing of pre-mRNAs, with early and late promoters present in multiple regions of the genome

Figure 10.6. The formation of a wart by cell proliferation caused by infection of basement epithelial cells with human papillomavirus (HPV). Early gene expression leads to stimulation of cell division and terminal differentiation. This results in late gene expression and virus replication in a terminally differentiated, dying cell, which produces large quantities of keratin

Figure 10.7. The genetic and transcription map of the 36-kbp adenovirus genome. There are three kinetic classes of transcripts. The E1 transcripts are controlled by enhanced promoters and require no modification of the host cell because some functions of their expression are similar to those of T antigen in SV40 virus replication. These functions include stimulating cellular transcriptional activity and cell replication. Early in infection, only early transcripts are expressed. These include mRNAs encoding viral DNA polymerase and terminal proteins. There are a number of early promoters and transcription units. The E2 transcription unit also has a 72-kd DNA-binding protein (72KDBP) that shuts off early transcription. Two primary transcripts, E2A and E2B, are expressed from the same E2 promoter. The mRNA for the DNA-binding protein continues to be expressed late because there is a second promoter upstream of the E2 promoter that is not shut off by the DNA-binding protein. The major late promoter at map position 15 is always "on," but polyadenylation and splicing patterns change markedly as infection proceeds. Late in infection, the late transcription unit extends to one of five polyadenylation signals, and differential splicing results in the generation of a myriad of late mRNAs encoding structural proteins as well as proteins involved in host cell modification and virus maturation

Figure 10.8. Adenovirus DNA replication. The 5′ ends of the viral genome have 50,000-dalton terminal proteins bound to them. Of major interest is the fact that there is no discontinuous strand synthesis. The process is marked by the accumulation of a large amount of single-stranded DNA (unusual in eukaryotic DNA replication). Further, the initial priming event requires the first nucleotide of the new DNA strand being covalently bound to the 80,000-dalton precursor of the 50,000-dalton terminal protein. Following complete second-strand synthesis, the precursor end proteins are proteolytically cleaved to form the mature terminal proteins. TBP = precursor to terminal binding protein

Figure 10.9. The 5000-nucleotide (NT) linear genome of adenoassociated virus (AAV). This ssDNA has repeated sequences on both ends that allow it to form a "hairpin" structure. This serves as the template for cellular enzymes to convert into dsDNA. Cellular enzymes also mediate replication of the viral genome. Three families of coterminal mRNAs are expressed from the three AAV promoters; the genome encodes replication proteins and a capsid protein but depends on cell replication for its ability to replicate its genome. This cellular replication is induced by a helper virus such as adenovirus in the animal, but the virus can replicate in cultures of some actively replicating cells. Other groups of parvoviruses, such as the minute virus of mice (MVM), are able to replicate in some actively replicating cells of their natural host

Figure 10.10. The capsid structure and compressed genome of bacteriophage ΦX174. The capsid is made up of three proteins: major capsid, major spike, and minor spike. In all, a total of 10 genes are compressed into 3.4 kb of ssDNA! This is accomplished by very short intergenic regions and two completely overlapping genes. The functions of the proteins encoded by these genes are listed

List of Tables

PREFACE

Viruses and viral diseases have profoundly influenced science, agriculture, and medicine throughout history. A notable example is smallpox, once humanity's deadliest scourge, which played a pivotal role in altering the course of history during the European colonization of the Americas. However, it stands out as the only disease ever completely eradicated worldwide. This achievement began with Edward Jenner's groundbreaking experiment in 1796, demonstrating that inoculation with cowpox could protect against the much more dangerous variola major virus.

Following Jenner's discovery, a concerted global vaccination campaign led by the World Health Organization ultimately succeeded in eradicating smallpox by 1979. This breakthrough marked just the beginning of a series of vital investigations and breakthroughs catalyzed by the study of viruses. Initially, virologists focused on viruses as agents of disease, leading to significant strides in preventing and managing many acute viral infections through vaccines and public health initiatives. As a result, diseases like measles, polio, rabies, and yellow fever are now uncommon in developed countries. Additionally, the development and widespread use of effective antiviral medications further underscore the ongoing progress in combating viral illnesses globally.

Content/Structure of the Book

This book is structured into ten comprehensive chapters, each focusing on a crucial aspect of virology to provide a detailed understanding of the subject.

Chapter 1 is an introductory chapter that lays the groundwork for the study of virology. It covers the history of virology, the fundamental principles of virus biology, and the significance of viruses in the broader context of microbiology and infectious diseases.

Chapter 2 covers the mechanisms of virus replication and the processes by which viruses cause disease in their hosts. It explores the cellular and molecular interactions involved in viral infection and the resulting pathogenesis.

In Chapter 3, the book examines various viral diseases that affect humans, discussing their epidemiology, clinical manifestations, and the patterns of disease progression. It provides case studies and examples of significant viral infections.

Chapter 4 focuses on the physical structure of viruses and their classification. It details the structural components of viruses, such as capsids and envelopes, and the criteria used to classify viruses into different families and genera.

In Chapter 5, the book explores the critical stages of the viral replication cycle, from the initial attachment and entry into the host cell to the assembly and release of new virions. It provides a detailed overview of the entire replication process.

Chapter 6 examines the vertebrate immune system's response to viral infections. It discusses the innate and adaptive immune responses, including the roles of antibodies, T-cells, and other immune components in combating viral infections.

Chapter 7 addresses the methods used to replicate viruses in laboratory settings and measure their biological activity. It covers techniques such as cell culture, plaque assays, and other experimental approaches used to study viruses.

Chapter 8 highlights the molecular genetics of viruses, exploring the genetic material of viruses, how it is organized, and how viral genes are expressed and regulated. It covers topics such as viral genome replication, transcription, and translation.

Focusing on a specific group of viruses, Chapter 9 examines the replication strategies of positive-sense RNA viruses. It explains how these viruses replicate their RNA genomes and produce proteins within the host cell.

The final chapter explores the replication mechanisms of small and medium-sized DNA viruses. It discusses the unique strategies these viruses use to replicate their DNA and produce progeny within the host cell.

Each chapter in this book is designed to build upon the previous one, providing a cohesive and thorough exploration of virology. This structured approach ensures that readers gain a comprehensive understanding of the field, from fundamental concepts to specific viral replication strategies.

Audience

The intended audience for this book includes students, researchers, and professionals across various disciplines within biology and medicine who seek a comprehensive understanding of viruses and viral diseases. It is tailored for undergraduate and graduate students studying virology, microbiology, molecular biology, and related fields, providing a foundational knowledge base supplemented with advanced concepts and current research trends. Researchers in academia and industry engaged in virology and infectious disease studies will find the book valuable for its detailed exploration of virus structure, replication mechanisms, and pathogenesis. It serves as a resource for staying updated on cutting-edge research and methodologies in virology, offering insights into fundamental virological principles as well as applied aspects relevant to public health and medical research. Healthcare professionals, including clinicians and epidemiologists, can benefit from the book's insights into viral diseases and their management. It provides a solid grounding in the principles of virology essential for understanding disease transmission, diagnosis, and treatment strategies. This book aims to bridge the gap between foundational knowledge and advanced research in virology, catering to a diverse audience interested in viruses and their profound impact on human health, agriculture, and the environment. Whether exploring the basics or delving into specialized topics, readers will find the book an indispensable resource for navigating the complexities of virological science.

Learning Features

- Start-of-chapter learning objectives
- Key terms in each chapter
- Summary Points at the end of each chapter
- Review Questions at the end of each chapter
- Case Studies and Activities in each chapter for solving real problems
- References at the end of each chapter

CHAPTER 1

The Impact of Viruses on Our View of Life

LEARNING OBJECTIVES

After studying this chapter, you will be able to:

- Understand the effect of virus infections on the host organism and populations—viral pathogenesis, virulence, and epidemiology;
- Explain the interaction between viruses and their hosts;
- Know the history of virology;
- Give examples of the impact of viral diseases on human history;
- Discuss examples of the evolutionary impact of the virus-host interaction.
- Understand the origin of viruses
- Describe how viruses have a constructive as well as destructive impact on society
- Recognize that viruses are not the smallest self-replicating pathogens

KEY TERMS FROM THIS CHAPTER

Algae	Cell growth	Cells
Epidemiology	Genetic variation	Genetics
Genotype	Inoculum	Mycoplasma
Organelle	Organism	Protein synthesis
Respiratory infection	Selfish genes	Vaccines
Virologists	Virulence	Viruses

INTRODUCTION

Many of our most basic concepts in contemporary biology, genetics, and medicine have their roots in the study of viruses, both historically and presently. The study of biological macromolecules, cellular gene expression processes, mechanisms for producing genetic diversity, processes governing cell growth and development, elements of molecular evolution, the mechanism of disease and the host's response to it, and the spread of disease in populations have all been impacted by virology (Anderson et al., 1998).

Essentially, viruses are just genetic data sets with one goal in mind: to replicate themselves. They represent the pinnacle and archetype of "selfish genes." The molecular machinery of the infected cell must decode the "blueprints" for virus replication encoded in the genetic code found in the viral genome in order for the virus to replicate. Thus, viruses are obligatory intracellular parasites that rely on the genetic and metabolic processes of living cells. Submicroscopic viruses come in a variety of sizes, from less than the tiniest organelle to slightly smaller than the mycoplasma and simple unicellular algae, which are the simplest cells with the ability to metabolize energy and synthesize proteins. Even though they are small in size, they have adapted and developed a mode of reproduction and propagation that allows them to survive in free-living organisms that are ten to ten million times larger and have more genetic complexity than they do.

1.1. THE EFFECT OF VIRUS INFECTIONS ON THE HOST ORGANISM AND POPULATIONS—VIRAL PATHOGENESIS, VIRULENCE, AND EPIDEMIOLOGY

An infectious disease that spreads from person to person is often (though not always) a sign of the replication and propagation of a particular virus in a population. Viral pathogenesis is a broad term used to describe the study of how viruses infect hosts. The pathogenicity of a particular virus is defined as the totality of its encoded functions that support virus propagation in the infected cell, the host organism, and the population. This term refers to a virus's ability to cause disease and propagate through a population based on its genetic makeup. Members of a particular virus population are thought to be genetically similar. Therefore, the genotype, or genetic composition of a virus, plays a significant role in determining how pathogenic it is.

Complex factors can contribute to the severity of a viral disease's symptoms in an organism or population. It is the outcome of a complex interplay between the expression of the viral genes that control pathogenicity, the infected person's body's reaction to these pathogenic factors, and the population's reaction to the virus's presence as it spreads. When combined, these elements establish the virus's virulence and the illness it causes. The interplay between particular viral genes and the genetically encoded defenses of the infected person is a fundamental factor contributing to virulence. However, it is crucial to realize that virulence is also influenced by the general health and genetic composition of the infected population, as well as by societal and economic factors that affect the type and intensity of the immune response to the infection (Bell, 2001). Understanding the differences and interpretations of the terms "pathogenesis" and "virulence" comes from considering the various aspects of disease severity and dissemination present in a population infected with a virus. Imagine a virus whose genotype renders it highly effective in causing a disease, the symptoms of which are crucial for transmission—such as a respiratory infection accompanied by coughing, sneezing, and other symptoms. When this optimal virus replicates and spreads throughout an individual and the population, it can incorporate a multitude of genetic modifications. Therefore, some viruses created during the disease may contain genes that are less effective at causing symptoms. This type of virus has lower virulence, and in the worst situation, it could accumulate so many pathogenic gene mutations that it is completely unable to cause disease (i.e., it becomes an apathogenic or avirulent strain). Even though an avirulent virus might not cause illness, an infected person may develop full or partial immunity against the most virulent genotypes as a result of the infection. This is how vaccinations work. However, the ability to elicit an immune response and the subsequent development of herd immunity also implies that a virus's virulence must alter or adapt genetically to the host as the infection spreads throughout a population.

Variations in the virulence of a pathogenic genotype are also caused by other factors that are not entirely related to the genetic composition of a virus. The same virus genotype infecting two immunologically naïve individuals (i.e., individuals who have never been exposed to any form of the virus) can cause very different outcomes. One person may exhibit minimal symptoms due to limited viral exposure, infection through an inefficient pathway, or a strong immune system and other defense mechanisms ingrained in their genetic composition. Another person may experience severe symptoms or even death if exposed to a high dose of the virus, has a weakened immune system, or is physically stressed from illness or malnutrition.

Furthermore, two genetically similar populations with different economic and technological resources may experience markedly different disease outcomes from the same virus genotype. This may occur as a result of disparities in a society's social safety net's capacity to offer medical care, isolate sick individuals, or provide the best treatment regimens. When considered as a whole, epidemiology is the study of infectious diseases in humans caused by viruses and other pathogens (in animals, it is called epizoology). A solid grasp of the nature of the disease being studied, available medical treatments, and an appreciation of the dynamics of the community or population where the disease is prevalent are essential in this field.

1.2. THE INTERACTION BETWEEN VIRUSES AND THEIR HOSTS

There is constant interaction between viruses and their hosts, as well as other infectious agents. Through biomedical research, effective physiological responses to infectious diseases have evolved in organisms and, more recently, in society (Bos, 1999). In response, viruses use their natural genetic variation to accumulate and select mutations that enable them to become resistant to these responses. Diseases believed to be under control have periodically reappeared due to this resistance.

Certain viruses now have more opportunities to "break out" and spread both established and newly discovered diseases due to the increasing rate at which humans exploit the environment and the rate at which agricultural populations are growing. The ongoing AIDS epidemic and occasional outbreaks of viral diseases like hemorrhagic fevers in Asia, Africa, and the Southwest of the United States provide evidence of this. Examining how a viral disease spreads and how society reacts to it offers insights into the function of social policies and the social behavior of illness in general.

The recent global spread of AIDS is a prime example of how human behavior, including economic factors, can contribute to the onset of disease. The introduction of the causative agent, HIV, into the human population may have been facilitated by agricultural encroachment on animal habitats in equatorial Africa. This demonstrates how financial need amplifies risk.

HIV is not a highly effective pathogen; for it to spread, infected blood or bodily fluids must be directly injected. A major contributing factor to the spread of HIV and AIDS among the male homosexual community in the Euro-American world was the urban concentration of homosexual males with high-risk sexual habits. Due to overlap between this population and those who use intravenous drugs and engage in commercial sex, the virus spread to other urban populations. As a result, AIDS became a double-edged sword, threatening two different urban populations: the relatively wealthy homosexual community and the impoverished heterosexual drug users, both highly concentrated in urban areas. Commercial sex in the latter group led to further spread in heterosexual communities, particularly among young, single men and women (Courtney, 2021).

Another factor is that whole blood transfusions and pooled blood fractions for therapeutic use became commonplace due to the strong medical and financial resources of the "economic first world." As a result, AIDS appeared suddenly in hemophiliacs and occasionally in those who received large transfusions after surgery. Fortunately, efficient blood screening methods have reduced the incidence of the disease in these populations.

The distribution of HIV and AIDS in equatorial Africa and Southeast Asia differs due to distinct societal factors. In these parts of the world, the disease primarily affects heterosexual populations, with women being disproportionately affected. This spread of AIDS was caused by a small number of urban commercial sex workers who infected working men living away from their families. The virus spread to isolated family units when these men returned to their homes sporadically. Infected women who left prostitution to start families in their villages contributed to further spread.

Technology is another dominant factor contributing to the spread of AIDS. A century ago, HIV could not have spread and posed the threat it does today. Lower concentrations of at-risk individuals and generally lower population densities would have prevented HIV from spreading throughout the population. Rapid communication, less restrictive travel, migration policies, and the use of blood products in therapy, which were unknown a century ago, have all contributed to its spread.

Viruses are not the only pathogens that exhibit this dynamic relationship with their host. Many evolutionary processes and population dynamics can be better understood by investigating the genetic adaptations viruses make in response to human-directed efforts to control disease. Many of the techniques developed to study interactions between organisms and their environment can also be applied to examine how pathogens and hosts interact (DeBuysscher et al., 2021).

1.3. THE HISTORY OF VIROLOGY

The history of virology can be traced through the theories and research findings of key figures. Even though infections like smallpox, yellow fever, and rabies were recognized in early human history, it was not until around 1898—only about 120 years ago—that individual viruses were first isolated and linked to specific diseases. A mere lifetime ago, Peter Medawar, awarded the 1960 Nobel Prize in Medicine and Physiology, described viruses as "a strand of nucleic acid encircled by unfortunate circumstances."

Early virology pioneers had few tools and could not categorize or recognize viruses as microbiological agents. There were no cell cultures to cultivate viruses, no microscopes powerful enough to view viral agents, and no accessible probes like antibodies or nucleic acids to identify infectious agents. Biophysical tools such as acrylamide gels or centrifuges to separate viral proteins or nucleic acids also did not exist. Eventually, these methods were developed due to the desire for discovery. These tools helped define modern virology and its classifications and contributed to the development of contemporary cell and molecular biology, immunology, epidemiology, and treatment options. Modern pharmacology has made significant strides in developing vaccines and, more recently, antiviral medications. Vaccines, in particular, transformed the field of virology.

While deadly infectious diseases like poliomyelitis, measles, rubella (German measles), yellow fever, mumps, and others have been eradicated, others like smallpox, which once killed millions, still exist but can now be effectively managed. Unfortunately, some diseases, like influenza, remain beyond control. The nearly universal death rate associated with AIDS has been significantly reduced by antiviral therapies against HIV. Furthermore, viruses are responsible for at least 20% of all cancer cases. As a result, vaccines are now widely used to prevent cancers of the liver, uterus, cervix, and penile regions, as well as hepatitis B virus. Vaccines have been one of the most cost-effective medical interventions.

In retrospect, assessing the accuracy of early theories developed by pioneering and contemporary virologists regarding the illnesses and causes they studied reveals the evolution of virological science. Comprehensive historical accounts of viruses that have caused or are causing plagues can be found in the works of McNeill (1976), Oldstone (2010), and Waterson and Wilkinson (1978).

The groundbreaking work of Louis Pasteur and colleagues in the middle of the 1800s laid the groundwork for the later identification of viruses as infectious agents. During that time, the process of culturing bacteria in a lab was developed. These bacteria could be grown on broths or enriched agar preparations, and they could be identified by fixing them on glass slides, staining them, and observing them under a microscope. Retention of bacteria on filters with particular pore sizes allowed for their isolation (Figure 1.1(a)). Koch's (Henle–Koch) postulates could be used to associate particular bacteria with specific diseases following their collection on filters, growth on agar, and identification (Figure 1.1(a)). The first viruses were discovered using this framework (Figure 1.1(b)). In 1898, researchers Dmitri Losifovich Ivanovski from Russia and Martinus Beijerinck (Beijerinck, 1999) from the Netherlands showed that the material causing a tobacco plant disease did not remain in the plant; rather, it passed through the smallest pore-sized filter of the Pasteur-Chamberland apparatus without losing its infectious properties. Therefore, when the material collected at the bottom of the filter-containing flask was given to an uninfected recipient plant of the same kind, it remained infectious because it was smaller than bacteria (Figure 1.1(b)).

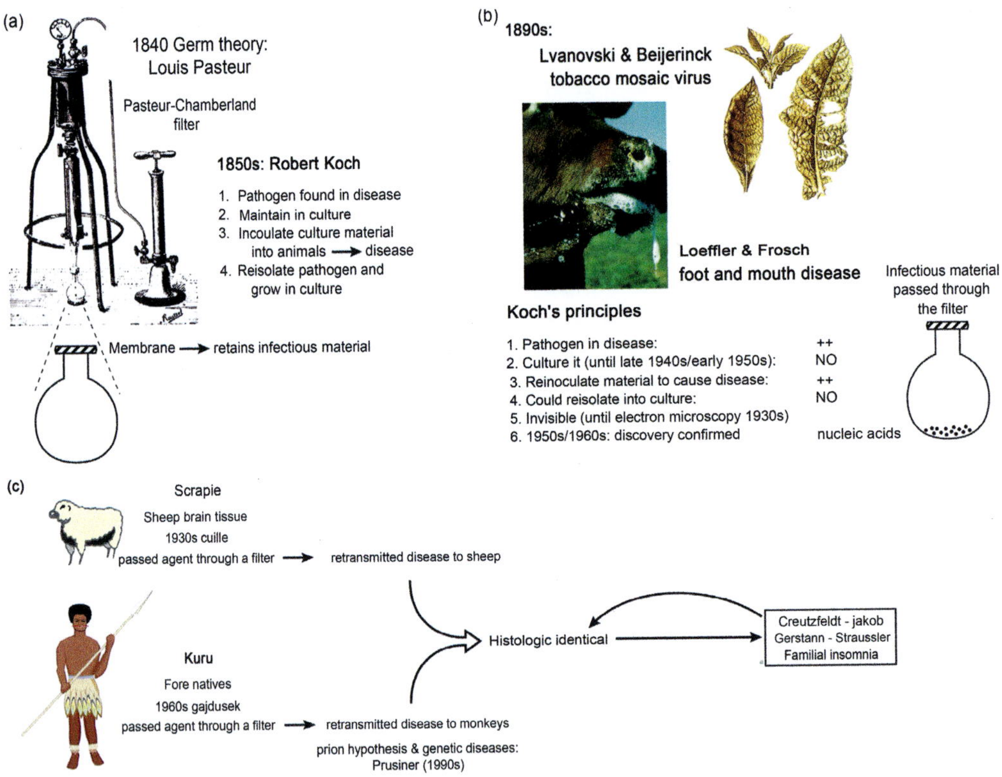

Figure 1.1. *The identification of viruses as infectious agents.*

Source: Oldstone MBA. History of Virology. Encyclopedia of Microbiology. 2014:608–12. doi: 10.1016/B978-0-12-801238-3.00078-7. Epub 2019 Aug 28. PMCID: PMC7150144.

The first virus was discovered by Ivanovsky and Beijerinck, who identified it as the tobacco mosaic virus in plants. Using a similar methodology, Friedrich Loeffler and Paul Frosch in Germany came to the almost simultaneous conclusion that the agent causing cattle foot-and-mouth disease also passed through porcelain filters and produced disease symptoms when reintroduced into healthy cattle (Figure 1.1(b)). These findings, which were very contentious at the time, served as the foundation for the definition of viruses as subcellular organisms that caused unique types of tissue destruction, later associated with specific diseases (Dolnik et al., 2021). When one considers that the infectious virus particles were too small to see and could not grow on the available culture media, the uniqueness of this detective work becomes even more striking (Table 1.1). It took the development of electron microscopy in the mid-1930s to visualize viruses, and it wasn't until the late 1940s or early 1950s that scientists were able to cultivate living cells, which is required for viral replication.

Figure 1.1(a) depicts the origins of microbiology, while Figure 1.1(b) lists the conceptual challenges these virologists faced while proving Koch's postulates and shows the first viral isolations. In theory, a toxin was differentiated from a viral replicating agent when the material that was diluted and passed through the Pasteur–Chamberland filter became less potent when injected into the appropriate host. On the other hand, the diluted substance was a replicating agent if it maintained its potency after injection (Iyer et al., 2006). However, not all material taken from an infected source, put through a filter similar to Pasteur-Chamberland, and then spread via infectious supernatant material turned out to be a virus (Figure 1.1(c)). For example, when scrapie-infected sheep were vaccinated against healthy sheep (their natural host), the infected material produced a disease similar to the one described. However, research did not reveal until the 1980s and 1990s that the disease was caused by misfolded abnormal prion protein (called PrPres or PrPsc) from infected sheep rather than nucleic acids, as Table 1.1 demonstrates.

Table 1.1. *Selected Milestones in Virology*

Date(s)	Virologists/Investigators	Discovery
400 BCE	Hippocrates	Greek physician, father of medicine, epidemiologic observations of many viral diseases
1796	E. Jenner	Application of cowpox virus for vaccination against smallpox, inflammation origins of virology and immunology
1846	P. Panum	Immunologic memory (protection) against reinfection (measles). The conceptual basis for vaccine strategy
1857	L. Pasteur, R. Koch	Founding of microbiology
1858	C. Darwin, (a) Wallace	Concepts of evolutionary progression, natural selection
1865	G. Mendel	Founding of genetics
1883	E. Metchnikoff, P. Ehrlich	Founding of immunology
1885	L. Pasteur, E. Roux	Development of rabies vaccine
1892–98	D. Ivanovsky, M. Beijerinck	First demonstrations of a filterable plant virus: tobacco mosaic virus
1898	F. Loeffler, P. Frosch	First demonstration of a filterable animal virus: foot-and-mouth disease virus

1900	J. Carroll, J. Lazear, (a) Agramonte, W. Reed, C. Finlay, W. Gorgas	First human virus: yellow fever virus, first use of consent form for human clinical investigation, identify mosquito as a transmitting agent, control of virus by elimination of mosquito breeding sites
1904–08	V. Ellermann, O. Bang, H. Vallee, H. Carre	First demonstration of a leukemia-causing virus, retrovirus
1908	C. von Pirquet	First report of virus causing immunosuppression: measles virus
1909	K. Landsteiner, E. Popper	Isolation of poliomyelitis virus
1911	J. Goldberger	Discovery of the measles virus
1911	P. Rous	First demonstration of a solid tumor virus: Rous sarcoma virus
1915	F. Twort	Discovery of bacteriophages
1917	F. d'Herelle	Bacteriophages, plaque assay
1919	A. Löwenstein	Discovery of herpes simplex virus
1923–28	A. Carrel, H. Maitland, M. Maitland	Tissue culture of embryo explants and first tissue culture cultivation of virus: Rous virus, vaccinia virus
1928	R. Lancefield, E. Lennette et al.	Beginning of viral disease diagnosis
1931	J. Furth	Use of mice as a host for viruses
1931	A. Woodruff, E. Goodpasture	Use of embryonated hen's eggs as a host for viruses
1933	W. Smith, C. Andrewes, P. Laidlaw	Isolation of human influenza virus
1933	R. Shope	Rabbit papillomavirus: First DNA tumor virus
1933	A. Tiselius	Development of electrophoresis
1936	J. Cuillé, P.-L. Chelle	Transmission of scrapie to normal sheep by cell-free material from diseased sheep
1936	P. Rous, J. Beard	Rabbit papillomavirus induces carcinomas in a different species
1936	C. Armstrong, T. Rivers, E. Traub	Discovery of lymphocytic choriomeningitis virus: The first arenavirus
1937–51	M. Theiler and others	Development of 17D strain human yellow fever vaccine: Made in animal and embryonic cultures
1939	G. Kausche, P. Ankuch, H. Ruska	First electron micrograph of a virus: Tobacco mosaic virus
1939	M. Debbruck, E. Ellis	Identification of the one-step virus growth curve
1941	G. Hirst	Discovery of influenza virus hemagglutination inhibition test
1941	M. Gregg	Discovery of rubella virus congenital abnormalities

1944	O. Avery, C. MacLeod, M. McCarty	Identification of DNA as the material of inheritance
1945	T. Francis, J. Salk, G. Hirst, F. Davenport, E. Kilbourne et al.	Development of inactivated influenza vaccines
1946	W. Stanley, J. Summer, J. Northrop	Preparation of a viral protein in a pure form: tobacco mosaic virus
1948–55	J. Enders, T. Weller, F. Robbins, H. Eagle	Routine use of tissue culture to grow and study viruses, development of optimal media for growing cells
1953	J. Watson, F. Crick, M. Wilkins, R. Franklin	Discovery of the structure of DNA
1953	A. Coons	Development of immunofluorescence
1954	J. Salk, J. Youngner, T. Francis et al.	Development of inactive poliomyelitis vaccine
1954	B. Sigurdsson	Development of the concept of slow viruses (maedi-visna virus)
1954	W. Rowe	Role of thymus in immune responses to virus
1954–61	J. Salk, (a) Sabin, H. Koprowski, J. Enders, S. Katz, S. Krugman	Development of poliomyelitis virus vaccine, development of measles virus vaccine: made in tissue culture
1956	H. Fraenkel-Conrat, (b) Singer	Discovery of the infectivity of viral RNA (tobacco mosaic virus)
1956	M. Smith, W. Rowe, T. Weller	Discovery of cytomegalovirus
1957	A. Isaacs, J. Lindenmann	Discovery of interferon
1957	J. Enders, M. Hilleman, (a) Gershon, S. Katz, S. Plotkin, M. Takahashi et al.	Development of vaccines against measles, mumps, rubella, hepatitis A, hepatitis B
1959	A. Sabin, H. Cox, H. Koprowski	Development of attenuated live-virus poliomyelitis vaccine
1959	S. Brenner, R. Horne	Invention of negative stain electron microscopy
1959	R. Porter, G. Edelman, (a) Nisonoff	Discovery of the structure and molecular function of antibodies
1959	R. Yalow, S. Berson, F. Dixon	Development and use of radioimmune assays
1950s–70s	S. Luria, M. Delbruck, J. Monod, (a) Lwoff, F. Jacob, S. Brenner, M. Nirenberg, S. Ochoa, H. Khorana, E. Wollman, (a) Hershey, S. Benzer, S. Cohen, D. Nathans, R. Dulbecco, D. Baltimore, H. Smith, W. Arber, J. Kates, H. Temin, P. Sharp et al.	Quantitative plaque assay, origins of molecular biology, and molecular virology
1950s–80s	B. Sigurdsson, (b) Blumberg, C. Gajdusek, S. Prusiner et al.	Persistent, latent, and slow virus infections, prions

1960s–70s	G. Palade, (a) Claude, K. Porter, C. DeDuve	Description and techniques for fine structure and biochemistry of cellular organelles
1962	A. Klug, D. Caspar	Discovery of the principles of icosahedral virus structure
1967	J. Maizel, U. Laemmi	Discovery of SDS polyacrylamide gel electrophoresis
1969–76	R. Huebner, P. Vogt, M. Bishop, H. Varmus, R. Weinberg et al.	Oncogenes
1970s–2000s	R. Zinkernagel, P. Doherty, M. Oldstone, (b) Fields, (b) Moss, (a) Notkins, R. Ahmed, F. Chisari, N. Nathanson et al.	Major histocompatibility restriction and cytotoxic T lymphocytes, immune-mediated viral diseases, molecular pathogenesis
1973	D. Nathans	Completion of the restriction map of a viral genome (SV40)
1974	G. Köhler, C. Milstein	Development of monoclonal antibodies
1975	B. Blumberg, (b) Larouze, W. London et al.	Discovery of the relationship of hepatitis B virus with hepatocellular carcinoma
1976	T. Diener	Discovery of viroids (infectious naked RNA molecules)
1976	K. Johnson, P. Webb, J. Lange, F. Murphy, J. McCormick et al.	Discovery of Ebola virus
1977	World Health Organization, D.A. Henderson, F. Fenner, and many health workers and virologists	Eradication of smallpox as a disease that killed over 300 million people in the twentieth century
1978–83	S. Harrison, (a) Olson, J. Hogle, M. Rossman, R. Rueckert	First atomic structure of a plant (tomato bushy stunt) and animal (poliovirus, rhinovirus) virus
1979	D. Lane, L. Crawford, D. Linzer, (a) Levine	SV40T-antigen-p53, virus-host cell interaction-tumor suppressors
1980s–90s	J. Strauss, E. Strauss, and others	Molecular analysis, structure/function of yellow fever
1980s–2000s	B. Roizman, J. Subak-Sharpe, E. Kieff, P. Spears, E. Wagner, J. Nelson, E. Goodpasture, J. Stevens, (a) Notkins	Genetic map and function of genes of herpes viruses, herpes latency
1980s–2000s	H. zur Hausen, D. Galloway, D. Lowy, I. Frazer et al.	Recognition of subtypes of papillomaviruses associated with cervical and penile cancer and development of papillomavirus vaccine
1981	J. Skehel, D. Wiley, I. Wilson et al.	Structure/function of influenza virus hemagglutinin, the first atomic structure of a glycoprotein
1981–84	R. Gallo, F. Barre-Sinoussi, L. Montagnier, Y. Hinuma, J. Chermann et al.	First human retrovirus

1984–2000	M. Hillemann et al.	First molecular recombinant virus vaccine: hepatitis B virus, first vaccine to successfully treat cancer: hepatitis B virus-induced liver cancer
1985	K. Mullis et al.	Invention of polymerase chain reaction (PCR)
1987	S. Broder, H. Mitsuya, M. Fischl, D. Richman et al.	Development of the first anti-HIV drug approved by the FDA
1987	M. Capecchi, M. Evans, O. Smithies	Development of knockout and other genetically manipulated mice
1988–2000s	M. Eigen, C. Weissmann, J. Holland, E. Domingo, J.C. de la Torre, R. Adino, E. Koonin, (a) Gibbs et al.	Development of modern concepts in viral evolution and quasispecies
1991	C. Venter, H. Smith	Invention of shotgun cloning methods
1993	S. Falkow	Molecular criteria for proof of viral disease causation: revisiting the Henle–Koch postulates
1994–2000s	K.-K. Conzelmann, M. Schnell, T. Mebatsion, P. Palese, J.C. de la Torre, Y. Kawaoka	Development of reverse genetics for negative-strand RNA virus
1990–present	R. Ahmed, M. Bevin, M. Slifka, and others	Kinetics, cell, and molecular basis of T cell and B cell generation and immunologic memory
1998–2006	R. Ahmed, (a) Zajac, E.J. Wherry, M. Oldstone, D. Brooks et al.	Persistent virus infections associated with T cell exhaustion and caused by negative immune response regulators PD-1 and IL-10
1990–present	W.I. Lipkin, J. diRossi, D. Ganem, H. Virgin, and others	Modern molecular detection of new viruses
2003	C. Urbani, J. Peiris, S. Lai, (a) Osterhaus et al.	Discovery of SARS coronavirus, the first viral pandemic of the 21st century
2005	J. Taubenberger, P. Palese, T. Tumpey, (a) Garcia-Sastre, Y. Kawaoka et al.	Molecular reconstruction, sequencing of the 1918–19 influenza virus; determining viral genes associated with pathogenesis
2005	T. Wakita, C. Rice, F. Chisari	Manipulation of virus and cells for first *in vitro* hepatitis C virus culture system
2000s	B. Beutler, J. Hoffmann, S. Akari, C. Janeway, R. Medzhitov	Discovery of cell-surface molecules activated by innate immunity and pathways involved
2000s	R. Steinman, Z. Cohn et al.	Discovery of the dendritic cell and its role in adaptive immunity
2000–2015	World Health Organization – many health workers and virologists	Planned eradication/elimination of poliomyelitis virus and measles virus (by 2015) as a disease

Source: Oldstone MBA. History of Virology. Encyclopedia of Microbiology. 2014:608–12. doi: 10.1016/B978-0-12-801238-3.00078-7. Epub 2019 Aug 28. PMCID: PMC7150144.

It is important to consider the effects of acute viral infections such as smallpox, measles, and yellow fever in the pre-vaccine era, even though today, the majority of life-threatening and crippling infections are prevented by vaccination (Kirui et al., 2021). Prior to its eradication in 1980, smallpox claimed the lives of over 300 million people in the twentieth century, with a mortality rate of about 33%. In the sixteenth century, when Native Americans were exposed to infectious viruses brought by European colonizers, the population in Mexico and Latin America dropped from 20 million to 2 million due to diseases like smallpox and measles that were previously unknown in the New World. While smallpox has been eliminated globally and measles controlled in developed countries, measles infection can still result in severe central nervous system damage for one in every thousand individuals.

Yellow fever spread along the Mississippi River in 1878, carried by infected travelers on riverboats from the Caribbean to New Orleans, Vicksburg, and Memphis, where the first case occurred by mid-August. Nobody was aware at the time that *Aedes aegypti* mosquito bites could spread the infection. After sucking blood from an infected person, the mosquito would bite a susceptible individual who had not yet contracted the virus. About 98,000 people were living in Memphis at the time of the yellow fever outbreak. Two months into the outbreak, twenty thousand people had fled the city. Of those who remained, 77% were infected, and 70% perished. Memphis's status as the American South's commercial hub was destroyed, and Atlanta took its place.

Writer Robert Blakeslee, dispatched from New York City to cover the yellow fever epidemic in Memphis, wrote in the *New York Herald*, "The city was almost deserted. We had not gone far; however, before the evidence of the terrible condition of things became apparent." He continued, "The first vehicle-like object I saw was a truck (wagon) that was transporting coffins and was circling the dead. You can imagine how quickly I realized how desolate this place was, given that it was only four blocks from the depot. We had to walk between the coffins arranged in tiers on the sidewalk in front of the undertaker's shop, two blocks away. A young doctor told me that everyone was extremely terrified. 'To stay here requires a man of great moral courage.' You speak with a man tonight, and the following day you find out he has passed away" (Oldstone, 2010).

For those who view this as archaic and possibly irrelevant to modern history, it is important to remember that more than 600,000 people lost their lives in less than two years during the 1918–19 influenza epidemic in the United States. The lack of coffins in Philadelphia and other locations resulted in the interment of the dead in mass graves. A similar situation occurred in parts of Africa as recently as the 1980s and 1990s, when the HIV-related AIDS epidemic led to widely reported mass deaths, and it continues to occur in some parts of Asia and Africa (Krupovic et al., 2019).

1.4. EXAMPLES OF THE IMPACT OF VIRAL DISEASE ON HUMAN HISTORY

Egyptian mummies and medical texts contain archeological evidence of easily recognized viral infections, such as poliomyelitis and genital papillomas (warts). Additionally, historical accounts of viral diseases that afflicted human populations in classical and medieval times are not entirely reliable. Because of the effectiveness of vaccines against the disease, the genetic stability of the virus, and a well-coordinated political and social effort to carry out the eradication, smallpox has been successfully eradicated from the human population. However, over thousands of years, the disease caused havoc and had a significant impact on human history.

Epidemics of smallpox during the Middle Ages and later in Europe caused large-scale population losses and profound changes in people's social, political, religious, and economic lives. While smallpox gradually declined in Europe and North America due to the success of vaccination campaigns, the disease still caused significant deaths and disruptions in other parts of the world until the end of World War II. Though the virus appears to be under control, concerns have recently surfaced regarding its potential use as a bioterrorism agent due to its high virulence and mode of transmission.

In human history, other viral epidemics have played equally significant roles. The introduction of infectious viral diseases like measles was a major factor in the social, economic, and political chaos that native populations experienced due to European conquest and expansion during the fifteenth through nineteenth centuries. These illnesses claimed the lives of large portions of the native population in the western hemisphere (La Scola et al., 2003).

Even today, there is still a chance that daily life could be seriously disrupted politically and socially. For instance, the 1918–19 "Spanish" influenza killed millions of people worldwide and, combined with the aftermath of World War I, nearly caused a collapse in global civilization. It is possible that the precise causes of this disease's virulence will never be fully understood. In the future, another illness may emerge with similar or even worse effects.

Several infectious diseases have the potential to spread throughout the general public due to drug resistance or human disturbance of natural ecosystems. Viral illnesses that may be involved include yellow fever, equine encephalitis, dengue fever, Ebola and Rift Valley fevers, Hantavirus pathologies, and the recently discovered coronavirus that causes severe acute respiratory syndrome (SARS).

Additional potential sources of disruptive viral infections include pathogens found in plants and animals. Sporadic viral outbreaks in domestic animals, such as avian influenza in chickens and vesicular stomatitis in cattle, cause significant financial and human losses. Over the past 50 years, rabies has spread among wild animal populations in the eastern United States. This illness poses a serious threat to household animals and, on rare occasions, to people as well.

An agricultural virus example that caused significant economic disruption is the increasing spread of the Cadang-Cadang virus in coconut palms in the Philippines and other parts of Oceania. The loss of coconut palms led to severe financial hardship for the local population.

1.5. EXAMPLES OF THE EVOLUTIONARY IMPACT OF THE VIRUS–HOST INTERACTION

Numerous genetic studies have demonstrated that the interactions between viruses and their hosts have influenced host evolution in significant ways (Lander et al., 2001). Viral environmental stressors cause organisms to evolve defense mechanisms. Furthermore, viruses' capacity to acquire and transfer genes between organisms may provide a mechanism for gene transfer between lineages.

Successful genetic adaptation to infection leads to the development of the immune system, the cellular-based antiviral interferon (IFN) response, and many other defensive and inflammatory responses that multicellular organisms can mount to fend off infection. Viral infection may also offer a fundamental—yet little-known—mechanism for influencing the course of evolution.

Strong evidence suggests that the unique ancestry of placental mammals can be traced back to an ancestral species that contracted an immune-suppressive proto-retrovirus. It is hypothesized that during a protracted gestational period, this immunosuppression enabled the mother to immunologically accommodate the development of a genetically distinct individual in the placenta (Mahy and Regenmortel, 2009).

Two recent examples provide compelling evidence for the ongoing role of viruses in the evolution of plants and animals. Some parasitic wasps deposit their eggs inside caterpillars of other insects. Undoubtedly, the host does not welcome this attack and mounts an immune defense against the invader—especially during the early stages of the wasp's embryonic development. The wasp larvae develop and devour the host, saving the most edible parts for last to keep the food supply fresh.

In wasps that are not infected with a polydnavirus, the larvae frequently die, and parasitism has a low success rate. However, when a polydnavirus infects the same species of wasp and becomes a persistent genetic passenger in the wasps' ovaries and egg cells, the situation changes. The polydnavirus, which remains in the developing female wasps' ovaries, is inserted into the caterpillar along with the wasp egg. This causes a systemic, immunosuppressive infection that prevents the caterpillar from eliminating the embryonic tissue at an early stage.

Another example involves the Chlorella virus, which illustrates how a virus contributes to the development of a symbiotic relationship between its host and another organism.

These viruses are found in freshwater across the United States, China, and likely other parts of the world at concentrations as high as 4×10⊠ infectious units/ml. Despite its effectiveness as a pathogen, the virus is limited to infecting free-living algae and cannot infect the same algae when they are in a semisymbiotic relationship with a paramecium species. The virus's presence creates a strong selective pressure to establish or maintain the symbiotic relationship, shielding the algae cells within their symbiotes from infection.

1.6. THE ORIGIN OF VIRUSES

Viruses are present in all environments and have likely existed since the earliest living cells (Iyer et al., 2006). Since viruses do not fossilize, their origin remains unknown. However, molecular methods provide insight into their ancient existence (Sanjuán et al., 2010). Occasionally, viral genetic material integrates into the host organism's germline, allowing the virus to be vertically passed down to the host's offspring for many generations. Paleovirologists use this as a crucial source of information to track ancient viruses that date back millions of years (Figure 1.2).

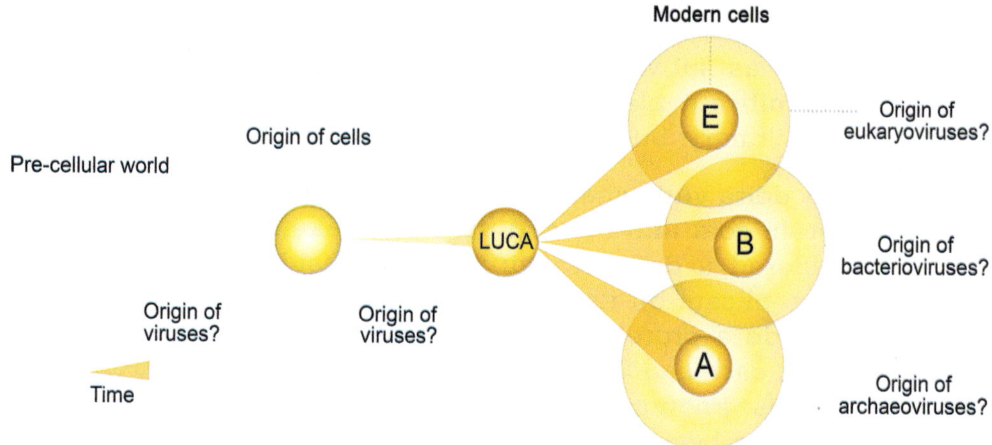

Figure 1.2. *Different scenarios for the origin of viruses.*

Source: Nasir A, Romero-Severson E, Claverie JM. Investigating the Concept and Origin of Viruses. Trends Microbiol. 2020 Dec;28(12):959-967. doi: 10.1016/j.tim.2020.08.003. PMID: 33158732; PMCID: PMC7609044.

There are three main hypotheses that aim to explain the origins of viruses (Krupovic et al., 2019).

1.6.1. Regressive Hypothesis

It's possible that viruses were once tiny cells that parasitized larger cells. Genes not needed for their parasitism disappeared over time. Like viruses, the bacteria *Chlamydia* and *Rickettsia* are living entities that can only multiply inside their host cells. They provide credence to this theory because their reliance on parasitism likely resulted in the loss of genes necessary for survival outside of cells. This is also referred to as the "degeneracy hypothesis" or "reduction hypothesis" (Mahy and Regenmortel, 2009).

1.6.2. Cellular Origin Hypothesis

Certain viruses might have developed from fragments of RNA or DNA that "escaped" from a larger organism's genes. The DNA that escaped could have originated from transposons, which are DNA molecules that replicate and move to different locations within a cell's genes, or from plasmids, which are bare DNA fragments that can travel between cells. Transposons, formerly known as "jumping genes," are movable genetic elements that may have originated from certain viruses. Barbara McClintock discovered them in maize in 1950. This hypothesis is occasionally referred to as the "escape hypothesis" or the "vagrancy hypothesis."

1.6.3. Co-evolution Hypothesis

The "virus-first hypothesis" postulates that viruses may have originated from complex protein and nucleic acid molecules concurrent with the emergence of cells on Earth and that they have been reliant on cellular life for billions of years. Since they lack a protein coat, RNA molecules known as viroids are not categorized as viruses. They are frequently referred to as subviral agents because they share traits with multiple viruses. Viroids are a significant plant pathogen. They interact with the host cell and replicate by using its machinery, but they do not encode proteins (Tsagris et al., 2008).

The human hepatitis delta virus shares an RNA genome with viroids, but it is unable to produce its own protein coat and instead uses one taken from the hepatitis B virus. As a result, it is considered a defective virus. Once inside a host cell, the hepatitis delta virus genome can replicate on its own, but to spread to new cells, it requires the hepatitis B virus to provide a protein coat. Similarly, the mimivirus, which infects the protozoan *Acanthamoeba castellanii*, is the source of the sputnik virophage (La Scola et al., 2008). These viruses, known as "satellites," may be evolutionary intermediates between viroids and viruses because they rely on the presence of other virus species in the host cell.

All these theories had past flaws. For example, the regressive hypothesis failed to explain why even the smallest cellular parasites are completely unlike viruses. The escape theory did not explain the intricate capsids and other structures of virus particles. Since viruses require host cells to reproduce, the virus-first hypothesis contradicts this definition. It is now acknowledged that viruses are ancient and their origins predate the division of life into the three domains. Contemporary virologists have reexamined and reconsidered these traditional theories in light of new discoveries.

The identification of the progenitors of contemporary viruses may be possible through evidence for an ancestral world of RNA cells and the computer analysis of viral and host DNA sequences. These findings enhance our understanding of the evolutionary relationships between various viruses. However, these analyzes have not definitively proven which of these theories is true. It appears unlikely that all known viruses descended from a single ancestor, and viruses have most likely emerged repeatedly in the past through one or more mechanisms (McNeill, 1976).

1.7. VIRUSES HAVE A CONSTRUCTIVE AS WELL AS DESTRUCTIVE IMPACT ON SOCIETY

Politicians and the media frequently portray viruses and infectious diseases as entirely negative forces, but as Sportin' Life from Gershwin's *Porgy and Bess* says, this "ain't necessarily so." Our ever-deepening knowledge of biology is unlikely to have advanced as it has without the influence of infectious diseases. As previously mentioned, a significant portion of our understanding of biological processes comes from studies on viruses.

While unbridled human curiosity led to the discovery of many fundamental biological patterns and supported Darwin's theory in *The Origin of Species*, the impetus for focused research on the microscopic world of pathogens came from medical necessity. The most illustrious figures in European microbiology during the 1800s and early 1900s—such as Pasteur, Koch, Ehrlich, and Fleming—were medical microbiologists. Many of today's booming biotechnology sector and research infrastructure initiatives are driven by medical and financial reasons (Moreira et al., 2021).

In the modern era, many of the basic biochemical processes that viruses encode can be modified to benefit humans. For example, using viral diseases to target plant and animal pests could offer a practical and regulated means of pest control. The introduction of the myxoma virus, a pathogen of South American rabbits, successfully reduced the European rabbit population in Australia, though the effect was only temporary. The dynamics of this adaptation between host and parasite taught researchers valuable lessons on coadaptation.

Scientists are also leveraging the precise targeting abilities of viruses to create tools for delivering therapeutic genes to the cells and organs of individuals with genetic and degenerative diseases. By modifying viral proteins and manipulating viral genomes, they are developing prophylactic vaccines and other therapeutic agents with potentially high specificity. The range of possibilities continues to grow each month (Muralidharan and Reid, 2021).

1.8. VIRUSES ARE NOT THE SMALLEST SELF-REPLICATING PATHOGENS

Prions are smaller and more complex pathogens than viruses, capable of regulating their own self-replication within a host cell. Despite this, the fundamental methodology for studying all subcellular pathogens stems from the study of viruses and the diseases they cause. Viruses are defined as consisting of a genome and one or more proteins that cover the genome. The genome contains the genetic information required to produce the protein coat and replicate the genome. Some virus variants lack the information needed for coat proteins or genome replication. Since these virus-derived entities can be traced back to a parental form with complete genetic information, the mutant forms are commonly referred to as defective virus particles.

Defective viruses can only reproduce if a helper virus is also present. Hepatitis delta virus is a prime example, as it can only spread through coinfection with hepatitis B virus. The hepatitis delta virus shares certain characteristics with a class of RNA pathogens that replicate in plants through as-yet-unidentified mechanisms. These RNA molecules, known as viroids, are economically significant pathogens that do not encode any proteins but can be mechanically transferred between plants (Nelson and Holmes, 2007).

Some pathogens, known as prions, appear to consist entirely of proteins. These prions resemble proteins found in cells but fold abnormally. When prions interact with normally folded proteins in neural tissue, they cause the normal proteins to fold incorrectly, impairing neural cell function and resulting in disease. Although much research remains to be done on prions, it is known that they are difficult to inactivate and can be transmitted efficiently among hosts. Several prion diseases, including kuru and Creutzfeldt-Jakob disease (CJD), infect humans. Prion diseases of sheep and cattle, such as scrapie and mad cow disease, have had a significant negative economic impact on British agriculture. Alarmingly, the spread of sheep scrapie among cattle in England has given rise to a new human illness related to but distinct from CJD.

The existence of these pathogens provides additional circumstantial evidence that viruses originated from cells and that viruses evolved from multiple sources over the course of evolution (Nguyen-Dinh and Herker, 2021).

CASE STUDY

PREVENTING EBOLA IN UGANDA

Case Study From The Makerere University School Of Public Health And The Johns Hopkins Center For Health Security

Introduction

On August 1, 2018, four cases in the eastern Democratic Republic of the Congo's North Kivu Province were confirmed, marking the start of the Kivu Ebola epidemic (DRC). Over 3,400 confirmed and suspected cases as well as over 2,200 deaths have resulted from the epidemic in the DRC to date. The Ugandan government and its partners launched a high alert and mobilized resources to prevent the importation of cases, detect imported disease quickly, contain the spread of imported disease, and treat sick people appropriately due to the frequent movement from North Kivu across the border into Uganda, including a regular influx of refugees.

Uganda was exposed to the virus in June 2019, but as of May 2020, there had only been four reported cases of importation and no instances of in-country transmission. Uganda has concentrated on finding cases and stopping an outbreak. Creating a multisectoral team and strong coordinating mechanisms, conducting simulation exercises to gauge the nation's preparedness, and launching a preventative immunization campaign among frontline staff and contacts of suspected cases were some of the prevention activities. In cooperation with DRC officials, detection operations included cross-border surveillance and health messaging and communications (Nickol et al., 2009).

The Makerere University School of Public Health and the Johns Hopkins Center for Health Security conducted this study based on a thorough review of the gray literature and key informant interviews, in response to these achievements, despite the fundamental problems that Uganda's health system faces.

Context

Uganda is a landlocked nation surrounded by Tanzania, Rwanda, South Sudan, Kenya, and the Democratic Republic of the Congo. In terms of GDP, it is one of the poorest 15 nations in the world. With a median age of 15, it has the lowest population median worldwide, with an estimated 42 million people. Following six years of guerilla warfare, Yoweri Kaguta Museveni, the current president, took office in 1986. He abolished the president's term limits and won bitter elections in 2011 and 2016 to retake the presidency. In terms of corruption, the nation ranks 151st out of 176, and a 2012 US Department of State report estimated that corruption costs the nation US$286 million annually. The NGO Amendment Act of 2006 severely restricted the output of nongovernmental organizations in the nation and stifled their activities. There is currently fighting between government and rebel forces in northern Uganda, and there are an estimated 1.4 million internally displaced people in the nation. Similar insecurity issues plague the neighboring Democratic Republic of the Congo (DRC), where there is continuous ethnic conflict in the border provinces of North Kivu and Ituri, leading to an estimated 1 million internally displaced people, many of whom cross into Uganda. The risk of disease transmission, including Ebola, is significantly increased by this instability close to the border (Paparisto et al., 2021).

Since 2000, Uganda has responded to Ebola outbreaks in the past. The government has made significant investments in its epidemic readiness as a result of learning from previous outbreaks. Uganda's reaction to the Ebola outbreak has significantly improved over time.

History of Ebola Outbreaks in Uganda

Year(s)	Species	Number of Cases	Number of Deaths	**Case Fatality Rate**
2000–2001	Sudan Ebolavirus	425	224	53%
2007–2008	Bundibugyo Ebolavirus	131	42	32%
2011	Sudan Ebolavirus	1	1	100%
2012	Sudan Ebolavirus	11	4	36%
2012-2013	Sudan Ebolavirus	6	3	50%
2019–Current	Zaire Ebolavirus	4	4	100%

Source: US Centers for Disease Control and Prevention (CDC). Years of Ebola virus disease outbreaks. https://www.cdc.gov/vhf/ebola/history/chronology.html. Last updated October 15, 2019. Accessed May 6, 2020.

Prevention

Coordination and Planning

In August 2018, Uganda's Ministry of Health received reports of an outbreak in the Democratic Republic of the Congo. To oversee preparation and response efforts, the ministry established a national task force, an emergency operations center, and a national rapid response team (Raoult and Forterre, 2008). The national task force is co-chaired by the World Health Organization (WHO) and consists of representatives from partner agencies as well as ministries, agencies, and departments from a variety of industries. Based on the WHO's six preparedness pillars, it is divided into the following six committees:

- Epidemiological surveillance, including contact tracing, points of entry, and laboratory testing;
- Case management, including psychosocial support, waste management, and safe burials;
- Risk communication and community engagement;
- Vaccination, therapeutics, and research;
- Emergency coordination including budgeting and resource mobilization;
- Logistics.

The national task force, shortly after its formation, developed a partner coordination matrix to try and prevent efforts from being duplicated or significant gaps from being unfilled among the numerous external partners providing financial and technical support. The US Centers for Disease Control and Prevention, the Uganda Red Cross, Doctors Without Borders (Médecins Sans Frontières), the United Nations Children's Fund, and the United Nations High Commissioner for Refugees are some of these partners. Interviewees concur that the matrix was helpful, but the challenges of managing so many partners persisted. Interviewees stated that some priorities were given too much attention by too many partners, while other priorities were given too little attention by too few (Raoult et al., 2004).

In order to manage daily operations, the national task force also created an incident management system that followed the same six-part committee structure. Twenty high-risk and ten moderate-risk districts were identified by the system, and district task forces with the same six-part structure were formed in each of them. Facilitating communication between the district task forces and the national task force was the responsibility of the incident management system. A decentralized structure has benefits and drawbacks as usual. The district task forces were more adaptable and had a deeper understanding of the communities they were serving because they were based closer to the ground. They were asked to work with the national task force on operational details, and their local expertise—including the understanding of frequent border crossings and the causes of community opposition to public health advisories—brought strength to the response. The national task force was in charge of budgeting and other major decisions that impacted the districts, even though the district task forces were independent. This could cause conflict. Sometimes there was a lack of clarity in the duties between the two, which caused epidemiological investigations to be delayed.

Following the identification of the districts that were at risk, each one had a national rapid response team dispatched to carry out a readiness assessment in accordance with WHO guidelines. On that evaluation, no district received a score greater than 50%. The National Ebola Contingency Plan was created and implemented by the national task force in collaboration with partners based on the readiness gaps that were identified. The plan was updated on a regular basis through formal assessments, simulation exercises, training, and input from system employees at all levels. All 30 at-risk districts had an average readiness score of 92% by the end of the year, a notable improvement in just four months.

Preparedness Simulation and Ongoing Training

In December 2018, the Ministry of Health proposed a full-scale simulation exercise to evaluate the preparedness and response plan, citing the significant improvement in districts' readiness scores. In April 2019, a system-testing exercise was carried out at the community, district, and national levels. In one scenario, community monitoring and the handling of a possible case in a Kasese District facility other than an Ebola Treatment Unit were tested, as well as the safe and respectable burial of the patient after their death. The response systems at the border crossing point of Mpondwe, a port of entry designated in the Kasese District, and the referral to an Ebola treatment unit were tested in a different scenario. The Entebbe International Airport's response systems were put to the test in a third scenario.

The case was found quickly, but the community's reporting to the federal government took some time. In order to quickly spread knowledge and skills, the Ministry of Health used the data from the simulation exercise to design training using a train-the-trainer model.

Prophylactic Vaccination

In order to protect the well-being of essential health care workers, the national task force requested authorization to administer an experimental vaccine, the rVSV-ZEBOV vaccine created by Merck. This vaccine had demonstrated success in combating the specific strain of Ebola in the DRC. It was initially granted a compassionate use exemption until it obtained formal approval from the appropriate regulatory bodies (Villarreal and DeFilippis, 2000).

The national task force obtained 3,000 doses of the experimental vaccine, trained and equipped vaccination teams, coordinated cold-chain logistics, determined which workers were most vulnerable, and started the voluntary vaccination program as soon as permission to use it was granted. To increase vaccination, 3,000 more doses of the vaccine were subsequently obtained. 4,699 healthcare workers from 165 facilities had received vaccinations as of June 2019. Furthermore, vaccinations were administered to 78 individuals who had come into contact with a DRC-infected person, as well as over 700 contacts of those individuals. Since this was the first Ebola outbreak where vaccinations were available, it makes sense that ethics and testing approval took some time to come together. As long as protocols are in place, regulatory bodies feel more at ease authorizing operational research during an outbreak, and responders and communities recognize the importance of implementing these medical countermeasures as soon as products are licensed for use, the process should proceed more quickly in subsequent outbreaks.

An interviewee stated that certain Ugandan communities were hesitant to accept a vaccine believed to be from the DRC but became more receptive when they learned it originated from Entebbe. District task forces adjusted their protocols for health care workers administering the vaccine based on the trust of local communities. Uganda's quick adaptability allowed them to implement public health strategies that gained community support.

Detection

Community Mobilization and Messaging

A key component of Uganda's approach has been open and honest communication. The national task force convened to report on its activities and made a media announcement in the evening of the day the first imported case of Ebola from the Democratic Republic of the Congo was discovered. Weekly or occasionally daily publicly accessible situational reports were made available by the WHO Uganda Country Office and the Ministry of Health.

Uganda gave top priority to ensuring that communities received important messages about symptoms, where to get treatment, and how to bury the dead safely. A toll-free phone number was frequently updated with pertinent information, authorities made the rounds of radio talk shows and posted on social media, educators attended community meetings and went door-to-door, and radio and television education spots were recorded. To counteract rumors and overcome community resistance, risk communication and educational awareness messages were carefully translated into local languages and faithfully delivered by trusted community and religious leaders. In order to prevent confusion when various groups engaged in different activities (like contract tracing or vaccination) visited the community on the same day, the district task forces also attempted to streamline operations. The intention was for the response to give the community a sense of "one voice."

Cross-Border Collaboration and Surveillance

Cross-border cooperation with the DRC was a major factor in Uganda's ability to prevent in-country transmission, as it facilitated the smooth operation of Uganda's point-of-entry screening procedure. In December 2018, the two nations signed a memorandum of understanding on cross-border collaboration for public health preparedness and response during a bilateral ministerial meeting in Goma, the Democratic Republic of the Congo. Through the establishment of four surveillance zones, Uganda and the DRC were able to prepare its ports of entry, hospitals, and local communities for potential cases that might escape isolation centers. The agreement made regular communication about potential cases possible. At refugee reception centers and ports of entry, volunteers received training on Ebola screening. Teams used infrared thermometers to screen refugees at reception centers as well as anyone entering Uganda from the Democratic Republic of the Congo. Following that, those with high body temperatures were checked for signs of

Ebola. Uganda also made improvements to its infrastructure and laboratory procedures to enable quicker and more precise diagnosis (Xiao et al., 2005). Point-of-entry screenings require a lot of work and resources, according to interviewees, but they have been a top priority for almost two years. Over 31,000 passengers had been screened at each of the 30 bus terminals and border crossings between the Democratic Republic of the Congo and Uganda as of May 2020.

COVID-19

As COVID-19 started to spread in recent months, it seems that the steps taken to prevent and treat Ebola were worthwhile. COVID-19 screening procedures were quickly added to the Ebola screening procedures at airports and border crossings. Additionally, laboratory and healthcare personnel received training on how to handle samples and patients, and community communications campaigns were updated to include information about COVID-19. Uganda made a mistake early on when it failed to identify the United Arab Emirates as a nation at high risk, which allowed multiple cases to be imported. Apart from that, the reaction has worked so far. As of May 7, 2020, 47,620 tests had been conducted, 101 confirmed cases, and more than 2,000 contacts had been located and examined in Uganda.

Conclusion

Due to its 20-year history with Ebola, Uganda has made investments in response and preparation plans for the virus. Despite governance challenges, this investment has paid off as the nation has progressively prevented or quickly contained outbreaks. Uganda has prioritized case detection at points of entry to combat the current Ebola epidemic in the Democratic Republic of the Congo, obviating the need for any kind of containment strategy inside Uganda. As an additional component of its prevention strategy, Uganda is administering the Ebola vaccine to certain health workers on a preventive basis.

The government's cooperative efforts with various partners constitute a fundamental aspect of its commitment. Task forces are a well-organized part of the government, both nationally and locally. The nation receives financial support for readiness and response from foreign donors, and outside partners—particularly the WHO—offer technical assistance. Ultimately, despite the possibility of tension, Uganda's relationship with the DRC has been positive. To help Uganda focus its screening efforts, the DRC keeps in regular contact with its authorities there to alert them to the possibility of infected individuals attempting to cross the border. Uganda's experience demonstrates that well-thought-out preparedness and response strategies are effective as long as a commitment, funding, and technical capacity available. That being said, maintaining a response that necessitates ongoing vigilance will be difficult as long as the Ebola epidemic in the Democratic Republic of the Congo persists (it will have been two years in August 2020) and as COVID-19 complicates matters even more. Failure to detect the virus will put the nation's capacity to contain it and care for the ill to the test.

CLASS ACTIVITY

THE SPREADING OF A VIRUS: A SIMULATION!

One of the simplest simulations to set up, it can be easily incorporated into a variety of classroom subjects, sparks interesting debates, and ends with a fun problem-solving exercise where students try to figure out who the original virus carriers were. With classes of at least 15, it functions best. Form groups with 20–28 pupils each. Some simulation and follow-up activities require going outside, and they both take about an hour.

Materials Needed:

- small paper cups (one per student – 3 oz. "bathroom cups");
- large opaque plastic cups (one for each pair of students – 18 oz. is good);
- plastic spoons (one per student);
- white flour;
- baking soda;
- vinegar;
- sticky notes (one per student);
- trash bag.

Setting It Up

Ascertain the quantity of cups infected with "the virus." The number ought to be as near to one-eighth of the total number as feasible. In a class of 24, contaminate three of the cups. In a class of 8–12, contaminate one.

While the students are not in the room, prepare the cups beforehand. Arrange the tiny cups and use a permanent marker to assign numbers to them. Select 1/8 of the cups to serve as the "carriers," or the ones that have the "virus" in them. Setting aside cups with the numbers 3, 7, and 18 out of a total of 22 to serve as the virus carriers is one way to do this. Keeping track of the cups that have been declared contaminated is crucial. Fill the "carrier" cups nearly two-thirds of the way with baking soda. Most of the remaining cups—roughly two-thirds full—should be filled with flour. Transfer flour onto the baking soda in the tainted cups, giving the impression that flour has been added to them as well. When the pupils arrive, every cup should have the same appearance.

The Activity

Discuss with the class what they already know about viruses and how they propagate before launching into the simulation. Inform them that they will be taking part in a simulated activity wherein students will be infected with a "fake virus." Nobody will be aware of who has the infection or who is at risk of getting it. Give each student a sticky note and instruct them to write their names on it. Show off a tiny cup and mention that each one has a powdered "drink" in it. Say, "You will share your 'drinks' in the following way: after we head outside, you'll find a sharing partner. You'll each receive a large eighteen ounces. cup and raise it so that neither of you can see what is inside by holding it high, above heads. You will both pour your "drinks" into the large cup while holding it high, then stir the contents while continuing to hold the cup high and avoid looking at the contents. After mixing everything well, divide the powder evenly between your small cups, carefully pouring half into each. Remind people to recall who they shared things with and in what sequence. They are told to share twice, with two distinct individuals. After sharing with two people, students should line up (e.g., in a row) as instructed by the teacher (e.g., a sidewalk, up against a wall).

Give a plastic spoon to every student. Next, give each student a numbered cup and instruct them to write and circle the number on a sticky note. As the students are led to an outdoor space, ask them to carry their cups delicately. Assign specific students to handle the large cups and trash bag. Give students the all-clear to start when you lay out the big cups so they are accessible. As students "share their drinks," keep an eye on them to make sure they are stirring the contents thoroughly and aren't staring into the cups as they pour. Remind them to keep in mind the two individuals they shared things with and the sequence in which they did so.

Once every student is in line, walk down the line and put a little vinegar into each cup, telling them it's time to find out if they have been infected. If the liquid appears to just settle on the contents of the cup, the student is not infected. If the contents of the cup bubble up, the student has the virus.

Students should return to the room after discarding their cups. After taking a seat, instruct students to write the numbers of the two individuals they shared with on their sticky notes, designating Person A for the first person and Person B for the second. Give them a little time to stand up and move around in order to gather their numbers; this should take no longer than two or three minutes, after which everyone should return to their seats

Lastly, make a chart to gather the information. Put it on the board or make a Google Form that students can fill out, and then show everyone the spreadsheet. Their mission is to identify the initial virus carriers. Permit students to solve problems using any approach they like, and allow them to write their thoughts on the board while they explain their solutions.

SUMMARY

- The replication and propagation of a given virus in a population frequently (but not always) manifest when an infectious disease spreads between individuals. The study of the effects of viral infection on the host is broadly defined as the study of viral pathogenesis.
- The identification of viruses as infectious agents followed pioneer work in the mid-1800s by Louis Pasteur and associates. During that period, the laboratory culturing process was developed for bacteria, which could be grown on enriched agar preparations or broths and were identifiable when the bacteria were fixed on glass slides, stained, and examined under the microscope.
- The interaction between viruses (and other infectious agents) and their hosts is dynamic. As effective physiological responses to infectious disease have evolved in the organism and (more recently) have developed societally through the application of biomedical research, viruses themselves respond by exploiting their naturally occurring genetic variation to accumulate and select mutations to become wholly or partially resistant to these responses.
- The early pioneers in virology had a very limited armamentarium on hand to classify or identify viruses as microbial agents. At the time of their initial discoveries, microscopes of sufficient power to see viral agents were not yet invented; cell cultures to grow viruses were unknown, and probes like antibodies or nucleic acids to mark an infectious agent were not available.
- There is archeological evidence in Egyptian mummies and medical texts of readily identifiable viral infections, including genital papillomas (warts) and poliomyelitis. There are also somewhat imperfect historical records of viral disease affecting human populations in classical and medieval times.
- Smallpox epidemics during the Middle Ages and later in Europe resulted in significant population losses as well as major changes in the economic, religious, political, and social life of individuals.
- Viruses provide environmental stresses to which organisms evolve responses. Also, the ability of viruses to acquire and move genes between organisms may provide a mechanism of gene transfer between lineages. Development of the immune system, the cellular-based antiviral interferon (IFN) response, and many of the inflammatory and other responses that multicellular organisms can mount to ward off infection result from successful genetic adaptation to infection.
- Viral genetic material occasionally integrates into the germline of the host organisms, by which they can be passed on vertically to the offspring of the host for many generations. This provides an invaluable source of information for paleovirologists to trace back ancient viruses that existed as far back as millions of years ago.

REVIEW QUESTIONS

1. What are the viral and host factors that affect viral pathogenesis?
2. What are the effects of virus infection on host cells?
3. What is the relationship between a virus and its host cell?
4. What do viruses do to their host?
5. What is the host response to a virus?
6. Who is the father of virology?
7. Do viruses have DNA?
8. Do viruses have ribosomes?
9. Which was the first virus, and who created viruses?
10. Why is it likely that viruses have not evolved from free-living organisms?

REFERENCES

1. Andersson, S. G. E., et al. (1998). The genome sequence of *Rickettsia prowazekii* and the origin of mitochondria. *Nature*, *396*(6707), 133–143. https://doi.org/10.1038/24094.
2. Bell, P. J. L. (2001). Viral eukaryogenesis: Was the ancestor of the nucleus a complex DNA virus? *Journal of Molecular Evolution*, *53*(3), 251–256. https://doi.org/10.1007/s002390010215.
3. Bos, L. (1999). Beijerinck's work on tobacco mosaic virus: Historical context and legacy. *Philosophical Transactions of the Royal Society of London B: Biological Sciences*, *354*(1383), 675–685. https://doi.org/10.1098/rstb.1999.0420.
4. Courtney, D. G. (2021). Post-transcriptional regulation of viral RNA through epitranscriptional modification. *Cells*, *10*(5), 1129. https://doi.org/10.3390/cells10051129.
5. DeBuysscher, B. L., Scott, D. P., Rosenke, R., Wahl, V., Feldmann, H., & Prescott, J. (2021). Nipah virus efficiently replicates in human smooth muscle cells without cytopathic effect. *Cells*, *10*(6), 1319. https://doi.org/10.3390/cells10061319.
6. Dolnik, O., Gerresheim, G. K., & Biedenkopf, N. (2021). New perspectives on the biogenesis of viral inclusion bodies in negative-sense RNA virus infections. *Cells*, *10*(6), 1460. https://doi.org/10.3390/cells10061460.
7. Iyer, L. M., Balaji, S., Koonin, E. V., & Aravind, L. (2006). Evolutionary genomics of nucleo-cytoplasmic large DNA viruses. *Virus Research*, *117*(1), 156–184. https://doi.org/10.1016/j.virusres.2006.01.009.
8. Kirui, J., Abidine, Y., Lenman, A., Islam, K., Gwon, Y. D., Lasswitz, L., Evander, M., Bally, M., & Gerold, G. (2021). The phosphatidylserine receptor TIM-1 enhances authentic Chikungunya virus cell entry. *Cells*, *10*(7), 1828. https://doi.org/10.3390/cells10071828.

9. Krupovic, M., Dolja, V. V., & Koonin, E. V. (2019). Origin of viruses: Primordial replicators recruiting capsids from hosts. *Nature Reviews Microbiology*, *17*(7), 449–458. https://doi.org/10.1038/s41579-019-0205-6.

10. La Scola, B., Desnues, C., Pagnier, I., Robert, C., Barrassi, L., & Fournous, G., et al. (2008). The virophage as a unique parasite of the giant mimivirus. *Nature*, *455*(7209), 100–104. https://doi.org/10.1038/nature07218.

11. La Scola, B., et al. (2003). A giant virus in amoebae. *Science*, *299*(5615), 2033. https://doi.org/10.1126/science.1081867.

12. Lander, E. S., et al. (2001). Initial sequencing and analysis of the human genome. *Nature*, *409*(6822), 860–921. https://doi.org/10.1038/35057062.

13. Mahy, W. J., & Regenmortel, M. H. V. (Eds.). (2009). *Desk encyclopedia of general virology* (p. 24). Academic Press.

14. McNeill, W. H. (1976). *Plagues and peoples*. Anchor Press. http://refhub.elsevier.com/B978-0-12-226980-6.00086-8/ref0025 (accessed on 1st October 2024).

15. Moreira, E. A., Yamauchi, Y., & Matthias, P. (2021). How influenza virus uses host cell pathways during uncoating. *Cells*, *10*(7), 1722. https://doi.org/10.3390/cells10071722.

16. Muralidharan, A., & Reid, S. P. (2021). Complex roles of neutrophils during arboviral infections. *Cells*, *10*(6), 1324. https://doi.org/10.3390/cells10061324.

17. Nelson, M. I., & Holmes, E. C. (2007). The evolution of epidemic influenza. *Nature Reviews Genetics*, *8*(3), 196–205. https://doi.org/10.1038/nrg2053.

18. Nguyen-Dinh, V., & Herker, E. (2021). Ultrastructural features of membranous replication organelles induced by positive-stranded RNA viruses. *Cells*, *10*(9), 2407. https://doi.org/10.3390/cells10092407.

19. Nickol, M. E., Lyle, S. M., Dennehy, B., & Kindrachuk, J. (2020). Dysregulated host responses underlie 2009 pandemic influenza-methicillin resistant *Staphylococcus aureus* coinfection pathogenesis at the alveolar-capillary barrier. *Cells*, *9*(11), 2472. https://doi.org/10.3390/cells9112472.

20. Oldstone, M. B. A. (2010). *Viruses, plagues, & history*. Oxford University Press. http://refhub.elsevier.com/B978-0-12-226980-6.00086-8/ref0030 (accessed on 1st October 2024).

21. Paparisto, E., Hunt, N. R., Labach, D. S., Coleman, M. D., Di Gravio, E. J., Dodge, M. J., Friesen, N. J., Cote, M., Muller, A., Hoenen, T., et al. (2021). Interferon-induced HERC5 inhibits Ebola virus particle production and is antagonized by Ebola glycoprotein. *Cells*, *10*(9), 2399. https://doi.org/10.3390/cells10092399.

22. Prangishvili, D., Forterre, P., & Garrett, R. A. (2006). Viruses of the archaea: A unifying view. *Nature Reviews Microbiology*, *4*(11), 837–848. https://doi.org/10.1038/nrmicro1527.

23. Raoult, D., & Forterre, P. (2008). Redefining viruses: Lessons from mimivirus. *Nature Reviews Microbiology*, *6*(4), 315–319. https://doi.org/10.1038/nrmicro1858.

24. Raoult, D., et al. (2004). The 1.2-megabase genome sequence of Mimivirus. *Science*, *306*(5700), 1344–1350. https://doi.org/10.1126/science.1101485.

25. Sanjuán, R., Nebot, M. R., Chirico, N., Mansky, L. M., & Belshaw, R. (2010). Viral mutation rates. *Journal of Virology*, *84*(19), 9733–9748. https://doi.org/10.1128/JVI.00694-10.

26. Tsagris, E. M., Martínez de Alba, A. E., Gozmanova, M., & Kalantidis, K. (2008). Viroids. *Cellular Microbiology*, *10*(11), 2168–2179. https://doi.org/10.1111/j.1462-5822.2008.01231.x.

27. Villarreal, L. P., & DeFilippis, V. R. (2000). A hypothesis for DNA viruses as the origin of eukaryotic replication proteins. *Journal of Virology*, *74*(15), 7079–7084.

28. Waterson, A. P., & Wilkinson, L. (1978). *An introduction to the history of virology*. Cambridge University Press. http://refhub.elsevier.com/B978-0-12-226980-6.00086-8/ref0035 (accessed on 1st October 2024).

29. Xiao, C., et al. (2005). Cryo-electron microscopy of the giant mimivirus. *Journal of Molecular Biology*, *353*(3), 493–496. https://doi.org/10.1016/j.jmb.2005.08.060.

CHAPTER 2

Virus Replication and Viral Pathogenesis

LEARNING OBJECTIVES

After studying this chapter, you will be able to:

- Explain the science of virus replication;
- Illustrate the virus replication cycle;
- Elaborate the general concepts of viral pathogenesis.

KEY TERMS FROM THIS CHAPTER

Cell culture systems
Disease symptoms
Influenza vaccines
Ribozyme function
Viral infection
Cell ribosomes
Genome
Liver inflammation
Satellite viruses
Cytoplasm
Immune system
Pneumonia
Viral diseases

INTRODUCTION

Virus replication refers to the process by which viruses produce new infectious particles, or virions, within a host cell. This process begins when a virus attaches to a specific receptor on the surface of a susceptible host cell. Following attachment, the virus penetrates the cell membrane and releases its genetic material, DNA or RNA, into the host cell cytoplasm. The viral genome then commandeers the host cell machinery to transcribe and translate viral proteins while also replicating its genetic material. Depending on the type of virus, replication can occur in the cytoplasm or the nucleus. For example, DNA viruses typically replicate in the nucleus using the host DNA-dependent DNA polymerase, whereas RNA viruses often replicate in the cytoplasm with their own RNA-dependent RNA polymerase. The newly synthesized viral components are then assembled into progeny virions, which are eventually released from the host cell through lysis or budding, ready to infect new cells (Bose et al., 2015).

Viral pathogenesis, on the other hand, refers to the mechanisms by which viruses cause disease in the host organism. This process is multifaceted, involving direct damage to infected cells, immune responses triggered by the infection, and the broader physiological effects of the virus on the host. Initially, the direct cytopathic effects of viral replication can lead to cell death or dysfunction, contributing to tissue damage. Moreover, the immune response to viral infection, while essential for clearing the virus, can also cause collateral damage to host tissues. Inflammatory responses and the activation of various immune cells can result in tissue inflammation and injury, exacerbating the disease symptoms. Additionally, certain viruses have evolved strategies to evade the immune system, such as mutating rapidly to avoid antibody recognition or hiding within immune-privileged sites, thereby prolonging their survival and pathogenicity within the host.

Furthermore, viral pathogenesis can vary significantly among different viruses and host species. For instance, the influenza virus causes respiratory illness by infecting epithelial cells in the respiratory tract, leading to symptoms such as coughing, fever, and pneumonia. In contrast, the hepatitis B virus primarily targets liver cells, causing liver inflammation and potentially leading to chronic conditions like cirrhosis and liver cancer. The severity of viral diseases also depends on factors such as the host's immune status, age, genetic background, and co-existing health conditions. In some cases, the immune response to the virus can be hyperactive, resulting in conditions like cytokine storms, which can cause severe tissue damage and increase mortality, as seen in severe cases of COVID-19.

2.1. THE SCIENCE OF VIRUS REPLICATION

Experimental virology has greatly emphasized on studying the molecular processes that follow virus replication since they are crucial to comprehending and managing all viral illnesses. Any replication cycle's biological goal is to create new viral genomes and proteins in sufficient amounts to guarantee the virus's propagation (a prime example of a selfish gene). To do this, the extracellular viral genome must be shielded from enzymatic degradation to be inserted into new target cells for additional replication cycles. Much has been learned about the early phases of attachment, and more thorough research reveals that the initial detection of the virus by the host is more intricate than previously thought.

In almost all human viruses, the host's innate immune response is suppressed in some way, making temporal modulation of intracellular activities crucial in all but the most basic viruses. In certain cases, the innate immune response has been known to facilitate viral replication into cells that the virus would not normally be able to infect. The lines dividing virus-directed events from cellular processes that regulate specialized cell functions are probably going to get even more blurred over the coming years, but comprehending these processes will provide a host of new targets for the development of immunotherapy and innovative antiviral medicine (Table 2.1) (Bose et al., 2015).

Table 2.1. *Characteristics of Virus Replication Among Different Families*

Family	Route of Uptake	Site of Nucleic Acid Replication	Eclipse Period (h[a])	Site of Maturation or Budding
Poxviridae	Variable	Cytoplasm	4	Golgi membrane
Herpesviridae	Variable	Nucleus	4	Nuclear membrane
Adenoviridae	Clathrin-mediated endocytosis	Nucleus	10	Nucleus
Papillomaviridae	Clathrin-mediated endocytosis	Nucleus	?	Nucleus
Polyomaviridae	Caveolar endocytosis	Nucleus	12	Nucleus
Parvoviridae	Clathrin-mediated endocytosis	Nucleus	6	Nucleus
Circoviridae	?	Nucleus	?	None
Hepadnaviridae	Clathrin-mediated endocytosis	Nucleus/cytoplasm	?	Endoplasmic reticulum

Retroviridae	Plasma membrane fusion or clathrin-mediated endocytosis	Cytoplasm/nucleus	10	Plasma membrane
Reoviridae	Clathrin-mediated endocytosis	Cytoplasm	5	Cytoplasm
Birnaviridae	Uncertain	Cytoplasm	4	None
Paramyxoviridae	Plasma membrane fusion	Cytoplasm	4	Plasma membrane
Rhabdoviridae	Plasma membrane fusion	Cytoplasm	3	Plasma membrane
Filoviridae	Plasma membrane fusion	Cytoplasm	2	Plasma membrane
Bornaviridae	Clathrin-mediated endocytosis	Nucleus	?	Plasma membrane
Orthomyxoviridae	Clathrin-mediated endocytosis	Nucleus	4	Plasma membrane
Bunyaviridae	Clathrin-mediated endocytosis	Cytoplasm	4	Golgi membrane
Arenaviridae	Clathrin-mediated endocytosis	Cytoplasm	5	Plasma membrane
Coronaviridae	Clathrin-mediated endocytosis/plasma membrane fusion	Cytoplasm	5	Endoplasmic reticulum
Arteriviridae	Clathrin-mediated endocytosis	Cytoplasm	5	Endoplasmic reticulum
Picornaviridae	Caveolar endocytosis/ plasma membrane insertion	Cytoplasm	2	Cytoplasm
Caliciviridae	Caveolar endocytosis/ plasma membrane insertion?	Cytoplasm	3	Cytoplasm
Astroviridae	Caveolar endocytosis/ plasma membrane insertion?	Cytoplasm	3	Cytoplasm
Togaviridae	Clathrin-mediated endocytosis	Cytoplasm	2	Plasma membrane
Flaviviridae	Clathrin-mediated endocytosis	Cytoplasm	3	Endoplasmic reticulum

[a]Differs with a multiplicity of infection, the strain of virus, cell type, and physiological conditions.

Source: https://www.ncbi.nlm.nih.gov/pmc/articles/PMC7173592/.

2.1.1. Growth of Viruses

Before the development of *in vitro* culture techniques, human viruses could only be propagated in laboratory animals. Much of our knowledge of virus growth and replication, therefore, comes primarily from the study of bacterial viruses. In 1931, vaccinia viruses and herpes simplex viruses could be cultured on the chorioallantoic membrane of embryonated chicken eggs. This system has become routine for the study of many mammalian viruses because virtually all virus families contain viruses that can be cultured in this manner. Although cell culture has long since replaced the use of eggs for this purpose, the technology is still used to prepare influenza virus stocks and produce influenza vaccines.

When *in vitro* cell culture systems were developed, virology underwent a sea change. Not only was it possible to analyze the intracellular processes that accompany virus replication in a way that was comparable to studying bacteriophages in bacterial cells, but it also gave researchers a way to measure the quantity of infectious viruses present in samples and virus stocks (Flint et al., 2016). An artificial medium was created to preserve the viability of cells, regardless of the original species. These cells might be primary cell cultures, organ cultures, explant cultures, monolayer cell cultures, or monolayer cell cultures that were immortalized into cell lines. For short-term research that require the preservation of fully differentiated cells, organ cultures can be helpful since they preserve the three-dimensional structure of the original tissue.

For instance, several human respiratory viruses were isolated thanks in large part to tracheal epithelial cells that were cultured and adhered to the trachea's cartilage matrix. Proteases like trypsin or collagenase are used in the creation of primary cell cultures to separate individual cells from a tissue, like foetal kidney or lung, and then the individual cells adhere to a cell culture substrate where they will divide only a limited number of times. Due to their short lifespan, it is necessary to repeatedly prepare new cultures from source tissue, which obviously poses reproducibility issues.

In contrast, continuous propagation of cells is possible with two types of long-term culture:

- "Semicontinuous," "diploid" cell strains, for example, human lung or foreskin fibroblasts (WI-38, MRC-5), in which cells eventually senesce after 20 to 30 divisions (the "Hayflick number") due to progressive shortening of telomeres; and
- HeLa and BHK-21 cells are two examples of continuous immortalized cell lines. They come from tumors or primary cells that spontaneously transform during cell culture. They can divide into practically unlimited numbers of cells, producing consistency, although this is frequently accompanied by a loss of differentiated cell activities. For example, HeLa cells were first isolated in cell culture from a patient who passed away from cervical cancer in 1951. Despite this, HeLa cells are still a crucial tool for culturing viruses and can even contaminate cell cultures from other sources due to their successful proliferation.

Initially, the identification of a virus had to rely on routinely examining cultures under a light microscope for indications of morphological alteration or cellular death, in contrast to uninfected cell cultures serving as controls. A particular family of viruses can be identified by the peculiar appearance of their cytopathic effect (CPE). For instance, herpesviruses can generate a unique cytopathology that is frequently accompanied by the fusing of dying cells. However, depending solely on cytopathology may yield inaccurate results, particularly if the sample is tainted with adventitious agents unrelated to the pathological process being studied, such as bacterial toxins (Flint et al., 2016).

These systems have helped to clarify gene expression and control techniques and characterize the fundamental mechanisms of transcription, translation, and nucleic acid replication for all major families of vertebrate viruses. Every virus family uses a different replicating strategy. As suggested by David Baltimore in 1971, classifying all viruses into six groups according to the structure of their genomes and the method by which they use host cell ribosomes to make messenger RNAs (mRNAs) for translation is a significant unifying and simplifying idea (Figure 2.1). This illustrates another aspect of all viruses' parasitic nature, which is their need for a host cell in order to synthesize viral proteins.

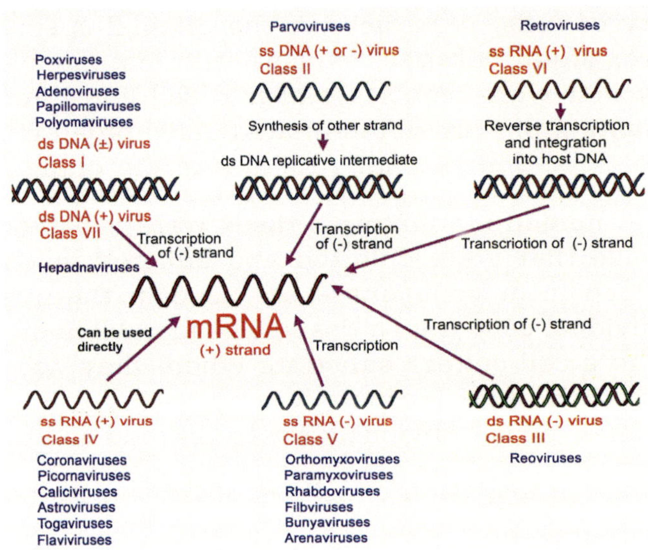

Figure 2.1. *Virus replication strategies based on genome composition and the pathways of viral mRNA synthesis.*

Source: https://www.ncbi.nlm.nih.gov/pmc/articles/PMC7173592/bin/f04-01-9780123751560.jpg.

2.1.2. Satellite Viruses and Viroids

Subviral particles known as satellite viruses have an RNA or DNA genome that codes for a capsid protein, However, they are completely dependent on the presence of another virus in order to replicate. Most are found in plants, although a few number affect public health and medical virology. For example, co-infection with another adenovirus is required for the replication of dependoviruses (family Parvoviridae, genus Dependovirus; adeno-associated viruses, now used as vectors for delivering heterologous genes of interest, e.g., vectored vaccines). The adeno-associated virus produces coat protein but not the enzymes required for genome replication. Although they have minimal to no nucleotide sequence homology with the helper virus, satellite viruses can occasionally prevent the helper virus from replicating.

Approximately 250 to 450 bases in size, viroids are short, rod-like, single-stranded RNA molecules with a high degree of secondary structure. They all share a conserved central genomic region necessary for replication as a structural characteristic. With the enzymatic characteristics of a ribozyme, an autocatalytic, self-cleaving molecule, the RNA produces a hammerhead shape. This ribozyme function is employed in the replication process to cleave multimeric RNA complexes. Most are plant infections, which are notable for lacking any protein coding. Since their initial description by Theodore Diener, viroids have come to be thought of as important transitional entities in the evolution of RNA-based life forms from inorganic antecedents.

2.1.2.1. Hepatitis Delta Virus

One rare example of a flawed virus is the hepatitis delta virus (HDV), which needs co-infection with the hepatitis B virus to produce its outer HBsAg-containing envelope. In contrast to other viroids, the HDV genome is a single-stranded RNA molecule that has a branching or rod-like structure. It codes for a single protein, known as the delta antigen. Two variants of this nuclear phosphoprotein were produced by RNA editing: the bigger form (214 amino acids) is required for assembly and release from infected liver cells, whereas the shorter form (195 amino acids) is essential for HDV genome replication.

(i) The host cell RNA polymerase (ii) enzyme replicates the HDV genome through a rolling circle mechanism, resulting in a succession of concatemers of antisense RNA first and subsequently RNA of the genome sense. This is the only instance of a mammalian virus using a ribozyme for genome replication; these are instantly cleaved by the ribozyme domain in the RNA genome to produce new viral genomes (Flint et al., 2016).

2.1.3. Generation of Genetic Diversity

In any sample of a virus, there exists a dynamic population of closely related genome sequences known as a "quasi-species." These sequences replicate concurrently, yet at different rates, influenced by various selection pressures. The mechanisms driving this diversity encompass nucleotide substitutions, insertions, deletions, sequence duplications or deletions, genetic recombination, and reassortment for viruses with segmented genomes. The error rate during nucleotide copying is approximately 1 in 10^5 for DNA viruses and 1 in 10^4 for RNA viruses. This ever-changing quasi-species population continually adapts its composition based on recent selection pressures, providing the foundation for antigenic drift and adaptation to new hosts.

Desirable selection traits *in vivo* involve high replication rates, efficient transmission between hosts, immune evasion, and resistance to host antiviral defenses or therapies. However, not all these traits are equally vital for successful replication in cell cultures compared to *in vivo* conditions. Virus stocks cultivated *in vitro* are likely to harbor different variants from those produced *in vivo*, such as in experimental animal tissues. Historically, serial passaging of wild-type viruses through cell cultures or foreign host species has been a commonly employed strategy to attenuate virulence for the original host, yielding "modified live-virus" vaccines.

Adaptations may involve various viral genes; for instance, mutations in viral polymerases can impact replicative efficiency or resistance to antiviral medications.

Alterations in antigenic epitopes facilitate immune evasion, while changes in receptor ligands can affect receptor preferences, thereby influencing multiple aspects of pathogenesis.

2.1.4. Quantitative Virus Assays

Progress in understanding virus replication, pathogenesis, diagnosis, vaccine development, and various other fields has been greatly facilitated by the ability to quantify viral load in samples or preparations. Especially with the increased use of antiviral medications, particularly in diseases transmitted through blood like HIV, hepatitis B virus, and hepatitis C virus, assessing disease stage and treatment effectiveness now commonly involves measuring viral load in the blood (Freed,2015).

Two main approaches exist for quantifying virus titers: biological and physical. Physical assays gauge actual virus particles, employing methods such as electron microscopy, hemagglutination, and serological assays to detect viral antigens. On the other hand, biological assays measure viral functions like infectivity or reverse transcriptase activity. Results from different techniques applied to the same sample can sometimes yield significantly different outcomes, underscoring the importance of understanding such discrepancies.

The variation between virus quantity detected via physical assays, like particle counting via electron microscopy, and biological assays, such as plaque assays, is often termed the particle-to-plaque-forming unit (pfu) ratio. In most cases, the number of particles surpasses that determined by biological assays, often by ratios exceeding 1000:1. Several factors contribute to this disparity: some virions may be incapable of replication due to inherent defects like incomplete or faulty genomes, empty particles, defective capsid assembly, or lethal mutations; environmental factors may have inactivated virions; the choice of cells for virus isolation and growth might not replicate the natural intracellular environment of the target organ optimally; and the successful initiation of infection between a viable virus particle and a permissive cell may not always occur, necessitating an excess of particles to enhance the likelihood of infection initiation (Figure 2.2).

Cell monolayers are typically cultivated in Petri dishes until reaching confluence, at which point the virus inoculum is introduced in a small volume to allow adsorption onto the monolayer. Agarose overlay is then applied to restrain the diffusion of virus particles, and the culture is left to incubate, initiating virus replication from individual cells and spreading as more cells become engaged. The addition of the vital stain neutral red follows; viable cells take up the stain, rendering the monolayer red and leaving transparent plaques. The size and development rate of these plaques vary based on the virus's replication cycle and yield. To ascertain the virus titer of the initial inoculum, serial 10-fold dilutions are applied to duplicate monolayers. Plaque numbers are counted on monolayers where discrete plaques are distinguishable (typically 10 to 50 plaques per dish), and the original virus titer is calculated by multiplying this count by the dilution factor. By isolating the virus from a single plaque, a genetically homogeneous virus stock can be obtained through plaque purification. Abnormal plaque sizes or morphologies may indicate spontaneous mutants within the virus inoculum, which can then be separated for further analysis.

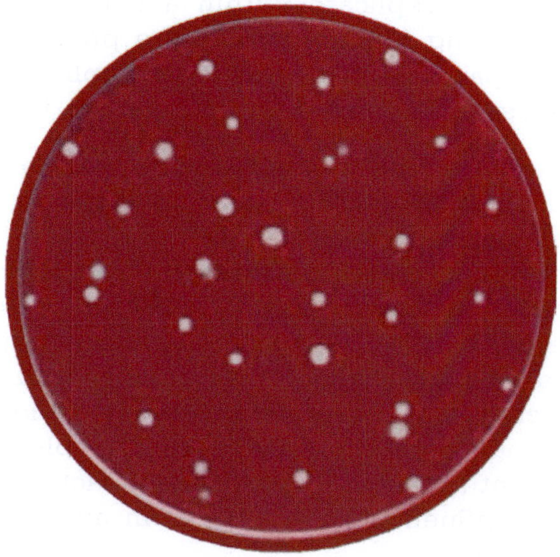

Figure 2.2. *Plaque assay for virus quantitation.*

Source: https://www.ncbi.nlm.nih.gov/pmc/articles/PMC7173592/bin/f04-10-9780123751560.jpg.

Arguably, no other technique has contributed as significantly to virology as the development of the plaque assay. Originally devised by Felix d'Herelle a century ago for studying bacteriophages, this method was later adapted to mammalian viruses in 1953 by Renato Dulbecco and Marguerite Vogt. The assay is elegantly straightforward: a virus sample undergoes serial 10-fold dilutions in a cell culture medium, which are then added to preformed monolayer cells and incubated under agar or carboxymethyl cellulose to prevent virus spread. After a period (typically 24 to 48 hours or longer, depending on the virus), necrotic cell plaques become visible under a light microscope. Fixation and staining, often with crystal violet, reveal distinct holes in the monolayer, each corresponding to a focus of necrotic cells initiated by a single viable virus particle. Serial dilution facilitates plaque counting, allowing the determination of the concentration (titer) in the original sample based on dilution, volume tested, and plaque count. Immunohistochemical staining with specific antisera offers an alternative method for visualizing plaques formed by non-cytopathic viruses.

The advent of real-time PCR assays enables the measurement of viral nucleic acid concentration in virtually any context. By comparing with copy number controls, the concentration of nucleic acid in the treated sample can be determined. However, this assay does not detect empty capsids (those lacking viral nucleic acid) and may not necessarily correlate with sample infectivity (Freed,2015).

2.1.5. Defective Interfering Viruses

During virus replication, non-infectious particles, akin to defective motor vehicles from a factory assembly line, are often generated. One notable example is defective interfering (DI) particles. These particles lack the full genetic information of the wild-type virus (deletion mutants) but can still replicate within a cell if co-infected by a wild-type virus, which provides the missing function through a process called complementation. Furthermore, these defective particles may impede the replication of the wild-type helper virus, possibly due to possessing a competitive advantage in replication.

High multiplicity infections, where the likelihood of co-infection with both full-length and defective viruses in the same cell is heightened, promote the production of DI particles. This phenomenon was first noted by Preben and Herdis von Magnus in 1944 with the influenza virus, where very high multiplicities resulted in poor virus replication and the generation of "interfering" virus stocks. A cyclical pattern emerges, with DI particles proliferating as long as there is sufficient replication of the wild-type helper virus. However, as DI particles begin to inhibit wild-type virus numbers, the overall virus population decreases. This reduction creates conditions for renewed rounds of wild-type virus replication, fostering a cycle of DI replication resurgence. Despite various hypotheses, the extent to which DI particles impact the pathogenesis of ongoing infection in a host, such as influencing recovery from an acute infection or promoting chronic infection, remains unclear.

2.2. THE VIRUS REPLICATION CYCLE

Most research on the replication of human viruses has used cultured mammalian cells either in suspension or as a monolayer adherent to a plastic surface. Classic research of this type defined the one-step growth curve, in which all cells in culture are infected simultaneously using a high multiplicity of infection, and the rise in the infectious virus over time is assessed by sequential sampling and viral titration (Figure 2.3). It is critical to recognize that, with simultaneous infection of the entire cell population, the events observed in the total culture can be utilized to examine events in a single cell.

The time required to produce maximum virus yield ranges from 6 hours (poliovirus) to more than 24 hours (adenovirus), whereas *E. coli* in broth culture only takes 15 to 20 minutes to generate.

Viruses released into the medium can be titrated separately from those that remain associated with cells. Shortly after infection, the inoculated virus "disappears"—infectious particles cannot be detected, even if the cells are disrupted. This eclipse period lasts until the first progeny virions are detected some hours later. As the infection progresses, non-enveloped viruses mature within the cell and can be detected as infectious, intracellular virions for several hours before being released through cell lysis. Many enveloped viruses, however, mature by budding from the plasma membrane of the host cell and are released into the medium. Virus release via cell lysis is sudden, whereas release through budding can continue over an extended period. The eclipse period generally ranges from 2 to 12 hours, varying among different virus families (Harrison, 2015).

Early studies using quantitative electron microscopy and infectious virion assays provided an overview of the replication cycle (attachment, penetration, maturation, and release) but lacked detailed insight into the processes involved. The introduction of biochemical methods for analyzing viral nucleic acids and proteins allowed for investigations viral genome expression and replication. Additionally, modern imaging methods enable tracking viral and cellular proteins throughout the viral replication cycle.

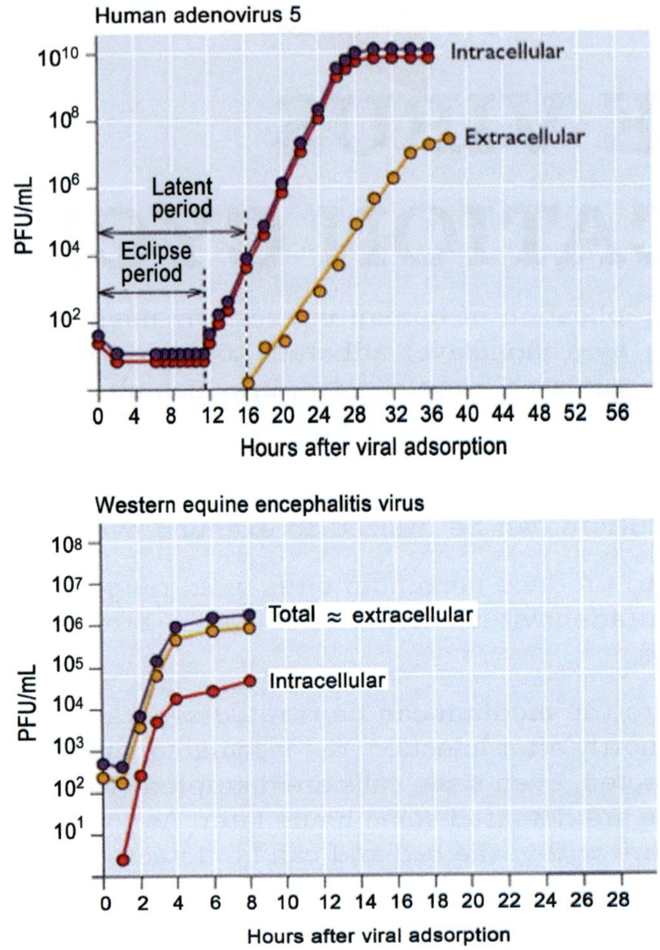

Figure 2.3. *One-step growth curve of an enveloped and a non-enveloped virus*

Source: https://www.ncbi.nlm.nih.gov/pmc/articles/PMC7173592/bin/f04-02-9780123751560.jpg.

2.2.1. Attachment

To initiate infection, virions must first bind to cells. This binding occurs between ligands on the virion's surface (viral attachment proteins) and receptors on the cell's plasma membrane (Table 2.2). The initial interaction between the virus and the host cell can be complex, and there is often a lack of correlation between attachment studies in cultured cells and intact hosts. For instance, several viral envelope glycoproteins of herpesviruses can act as attachment proteins, involving multiple cellular receptors in sequential order. This process may start with a loose attachment via one receptor, followed by an irreversible binding through a second receptor and ligand (Grieder et al., 1995). Utilizing more than one cellular ligand provides redundancy and may help herpesviruses invade both epithelial and neural tissues, supporting their latent and recrudescent infection cycles.

Table 2.2. *Examples of Cellular Receptors for Viruses of Medical Importance*

Virus	Family	Receptor
Human immunodeficiency virus	*Retroviridae*	CD4; coreceptors CCR5, CXCR4, CCR3
Vaccinia virus	*Poxviridae*	Various (heparan sulfate, chondroitin sulfate glycosaminoglycans, laminin)
Poliovirus	*Picornaviridae*	PVR (CD155)—Immunoglobulin (Ig) superfamily
Human rhinovirus 14	*Picornaviridae*	Intercellular adhesion molecule-1 (ICAM-1); Ig superfamily
Echovirus 1	*Picornaviridae*	α_2/β_1 integrin VLA-2
Hepatitis A virus	*Picornaviridae*	HAVCR1; member of TIM-1 family (transmembrane, Ig, mucin)
Adenovirus	*Adenoviridae*	CAR (Coxsackie and adenovirus receptor), member Ig superfamily; coreceptor α_v integrins
Herpes simplex virus 1	*Herpesviridae*	HveA (herpesvirus entry mediator A), heparan sulfate proteoglycan
Human cytomegalovirus	*Herpesviridae*	Heparan sulfate proteoglycan
Epstein–Barr virus	*Herpesviridae*	CD21, complement receptor 2 (CR2)
Adeno-associated virus 5	*Parvoviridae*	α(2,3)-linked sialic acid
Influenza A virus	*Orthomyxoviridae*	α(2,3)- or α(2,6)-αlinked sialic acid
Influenza C virus	*Orthomyxoviridae*	9-*O*-acetylsialic acid
Hendra virus	*Paramyxoviridae*	Ephrin-B2
Rotavirus	*Reoviridae*	Various integrins
Reovirus	*Reoviridae*	JAMs (junction adhesion molecules)
Dengue virus	*Flaviviridae*	Heparan sulfate proteoglycan
Rabies virus	*Rhabdoviridae*	Acetylcholine, NCAM (neural adhesion molecule)
Polyomaviruses	*Polyomaviridae*	Various gangliosides (glycosphingolipids with terminal sialic acid)

Source: https://www.ncbi.nlm.nih.gov/pmc/articles/PMC7173592/.

While there is a degree of specificity in how particular viruses recognize certain cellular receptors, different viruses (e.g., orthomyxoviruses and paramyxoviruses) can use the same receptor. Conversely, viruses within the same family or genus may use different receptors. Viruses have evolved to opportunistically utilize a variety of host cell surface proteins as receptors, typically employing host proteins essential for fundamental cellular functions, many of which are highly conserved throughout evolution. The receptor-ligand interaction between human cells and human immunodeficiency virus 1 (HIV-1) exemplifies the complexity of virus-host protein interactions. Initially, HIV-1 attaches to CD4 molecules on the surface of target cells, such as macrophages and T helper lymphocytes, via the viral gp120 envelope glycoprotein. This binding induces a conformational change that exposes a high-affinity chemokine receptor-binding site on gp120, which then binds to one of several chemokine receptors on the cell surface. This subsequent binding exposes a fusogenic domain of the viral transmembrane protein gp41. The direct contact between gp41's fusogenic domain and the cell membrane facilitates the fusion of the viral envelope with the cell's plasma membrane, allowing the viral nucleocapsid to enter the cell cytoplasm.

The cellular receptor for most orthomyxoviruses is the terminal sialic acid on an oligosaccharide side chain of a glycoprotein (or glycolipid) exposed on the cell surface. The viral ligand is located in a cleft at the distal tip of each monomer of the trimeric viral hemagglutinin glycoprotein. Viruses can adapt to new hosts by modifying the relevant attachment ligand. For example, influenza A viruses, which originate in aquatic birds, recognize sialic acid residues on the surface of cells in the respiratory tract. Avian influenza A viruses prefer sialic acid linked to the penultimate galactose moiety through an α2–3 bond, whereas human influenza A viruses prefer sialic acid linked via an α2–6 bond. This difference reflects the relative abundance of these linkages in the respective host species. These interactions are subtle, and a single amino acid change within a receptor site can significantly alter host range and epidemiology.

Proinflammatory reactions in the host can also influence receptor expression and indirectly modify interactions between the virus and potential host cells. For example, human adenovirus 5 recognizes both the Coxsackievirus and adenovirus receptor (CAR) and αvβ3/5 integrin to infect cells. Normally, neither receptor is exposed on the apical surfaces of cells. However, when macrophages respond to infection and secrete IL-8, the CAR and αvβ3/5 integrin receptors are moved from the tight junctions of polarized cells to the apical cell surface, making them available for virus attachment.

2.2.2. Penetration and Uncoating

The majority of mammalian cells are constantly engaged in receptor-mediated endocytosis, which allows macromolecules to be taken up via specialized receptors. This crucial cell function is used by many enveloped and non-enveloped viruses to commence penetration and uncoating (Figure 2.4). Virion attachment to receptors clustered around clathrin-coated pits is followed by endocytosis into clathrin-coated vesicles. Vesicles enter the cytoplasm and merge with endosomes once clathrin is removed. Virion proteins and surface structures are altered as a result of vesicle acidification. For example, the acidic pH of the endosome causes the hemagglutinin molecules of influenza viruses to alter shape (Hogle, 2002). This enables fusion to occur between the viral envelope and the endosomal membrane and results in the release of the viral nucleocapsid into the cytoplasm. Many other non-enveloped and enveloped viruses undergo comparable changes.

Endocytosis in animal cells can occur via four major mechanisms: clathrin-mediated endocytosis (CME), macropinocytosis, caveolae/lipid rafts, and phagocytosis (for larger particles, >0.75 μM). While pathways like IL-2, GEEC (GPI-enriched early endosomal compartment), Flotillin, and Arf6 (ADP-ribosylation factor 6) import specific cellular cargo, viruses are not currently known to use these routes. Examples of viruses include VSV (vesicular stomatitis virus), SFV (Semliki forest virus), HSV (herpes simplex virus), LCMV (lymphocytic choriomeningitis virus), SV40 (simian virus 40), mPy (mouse polyomavirus), and HPV (human papillomavirus).

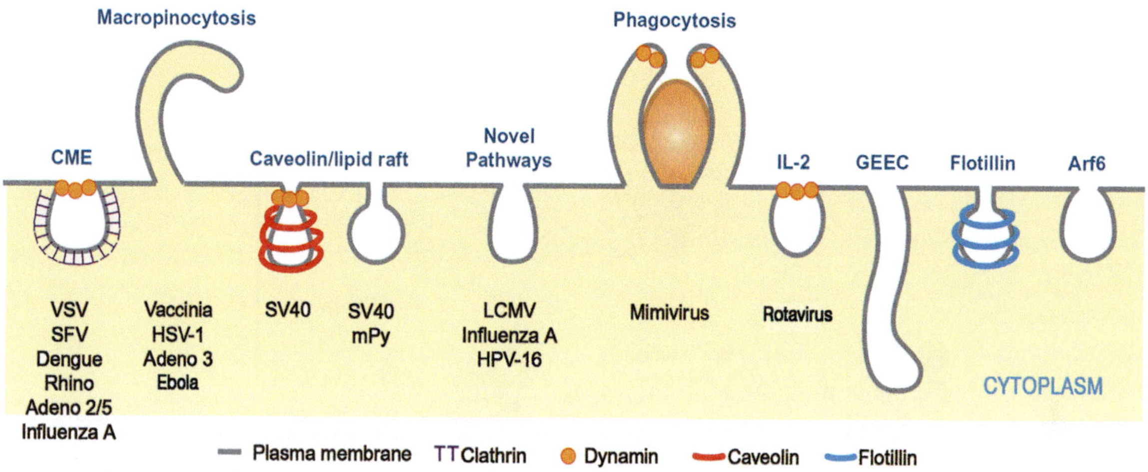

Figure 2.4. *Endocytic mechanisms for virus entry*

Source: https://www.ncbi.nlm.nih.gov/pmc/articles/PMC7173592/bin/f04-03-9780123751560.jpg.

Fusion at neutral pH is an alternative entry mechanism. The F (fusion) glycoprotein of paramyxoviruses enables the viral envelope to fuse directly with the cell's plasma membrane at pH 7, allowing the nucleocapsid to be released directly into the cytoplasm. Other important enveloped viruses, such as filoviruses, also fuse with the host cell plasma membrane, facilitating the entry of viral nucleic acid into the cytosol (Hogle, 2002).

A third entry mechanism, "direct entry," is used by some non-enveloped viruses, such as poliovirus and adenovirus. This process involves the virus binding to receptors at the endosomal membrane, which induces conformational changes in the viral capsid. These changes expose regions that interact with the endosomal membrane, creating channels that transport the viral genome across the plasma membrane (Figure 2.5).

The receptors induce conformational changes in the virion, converting the 160S infectious virion into a 135S particle. This transformation involves the loss of VP4 and the externalization of the N-terminus of VP1. The 135S particles are internalized by a mechanism that is independent of clathrin and caveolin but dependent on actin and tyrosine kinase. The viral genome is released only after internalization from an endocytic compartment located within 100 to 200 nm of the plasma membrane. Once the RNA genome is released, the empty capsid (80S) is transported away along microtubules.

For viral genes to be available for transcription, the virions must be at least partially uncoated. In enveloped RNA viruses that enter cells via fusion, the nucleocapsid is discharged directly into the cytoplasm, allowing transcription to commence from the viral nucleic acid while still associated with the nucleocapsid proteins. For non-enveloped icosahedral reoviruses, only specific capsid proteins are removed, and the viral genome expresses all its functions without being released from the virion core. However, for most other viruses, uncoating must proceed to completion for genome duplication to occur. In some cases, uncoating takes place in the nucleus, where the later stages of replication occur.

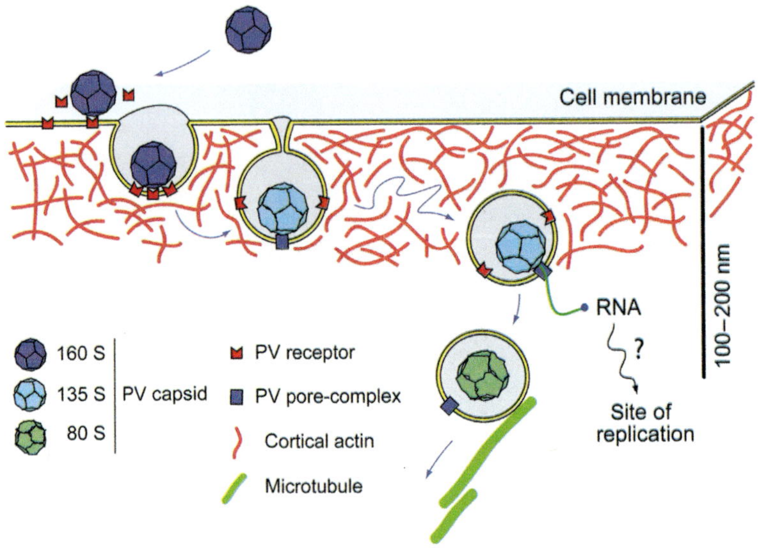

Figure 2.5. *Mechanism of poliovirus entry. The virus binds to specific receptors on the host cell plasma membrane (CD155 = poliovirus receptor PVR).*

Source: https://www.ncbi.nlm.nih.gov/pmc/articles/PMC7173592/bin/f04-04-9780123751560.jpg.

2.2.3. Replication of Viral Nucleic Acids

2.2.3.1. Replication of Viral DNA

Each DNA virus family has a different mechanism for DNA replication (Figure 2.6). These entail the synthesis of daughter strands via a replication fork (papillomaviruses, polyomaviruses, herpesviruses) or strand displacement (adenoviruses, parvoviruses, and poxviruses). Because cellular DNA polymerases cannot commence synthesis of a new DNA strand but can only prolong synthesis from a short (e.g., RNA) primer, one end of newly synthesized viral DNA molecules may retain a brief single-stranded segment. Various DNA viruses have developed different techniques to avoid this difficulty. Some viruses have a circular DNA genome, whereas others have a linear genome with complementary termini that function as DNA primers. Additionally, some viruses have a protein primer covalently bonded to the 5'-terminus of each DNA strand.

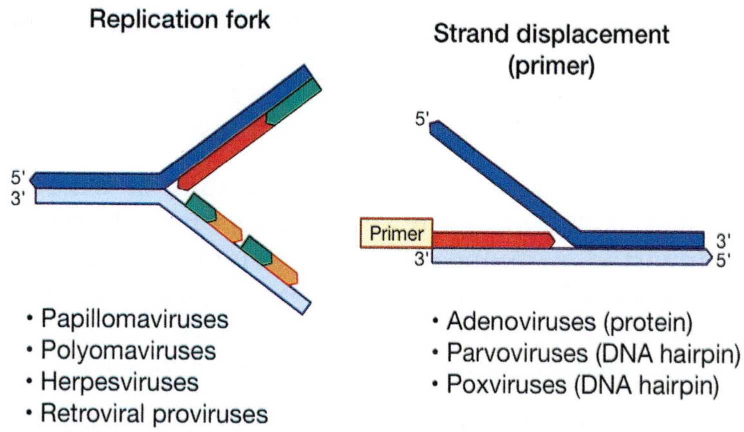

Figure 2.6. *Two mechanisms of synthesis of DNA virus genomes*

Source: https://www.ncbi.nlm.nih.gov/pmc/articles/PMC7173592/bin/f04-05-9780123751560.jpg

A replication fork is utilized by papillomaviruses, polyomaviruses, and herpesviruses for DNA replication. This process starts with an RNA primer, enabling continuous synthesis along the leading strand and discontinuous synthesis along the lagging strand, forming Okazaki fragments that are subsequently ligated. For circular DNA genomes, synthesis proceeds bidirectionally from the origin site (Knipe and Howley, 2013). Alternatively, adenoviruses, parvoviruses, and poxviruses replicate their DNA genomes through strand displacement, using protein primers (adenoviruses) or DNA hairpins (parvoviruses and poxviruses).

Several virus-encoded enzyme activities are typically necessary for viral DNA replication:

- **Helicase (with ATPase activity):** Unwinds the double helix.
- **Helix-Destabilizing Protein:** Keeps the separated strands apart until each is copied.
- **DNA Polymerase:** Copies each strand from the replication origin in a 5'- to 3'-direction.
- **RNase:** Degrades the RNA primer after its use.
- **DNA Ligase:** Joins the Okazaki fragments together.

Often, a single large enzyme complex performs multiple activities.

The genomes of papillomaviruses and polyomaviruses are structurally and functionally similar to cellular DNA, as they bind to cellular histones. These viral genomes use host DNA polymerase α-primase to synthesize the RNA primer necessary for genome replication. In polyomaviruses, an early viral antigen known as large T binds to the regulatory sequence, while in papillomaviruses, the E1 and E2 proteins initiate DNA replication. Replication of these circular double-stranded DNA molecules begins from a unique palindromic sequence, proceeding simultaneously with replication forks moving in both directions. As in mammalian DNA replication, both continuous and discontinuous

DNA synthesis occurs at the two growing forks. Discontinuous synthesis of the lagging strand involves the repeated creation of short oligoribonucleotide primers, which initiate short nascent DNA strands (Okazaki fragments). Each fragment is then covalently joined by DNA ligase to form one of the growing DNA strands.

The replication of adenoviruses is distinct from that of other DNA viruses. The adenovirus DNA genome is linear, with the 5'-end of each strand being identical (terminally repeated inverted sequences) and covalently linked to a protein, which serves as the primer for viral DNA synthesis. DNA replication proceeds continuously but asynchronously from both ends in a 5'- to 3'-direction, using a virus-encoded DNA polymerase. This process does not involve the synthesis of Okazaki fragments.

Herpesviruses encode many or all of the proteins necessary for DNA replication, including DNA polymerase, helicase, primase, single-stranded DNA binding protein, and a protein that recognizes the origin of replication. Poxviruses, which replicate entirely within the cytoplasm, are self-sufficient in their DNA replication machinery. Hepadnaviruses, similar to retroviruses, use positive-sense single-stranded RNA transcripts as intermediates for producing progeny DNA through reverse transcription. Single-stranded DNA parvoviruses utilize 3'-palindromic sequences to form a double-stranded hairpin structure, providing a primer for cellular DNA polymerase binding (Knipe and Howley, 2013).

2.2.3.2. Replication of Viral RNA

Transcription of RNA from an RNA template is a virus-specific event (Figure 2.7), requiring an RNA-dependent RNA polymerase, which is not found in uninfected cells. The replication of viral RNA begins with the synthesis of complementary RNA, which then serves as a template for the synthesis of further copies of viral RNA.

For viruses with negative-sense RNA genomes (e.g., orthomyxoviruses, paramyxoviruses, rhabdoviruses, filoviruses, bornaviruses, arenaviruses, and bunyaviruses), the complementary RNA is positive-sense. The RNA polymerase in these viruses performs the same function as the virion-associated transcriptase, which is used for the primary transcription of mRNAs. Most transcripts from negative-sense viral RNAs are subgenomic mRNA molecules; however, some full-length positive-sense strands must also be produced to serve as templates for the synthesis of full-length progeny viral RNA (replication). In some viruses, there is evidence that distinct RNA polymerases are used for transcription and replication, whereas in others, the same enzyme performs both functions.

For positive-sense RNA viruses (e.g., picornaviruses, caliciviruses, togaviruses, flaviviruses, coronaviruses, and arteriviruses), the complementary RNA is negative-sense, serving solely as a template for synthesizing more positive-sense RNA. Multiple viral RNA molecules can be transcribed simultaneously from a single complementary RNA template, with each RNA transcript being the product of a separately bound polymerase molecule. The resulting structure, known as the replicative intermediate, is partially double-stranded with single-stranded tails.

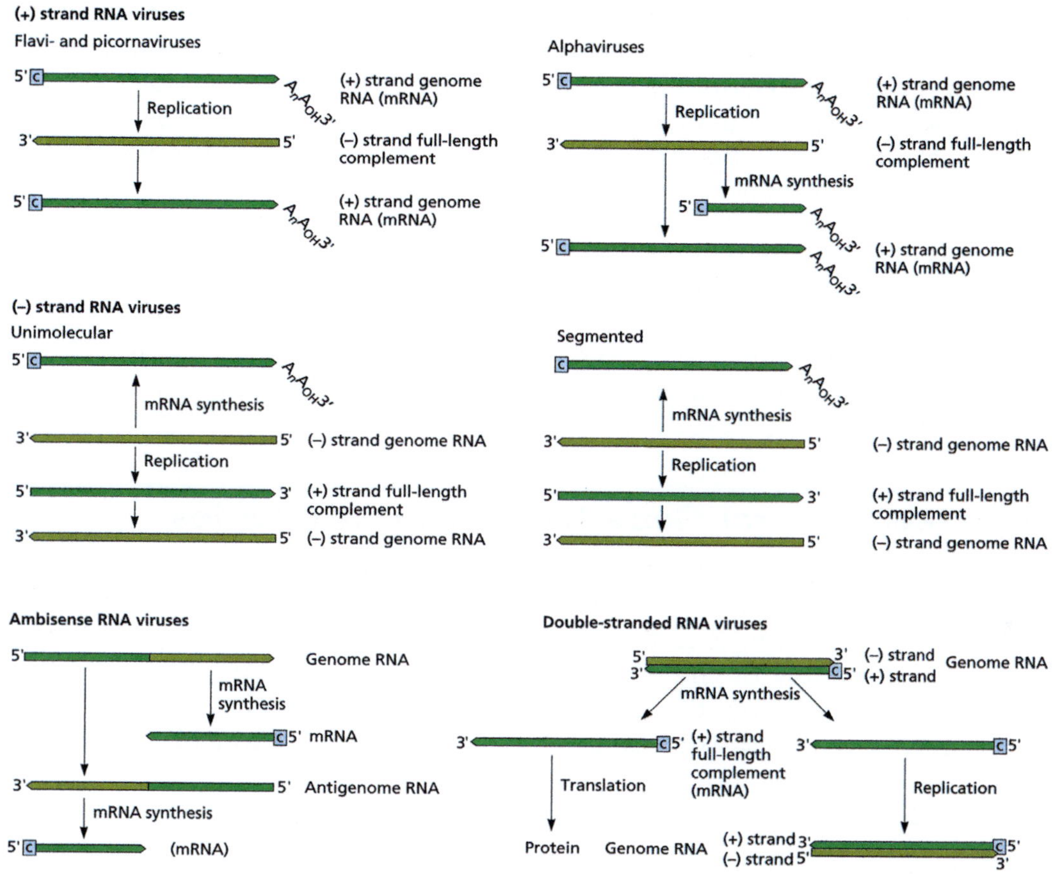

Figure 2.7. *Strategies for replication and mRNA synthesis of RNA virus genomes are shown for representative virus families. Picornaviral genomic RNA is linked to VPg at the 5′-end. The (+) genomic RNA of some flaviviruses does not contain poly(A). Only one RNA segment is shown for segmented (±) strand RNA viruses. Cap structures are shown as blue boxes containing "C."*

Source: https://www.ncbi.nlm.nih.gov/pmc/articles/PMC7173592/bin/f04-06-9780123751560.jpg.

Some RNA-dependent RNA polymerases can initiate RNA synthesis de novo, while others require a primer with a free 3'-OH group to which nucleotides can be added. For the replication of picornavirus and calicivirus RNA, initiation requires a bound protein as a primer, similar to adenovirus DNA replication, rather than an oligonucleotide. This small protein is covalently attached to the 5'-terminus of nascent positive and negative RNA strands and virion RNA, but not to molecules used as mRNA.

Little is known about what determines the fate of a given picornavirus positive-sense RNA molecule, whether it will:

- Be directed to a replication complex (a structure bound to smooth endoplasmic reticulum) to serve as a template for transcription by RNA-dependent RNA polymerase into negative-sense RNA.
- Be directed to a ribosome to serve as mRNA for translation into protein.
- Be directed to a procapsid, with which it associates to form a virion.

Retroviruses have a genome made up of positive-sense single-stranded RNA. Retroviruses, unlike other RNA viruses, replicate using DNA as an intermediary. The virion-associated reverse transcriptase creates a single-stranded DNA copy using a transfer RNA (tRNA) molecule as a primer, while also acting as a ribonuclease, removing the parental RNA molecule from the majority of the DNA: RNA hybrid, with the exception of several short RNA stretches known as poly-purine tracts.

The poly-purine tracts then serve as primers, duplicating the negative-sense single-stranded DNA strand to create a linear double-stranded DNA with an extra sequence known as the long terminal repeat (LTR) at either end. The viral integrase subsequently processes the double-stranded DNA and integrates it into the cellular chromosomal DNA. From this point on, transcription of viral RNA occurs from the integrated (proviral) DNA (Noack et al., 2014).

2.2.4. Overview of Viral Gene Expression Strategies

Understanding the different strategies viruses use to produce viral proteins is fundamental to comprehending virus replication (Figure 2.1). A summary of the properties of viral proteins is provided in Table 2.3. For many viruses, the production of viral proteins immediately after the entry of the viral genome is a crucial step. These early proteins play a key role in subverting the host's molecular machinery, facilitating the later production of new virus particles, and inhibiting the activation of the host's innate immune response, which would otherwise prevent virus replication and its spread to adjacent cells and tissues.

Table 2.3. *Properties of Viral Proteins*

Protein Category	Examples and Comments
Structural proteins of the virion	1. Capsid, and (for some viruses) core and/or envelope and matrix proteins 2. Virion-associated enzymes, especially polymerases (transcriptases)[a]
Non-structural proteins, mainly enzymes, required for transcription, replication of viral nucleic acid, and cleavage of proteins	DNA and RNA polymerases, helicases, proteases, etc. DNA viruses with large complex genomes, notably poxviruses, and herpesviruses, also encode numerous enzymes needed for nucleotide synthesis
Regulatory proteins that control the temporal sequence of expression of the viral genome	Site-specific DNA-binding proteins (transcription factors) bind to enhancer sequences in the viral genome or to another transcription factor. Some may act in trans (transactivators)
Proteins down-regulating expression of cellular genes	Usually by inhibiting transcription, sometimes translation

Oncogene products (oncoproteins) and inactivators of cellular tumor suppressor proteins	Upgrade expression of certain cellular genes; may lead to cell transformation and eventually to cancer. Observed with herpesviruses, adenoviruses, papovaviruses, and retroviruses
Proteins influencing viral virulence, host range, tissue tropism, etc.	Recorded so far mainly in the more complex DNA viruses (poxviruses, herpesviruses, adenoviruses) but may be more widespread
Virokines, which act on non-infected cells to modulate the progress of infection in the body as a whole	Mainly by subverting the immune response, e.g., by inhibiting cytokines, downregulating MHC expression, blocking the complement cascade, etc.

[a]RNA viruses of positive-sense and DNA viruses that replicate in the nucleus do not carry a transcriptase in the virion. Virions of some viruses, e.g., poxviruses, also contain many other enzymes.

Source: https://www.ncbi.nlm.nih.gov/pmc/articles/PMC7173592/.

Most families of DNA viruses conduct transcription and DNA replication within the cell nucleus, utilizing cellular RNA polymerase II and other cellular enzymes. Viruses like poxviruses and herpesviruses have identified a significant number of non-essential genes through gene deletion studies (more than 40%), albeit these genes likely play crucial roles for virus survival in natural environments where genetic economy is paramount.

In contrast, RNA viruses possess genetic information encoded in RNA, necessitating an understanding of their distinct replication and expression strategies, crucial for comprehending virus pathogenesis. RNA viruses, with diverse genome types (single-stranded or double-stranded, positive- or negative-sense, monopartite or segmented), have evolved unique mechanisms for mRNA production. Since eukaryotic cells lack RNA-dependent RNA polymerase, all RNA viruses encode their own RNA-dependent RNA polymerase (Noack et al., 2014).

Positive-sense single-stranded RNA viruses can directly bind to ribosomes for translation upon entry, without prior transcription. Conversely, other forms of viral RNA must first undergo transcription to generate mRNA, initiating the expression of the infecting viral genome. Thus, both negative-sense single-stranded RNA viruses and double-stranded RNA viruses must package an RNA-dependent RNA polymerase within the virion, derived from the previous infection cycle.

Cellular DNA-dependent RNA polymerase II is utilized for DNA viruses relying on the nucleus for replication. In contrast, double-stranded DNA viruses replicating in the cytoplasm encode their own DNA-dependent RNA polymerase.

Many viruses exhibit differential gene expression across various stages of their replication cycle. Early viral genes are initially transcribed into RNA, which can undergo several processing steps, including splicing. The proteins synthesized from this mRNA can be categorized into three main types, including those: that inhibit cellular nucleic acid and protein synthesis, regulate viral genome expression, and enzymes essential for viral nucleic acid replication. Once viral nucleic acid replication occurs, late viral genes are transcribed. Late viral proteins primarily consist of structural proteins crucial for assembling new virions. Some of these proteins undergo post-translational modifications and are often synthesized in large quantities. Structural proteins for coating new viral genomes are needed in multiple copies for each new nucleic acid molecule destined for encapsulation. Consequently, many viral strategies have evolved to ensure that newly replicated viral genomes can serve as templates for specific transcription and translation of structural proteins required for generating new virus particles.

The presence of overlapping reading frames, multiple splicing patterns of RNA transcripts, and post-translational cleavage of polyproteins complicates the straightforward assignment of one gene to one protein. Therefore, it is more accurate to refer to a transcription unit, defined as a genomic region starting from the transcription initiation site to the termination site (including all introns and exons in between), whose expression is regulated by a specific promoter. This approach acknowledges the complexity of viral gene expression and the diverse mechanisms viruses employ to ensure efficient replication and assembly of progeny virions.

Some viral proteins act as regulatory factors, influencing the transcription or translation of cellular genes and regulating early viral gene expression. Large DNA viruses also encode virokines, additional proteins that do not directly regulate the viral replication cycle but instead manipulate the host response to infection. These virokines can include viral protein analogs of cellular cytokines.

In 1978, Walter Fiers and colleagues provided the first comprehensive description of the genome of an animal virus, specifically the polyomavirus SV40. Their analysis of the circular double-stranded DNA molecule and its transcription yielded significant insights applicable to other double-stranded DNA viruses. Key findings included:

- Early and late genes are transcribed in opposite directions from different strands of the DNA.
- Some genes overlap, resulting in protein products with shared amino acid sequences.
- Certain regions of the viral DNA can be read in different reading frames (ORFs), producing distinct amino acid sequences from the same nucleotide sequence.
- Long stretches of viral DNA contain transcribed introns that are spliced out of the primary RNA transcript and not translated into protein.

Adenoviruses illustrate mechanisms that regulate viral genome expression at the transcriptional level. Adenovirus genomes are organized into several transcription units: "pre-early," "early," "intermediate," and "late," transcribed in a precise temporal sequence during the viral replication cycle. For instance, the early region E1A product stimulates transcription from other early regions like E1B. After viral DNA replication, there is a substantial increase in transcription from the major late promoter compared to early promoters such as E1B, accompanied by a decrease in E1A mRNA levels. Another control mechanism involves transcription termination. Early in infection, transcripts terminate at specific points, but transcription reads through these points later in infection, generating longer transcripts with different polyadenylation sites and proteins with diverse functions (Ortin and Martin-Benito, 2015). These examples underscore the efficiency of viral genomes in coding for complex functions using minimal nucleic acid sequences, highlighting the intricate regulatory strategies employed by viruses to ensure successful replication and interaction with host cells.

RNA viruses generally exhibit less complex transcriptional regulation compared to DNA viruses. In RNA viruses, particularly those with non-segmented negative-sense genomes, the temporal separation of early and late gene expression is not as distinctly defined as in DNA viruses. A notable example is found in picornaviruses, where a single polycistronic mRNA is translated into a polyprotein that is subsequently cleaved to yield equimolar amounts of all viral proteins. Togaviruses, on the other hand, produce excess structural proteins from a separate subgenomic RNA.

A distinct transcriptional mechanism is employed once the nucleocapsid enters the host cell cytoplasm among viruses with non-segmented negative-sense RNA genomes like paramyxoviruses. RNA polymerase initiates transcription from the 3' end of the genome. In paramyxoviruses, individual genes along the viral RNA are separated by consensus sequences that include termination and start signals, as well as short intergenic sequences rich in U residues. This arrangement facilitates the polymerase to generate a long poly(A) tail for each mRNA through a process known as reiterative copying or stuttering. Each mRNA is released sequentially as the polymerase transcribes each gene in turn. The efficiency of this process can vary, leading to differential synthesis of mRNAs and thus proteins, influenced by the order of genes along the viral genome.

Paramyxoviruses also employ RNA editing as a regulatory mechanism. For instance, the P gene of paramyxoviruses codes for two proteins, P and V, which share a common N-terminal sequence but have entirely different C-terminal sequences. This difference arises from a frameshift caused by the insertion of two non-template G residues into the RNA transcript during stuttering by the transcriptase.

These mechanisms illustrate the diverse strategies RNA viruses have evolved to regulate their gene expression, reflecting their efficient adaptation to replicate within host cells despite their simpler genomic structure compared to DNA viruses.

Eukaryotic cells lack the ability to translate polycistronic mRNA into multiple individual proteins because they typically cannot reinitiate translation along an RNA molecule. DNA viruses circumvent this limitation by utilizing cellular mechanisms to cleave and sometimes splice viral polycistronic RNA transcripts into monocistronic mRNA molecules.

In contrast, RNA viruses, which predominantly replicate in the cytoplasm, do not have access to the RNA processing and splicing enzymes found in the host cell nucleus. Instead, they have evolved diverse strategies to overcome the challenge of producing multiple gene products from a single, often compact genome. Orthomyxoviruses, for instance, have segmented RNA genomes where each gene is expressed and replicated as a separate molecule. Paramyxoviruses use a polycistronic genome but generate monocistronic RNA transcripts through transcription termination and reinitiation. Coronaviruses employ a nested set of overlapping RNA transcripts, each translated into a single gene product. Picornaviruses utilize a polycistronic genomic RNA that is translated into a polyprotein and subsequently cleaved into individual proteins. Certain viruses also employ mechanisms like internal ribosomal entry sites (IRES) to initiate translation internally at AUG start codons within polycistronic mRNA. This allows for the production of several gene products from a single transcript. Additionally, ribosomal frameshifting is utilized by retroviruses, where a ribosome slips along the RNA template, resulting in a shift in the reading frame and the synthesis of extended polyproteins (Ortin and Martin-Benito, 2015).

Furthermore, RNA splicing and RNA editing contribute to maximizing the coding potential of viruses with small genomes by exploiting overlapping reading frames. These mechanisms collectively illustrate the ingenious strategies employed by RNA viruses to efficiently express their genetic information despite the constraints imposed by their genomic size.

2.2.5. Regulatory Genes and Post-transcriptional Processing

The untranslated regions (UTRs) of viral genomes are of significant interest due to the presence of conserved sequences or motifs known as responsive elements. These elements play crucial roles in the regulation of transcription. For instance, each transcription unit within a viral genome contains an mRNA transcription initiation site, typically designated as nucleotide +1, located near its 3'-end. The promoter region lies upstream of this start site, within approximately one hundred nucleotides. This promoter is responsible for enhancing the transcription of specific genes or series of genes within the viral genome.

Adjacent to or distant from the start site, there may be extended sequences containing multiple, sometimes repeated, elements referred to as enhancers. These enhancers further amplify the transcription process. The activation of these regulatory regions occurs through binding proteins that are either viral or cellular in origin and capable of interacting with DNA. Often, several of these proteins bind to adjacent responsive elements, forming a complex structure that facilitates the attachment and activity of the viral RNA polymerase.

Viral genes encoding these regulatory proteins can function not only in cis (acting on the same molecule of nucleic acid from which they were transcribed) but also in trans. This means that these proteins have the capability to trans-activate genes located on physically separate nucleic acid molecules. This ability underscores their importance in regulating viral gene expression and replication, ensuring efficient viral propagation within host cells.

The regulatory mechanisms of human immunodeficiency virus 1 (HIV-1) highlight the sophisticated strategies employed by lentiviruses to control their gene expression. After the viral genome integrates into a chromosome of a resting T cell, it remains latent until external signals, such as T cell mitogens or cytokines, activate the cellular NF-κB family of DNA-binding proteins. NF-κB then binds to an enhancer sequence within the integrated provirus, initiating the transcription of HIV-1 regulatory genes.

One crucial regulatory gene is tat, which is present in all lentiviruses. Tat codes for a protein that specifically interacts with a responsive element called TAR (trans-activation response element), located within the provirus. Tat greatly enhances (trans-activates) the transcription of all viral genes, including its own. This process establishes a positive feedback loop that stimulates the production of HIV RNA transcripts. Furthermore, Tat enhances translation by interacting with TAR sequences in both viral mRNAs and the proviral DNA. Another key regulatory protein, Rev, functions differently by modulating gene expression through the regulation of nuclear export of viral mRNAs. Rev ensures that fully spliced and unspliced viral RNAs are efficiently exported from the nucleus to the cytoplasm for translation and packaging into new virions.

Despite the complexity of lentivirus transcriptional control mechanisms, HIV-1 and other lentiviruses possess only 9 genes in total, highlighting their ability to achieve regulatory sophistication with a relatively small genetic repertoire compared to DNA viruses, some of which may contain hundreds of genes. This efficient use of genetic material underscores the adaptability and evolutionarily refined strategies of HIV-1 in ensuring its replication and survival within host cells.

2.2.6. Post-transcriptional Processing

Primary RNA transcripts synthesized from eukaryotic DNA undergo several post-transcriptional modifications within the nucleus before being exported to the cytoplasm as mature mRNA. Firstly, a 7-methylguanosine cap (m7Gppp) is added to the 5'-end of the primary transcript. This cap structure facilitates stable complex formation with the host 40S ribosomal subunit, which is crucial for translation initiation. Secondly, a poly(A) tail composed of 50 to 200 adenylate residues is added to the 3'-end. This poly(A) tail serves as a recognition signal for mRNA processing and transport from the nucleus to the cytoplasm, and it also helps in mRNA stabilization against degradation in the cytoplasm. Thirdly, approximately 1% of adenylate residues throughout the RNA transcript undergo methylation at the sixth position (Albrecht et al., 1990).

Fourthly, introns are excised from the primary transcript through a process known as splicing, where exons are joined together. Splicing is a vital mechanism for regulating gene expression in nuclear DNA viruses. A single RNA transcript can contain multiple splicing sites and can be alternatively spliced in various ways to produce different mRNA species encoding distinct proteins. The preferred poly(A) site and the splicing pattern can undergo regulated changes during infection. For instance, in the case of HIV-1, the Rev protein facilitates the nuclear export of unspliced or singly spliced viral mRNA. Consequently, early in the replication cycle, cytoplasmic mRNA primarily consists of doubly spliced variants, whereas later, as Rev accumulates, there is a temporal shift towards different mRNA species.

An exceptional process known as cap-snatching deserves special mention. In the influenza virus, the transcriptase enzyme possesses endonuclease activity that hijacks 5'-methylated caps from newly synthesized cellular RNA transcripts in the nucleus. These stolen caps are then used as primers by the viral genome to initiate transcription.

The degradation rate of mRNA represents another potential regulatory level. Different mRNA species exhibit varying half-lives, and the half-life of a specific mRNA species may alter throughout the replication cycle.

Capped, polyadenylated, and processed monocistronic viral mRNAs adhere to ribosomes and undergo translation akin to cellular mRNAs. The sequence of events has been extensively analyzed in reovirus-infected cells. Each monocistronic mRNA molecule attaches via its capped 5'-end to the 40S ribosomal subunit, which subsequently moves along the mRNA until the initiation codon is reached. The 60S ribosomal subunit then joins the complex along with methionyl-transfer RNA and various initiation factors, facilitating the continuation of translation.

2.2.7. Post-translational Modifications

Most viral proteins undergo various forms of post-translational modification such as phosphorylation (for nucleic acid binding), fatty acid acylation (for membrane insertion), glycosylation, myristylation, or proteolytic cleavage. Newly generated viral proteins must also be carried to the various places in the cell where they are needed, for example, into the nucleus, which is the principal site of reproduction for viruses. The sorting signals that direct intracellular trafficking are only now being understood, as are the polypeptide chain-binding proteins (molecular chaperones) that govern the folding, translocation, and assembly of viral and cellular protein oligomers.

2.2.7.1. Post-Translational Cleavage

The polycistronic viral RNA is directly translated into a single polyprotein in positive-sense RNA such as picornaviruses and flaviviruses. This polyprotein is a large molecule that includes protease activity capable of cleaving itself at specific recognition sites into smaller functional proteins. The initial cleavage events occur while the polyprotein is still associated with the ribosome. Some intermediate products exist briefly, while others persist for a short period before being further cleaved by additional virus-encoded proteases into smaller proteins with distinct functions.

Post-translational cleavage is also observed in other RNA virus families like togaviruses and caliciviruses, where polyproteins spanning large portions of the genome undergo cleavage. Viruses often encode multiple types of proteases, typically categorized as trypsin-like (serine or cysteine proteases), pepsin-like (aspartyl proteases), or papain-like (thiol proteases).

Furthermore, cellular proteases located in organelles such as the Golgi complex or transport vesicles play critical roles in the maturation and assembly of many viruses. For instance, cleavage of viral glycoproteins like hemagglutinin in orthomyxoviruses or fusion glycoproteins in paramyxoviruses is indispensable for virion infectivity (Bose et al., 2015).

2.2.7.2. Glycosylation

Viruses often utilize cellular pathways intended for the synthesis of host cell secretory glycoproteins. This process begins with the amino terminus of viral envelope proteins, which contain a hydrophobic sequence of 15 to 30 amino acids known as a signal sequence. This sequence facilitates the binding of the nascent polypeptide chain to a receptor site on the cytoplasmic side of the rough endoplasmic reticulum, allowing the protein to pass through the lipid bilayer to the luminal side.

During translation in the rough endoplasmic reticulum, oligosaccharides are added in N-linkage to specific asparagine residues of the polypeptide chain. This occurs through the en bloc transfer of a mannose-rich core of preformed oligosaccharides. Subsequently, glucose residues are removed from these oligosaccharides by glycosidases in a process known as trimming.

After modification in the rough endoplasmic reticulum, the viral glycoprotein is transported to the Golgi complex. Here, further modifications of the core carbohydrate occur, involving the removal of additional mannose residues and the addition of N-acetyl-glucosamine, galactose, and terminal sugars such as sialic acid or fucose. These modifications result in the formation of complex oligosaccharides, which can be either N-linked (to asparagine) or, less commonly, O-linked (to serine or threonine). The completed oligosaccharide side chains consist of a mix of simple oligosaccharides, often referred to as high mannose oligosaccharides, and complex oligosaccharides.

Once fully modified, the glycoprotein is transported from the Golgi complex to the cellular membrane in coated vesicles. It is from here that the virus buds, acquiring its envelope. The precise composition of these oligosaccharides is determined not only by the amino acid sequence and tertiary structure of the viral proteins but also by the specific cellular glycosyl transferases that are active in the particular type of cell where the virus is replicating. These enzymes play a crucial role in modifying the glycoproteins to their final form before they are incorporated into newly formed viral particles.

2.2.7.3. Assembly and Release

All non-enveloped viruses that infect vertebrates exhibit an icosahedral structure. The structural proteins of these simple icosahedral viruses spontaneously associate to form units known as capsomers. These capsomers then self-assemble to construct capsids enclosing the viral nucleic acid. During this assembly process, the nascent capsid structure often undergoes conformational changes. Additionally, the formation of complete virions may involve the proteolytic cleavage of one or more types of capsid protein.

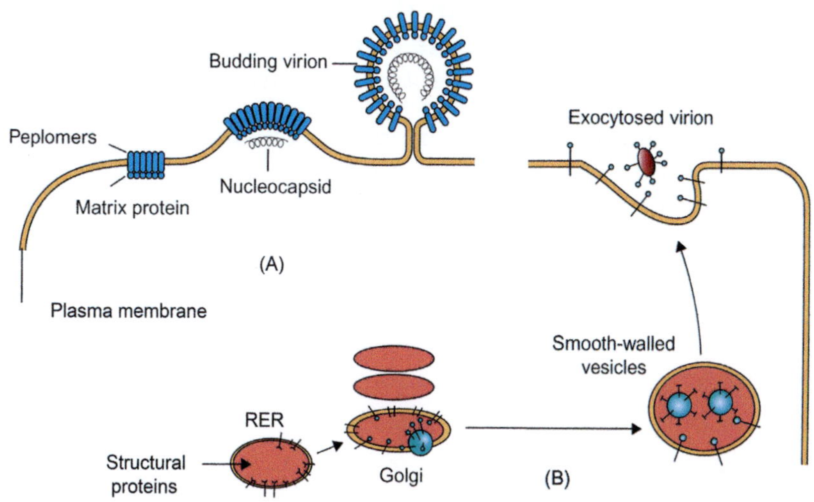

Figure 2.8. *Maturation of enveloped viruses. (A) Viruses with a matrix protein (and some viruses without a matrix protein) bud through a patch of the plasma membrane in which glycoprotein peplomers have accumulated over matrix protein molecules. (B) Most enveloped viruses that do not have a matrix protein bud into cytoplasmic vesicles (rough endoplasmic reticulum [RER] or Golgi), then pass through the cytoplasm in smooth-walled vesicles and are released by exocytosis.*

Source: https://www.ncbi.nlm.nih.gov/pmc/articles/PMC7173592/bin/f04-07-9780123751560.jpg.

The mechanism by which viral nucleic acid is packaged into pre-assembled empty procapsids has been well studied in adenoviruses. A specific protein binds to a nucleotide sequence located at one end of the viral DNA, known as the packaging sequenceÿÿ. This binding allows the viral DNA to enter the procapsid, where it binds to basic core proteins. Subsequently, some capsid proteins are cleaved to finalize the maturation of the virion (Albrecht et al., 1990).

Most non-enveloped viruses accumulate within the cytoplasm or the nucleus of infected cells. These viruses are typically released only when the host cell undergoes lysis (*see* Figure 2.8), which occurs when the cell membrane ruptures, releasing the mature virions into the surrounding environment. All helical nucleocapsid mammalian viruses mature by budding through cellular membranes to acquire an envelope, as do some icosahedral nucleocapsid viruses (such as herpesviruses, togaviruses, and retroviruses). The nuclear membrane, internal cytoplasmic membranes, or plasma membrane can all give rise to enveloped viruses, which are subsequently carried to the cell surface by vesicles once they have acquired an envelope inside the cell. Typically, viral glycoprotein(s) injected into the membrane's lipid bilayer cause membrane sections to budding. Cellular proteins are displaced laterally from that membrane patch, causing this to happen (Figure 2.9).

Viral glycoprotein-cleaved single molecules form oligomers that aggregate to form typical rod- or club-shaped peplomers. These oligomers comprise a short hydrophilic cytoplasmic domain that projects slightly into the cytoplasm, a hydrophobic trans-membrane anchor domain, and a hydrophilic domain that projects from the membrane's exterior. When it comes to icosahedral viruses, like togaviruses, the envelope is shaped around the nucleocapsid by the direct binding of each protein molecule to the cytoplasmic domain of the membrane glycoprotein oligomer. In the more common scenario of viruses with helical nucleocapsids, a matrix protein binds to the glycoprotein peplomer's cytoplasmic domain, and the nucleocapsid protein detects the matrix protein to start budding.

The plasma membrane remains intact during the release of enveloped virions, allowing for the shedding of numerous virus particles without causing significant cell damage. While most viruses that bud from the plasma membrane have slow cytopathic effects, some may not cause damage and can lead to persistent infections.

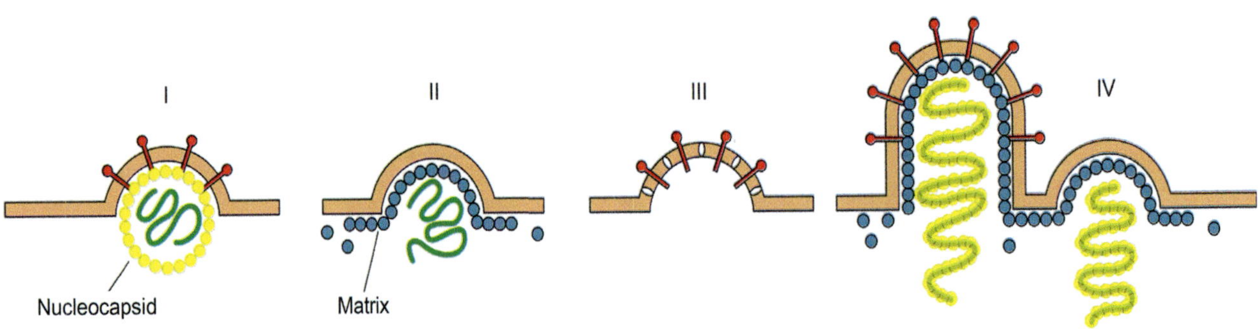

Figure 2.9. *Four distinct strategies used by different viruses for budding*

Source: https://www.ncbi.nlm.nih.gov/pmc/articles/PMC7173592/bin/f04-08-9780123751560.jpg.

- **In type I Budding,** exemplified by alphaviruses, both the envelope glycoproteins and the internal capsid are indispensable. Mutated or chimeric envelope proteins fail to support budding when they reach the plasma membrane. This indicates that the quaternary structures of the envelope heterodimers and capsid play crucial roles in driving this process.

- **Type II Budding,** such as Gag-dependent budding seen in many retroviruses, requires only the internal capsid or matrix proteins.
- **Type III Budding,** can be solely driven by the envelope proteins.
- **Type IV Budding,** is facilitated by viral matrix proteins but necessitates additional components. For instance, internal matrix proteins alone can induce budding in rhabdoviruses or orthomyxoviruses, albeit inefficiently or resulting in deformed particles, unless envelope glycoproteins or internal ribonucleoprotein are also present (Coen,1994).

Epithelial cells exhibit polarity, with an apical surface facing outward towards the external environment and a basolateral surface facing inward towards the interior of the body, separated by lateral cell-cell tight junctions. These surfaces possess distinct chemical and physiological characteristics. Viruses shed to the exterior (e.g., influenza viruses) typically bud from the apical plasma membrane. In contrast, viruses such as lentiviruses (e.g., human immunodeficiency virus) bud through the basolateral membrane (*see* Figure 2.10).

Flaviviruses, coronaviruses, arteriviruses, and bunyaviruses undergo maturation by budding through membranes of the Golgi complex or the rough endoplasmic reticulum. Vesicles containing the virus then migrate and fuse with the plasma membrane, leading to the release of virions by exocytosis (*see* Figure 2.9).

In contrast, herpesviruses acquire their envelope by budding through the inner lamella of the nuclear membrane. The enveloped virions subsequently pass directly from the space between the two lamellae of the nuclear membrane to the exterior of the cell via the cisternae of the endoplasmic reticulum. This unique process distinguishes the mechanism of viral maturation for herpesviruses from other enveloped viruses.

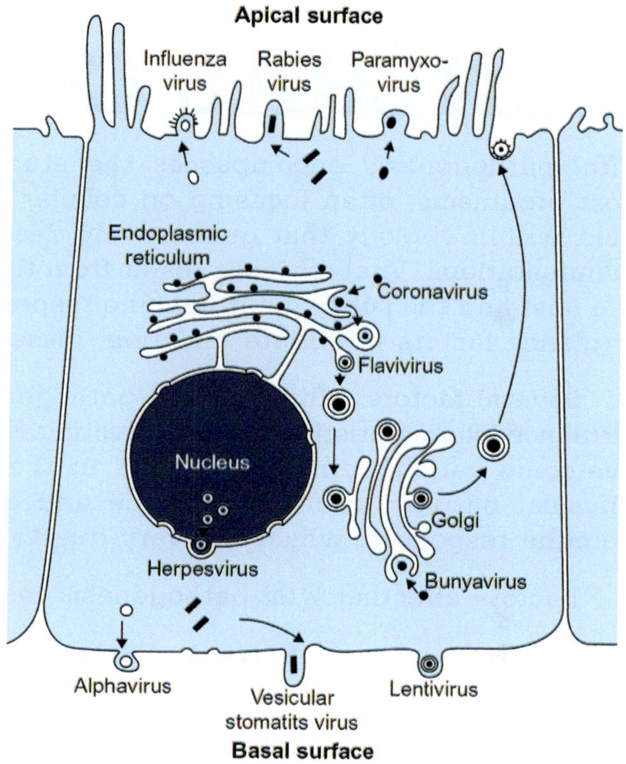

Figure 2.10. *Polarity of virus release from infected cells. Viruses that bud from apical surfaces can be shed in respiratory or genital secretions or intestinal contents. Viruses budding from basal surfaces are available for systemic spread via viremia or lymphatics. Some viruses, for example, flaviviruses, bunyaviruses, and coronaviruses, take a more circuitous route in exiting the cell. Viruses that do not bud are usually released by cell lysis.*

Source: https://www.ncbi.nlm.nih.gov/pmc/articles/PMC7173592/bin/f04-09-9780123751560.jpg.

2.3. VIRAL PATHOGENESIS: GENERAL CONCEPTS

Viral pathogenesis encompasses the study of how viruses cause diseases in their host organisms, often focusing on cellular and molecular processes. It is a specialized field within virology that qualitatively describes how initial infections lead to disease manifestations. Viral diseases result from the combined effects of viral replication within the host and the subsequent immune response against the virus. Viruses exploit specific virulence factors to initiate infection, disseminate throughout the body, and replicate.

Several factors influence viral pathogenesis (Coen,1994). Key among these are the virulence characteristics of the infecting virus. For a virus to cause disease, it must overcome various inhibitory factors in the host. These inhibitory factors can include physical barriers, such as distance and anatomical defenses, as well as the host's immune responses, which can vary due to genetic factors among individuals.

Factors affecting viral pathogenesis include:

- **Transmission, Entry, and Spread:** How the virus enters the host, spreads within the body, and reaches target organs.
- **Tropism:** The specific cells or tissues the virus targets for infection and replication.
- **Virus Virulence and Disease Mechanisms:** Characteristics of the virus that contribute to its ability to cause disease, such as its replication rate, ability to evade immune responses, and capacity to damage host cells.
- **Host Factors and Host Defense:** Genetic and immunological factors influence the host's susceptibility to viral infection and ability to mount an effective immune response.

Pathogenesis involves the sequential steps of virus implantation at the portal of entry in the body, replication at that site, and subsequent spread to target organs where disease may occur or where the virus is shed into the environment. Interestingly, many viral infections are subclinical, meaning they do not produce noticeable symptoms. Serological studies have revealed that a significant portion of the population may have antibodies against specific viruses despite never showing symptoms of the disease. These asymptomatic infections are epidemiologically important as they contribute significantly to the spread of viruses within populations and can confer immunity.

Several factors significantly influence the pathogenic mechanisms of viruses. An important early determinant is the accessibility of body tissues and organs to the virus. This accessibility is affected by physical barriers such as mucus and tissue barriers, the distance the virus must travel within the body, and natural defense mechanisms. For infection to occur in a particular organ, there must be susceptible cells capable of supporting virus replication. Cellular susceptibility depends on the presence of cell surface attachment sites (receptors) for the virus and an intracellular environment that supports virus replication and release. Even if the virus successfully initiates infection in a susceptible organ, the replication of sufficient virus to cause disease can be prevented by host defenses.

Other critical factors that influence whether infection and disease occur are the virulence characteristics of the infecting virus. Viruses must overcome inhibitory effects such as physical barriers, distance, host defenses, and variations in cellular susceptibility to infection to cause disease. These inhibitory effects are genetically controlled and can vary among individuals and different populations. Virulence characteristics enable viruses to initiate infection, spread within the body, and replicate to levels that can damage the target organ. These characteristics include the ability to replicate under specific conditions, such as inflammation or febrile responses, within migratory cells and in the presence of natural inhibitors and interferon (Fields,1983).

Viruses with extremely virulent strains may emerge within virus populations. Occasionally, these highly virulent strains become dominant due to unusual selective pressures. The specific viral proteins and genes responsible for these virulence functions are still being identified and studied. Fortunately for the survival of humans and animals, natural selective pressures typically favor the dominance of less virulent strains of viruses. These strains do not cause severe disease or death, allowing them to replicate and transmit efficiently without incapacitating the host. Inapparent or mild infections can result from the absence of one or more virulence factors. For instance, a virus lacking the ability to replicate at elevated temperatures may cause a milder disease than its fully virulent counterpart, which is halted during the febrile stage of infection. Live virus vaccines are often composed of strains deficient in virulence factors, capable of inducing immunity while causing only mild or asymptomatic infections.

The pathogenesis of diseases induced by viruses can be altered by spontaneous or induced mutations in viral genetic material. This phenomenon is particularly significant in viruses like HIV, where mutations can lead to the development of drug-resistant strains.

However, successful virus replication in a target organ does not always result in disease. Disease occurs only when the virus replicates sufficiently to directly damage essential cells, release toxic substances from infected tissues, affect cellular genes, or impair organ function indirectly due to the host immune response against viral antigens. Viruses as a group utilize various portals of entry, mechanisms of spread, target organs, and sites of excretion. This diversity is expected, given the vast numbers of viruses and their numerous variants.

2.3.1. Cellular Pathogenesis

Direct cell damage and death can occur due to the disruption of cellular macromolecular synthesis by infecting viruses. Viruses rely almost entirely on host cells to synthesize their genetic and structural components, making their replication a parasitic process that depletes the host cell's energy and macromolecular resources. This depletion severely impairs the host cell's function, often leading to cell death and contributing to disease.

At the cellular level, viral pathogenesis progresses through stages resulting in cellular dysfunction and disease. A fundamental aspect of viral pathogenesis is the competition between the synthetic requirements of the virus and those of the host cell. Viruses must hijack the cellular machinery to produce their own nucleic acids and proteins, necessitating mechanisms that alter the cell's normal functions to favor viral replication and progeny production. Viral genetic elements associated with virulence often play roles in creating conditions where the virus competes effectively for limited cellular macromolecular components and synthetic machinery, such as ribosomes (Fields,1983).

2.3.2. Tissue Tropism

Viruses often exhibit a preference for specific tissues, a characteristic known as tissue tropism or tissue specificity. This specificity is influenced by several factors including the selective susceptibility of cells, physical barriers within tissues, local environmental conditions such as temperature and pH, and the host's immune defenses. Numerous examples illustrate the diverse ways in which viruses target particular tissues.

For instance, polioviruses are known to infect and damage specific nerve cells that possess a higher density of surface receptors for polioviruses compared to resistant cells. Rhinoviruses replicate primarily in the upper respiratory tract, where conditions such as lower temperature and pH and higher oxygen levels are favorable for their multiplication. Enteroviruses thrive in the intestine, partly due to their resistance to digestive enzymes, bile, and acidic environments.

The receptors used by viruses to enter cells have been identified for several viruses. Rabies virus, for instance, binds to acetylcholine receptors present on neurons, while hepatitis B virus attaches to polymerized albumin receptors found on liver cells. Epstein-Barr virus targets B lymphocytes using complement CD21 receptors, and human immunodeficiency virus (HIV) utilizes CD4 molecules on T lymphocytes as its specific receptors.

Additionally, viral tropism is influenced by specific cellular transcription factors that interact with enhancer sequences within the viral genome. These enhancer sequences play roles in the pathogenesis of various viral infections. For example, the Moloney murine leukemia retrovirus has active enhancer sequences within its LTR regions that dictate tissue-specific viral activity. Similarly, JV papovavirus demonstrates enhancer activity specific to oligodendroglial cells, and hepatitis B virus exhibits enhanced activity in hepatocytes due to its specific enhancer sequences.

2.3.3. Sequence of Virus Spread in the Host

2.3.3.1. Implantation at Portal of Entry

All potential pathways—including the air, food, bites, and contaminated objects—allow viruses to enter the body. The same goes for all potential insertion sites, which include all body surfaces and interior locations accessible through mechanical penetration. Where the virus comes into direct touch with living cells (in the respiratory, alimentary, genital, and subcutaneous tracts), the frequency of implantation is highest. Certain viruses can implant themselves in the fetus at the time of fertilization through germ cells that are infected, later in the gestation period through the placenta, or even at birth.

Certain factors can affect the infection's course even in the pathogenesis's early stages (implantation). The site of the implantation and the virus's infectivity, dosage, and virulence, for instance, might influence whether the infection results in a moderate, severe, or fatal illness or is indiscernible (subclinical).

Even at the earliest stage of pathogenesis (implantation), certain variables may influence the final outcome of the infection. For example, the dose, infectivity, and virulence of virus implanted and the location of implantation may determine whether the infection will be inapparent (subclinical) or cause mild, severe, or lethal disease.

2.3.3.2. Local Replication and Local Spread

After successful implantation, viruses may replicate locally and spread within the infected tissue (Figure 2.11). Once inside the initially infected cell, the virus can propagate and subsequently disseminate to neighboring cells via two main routes: extracellular and intracellular spread.

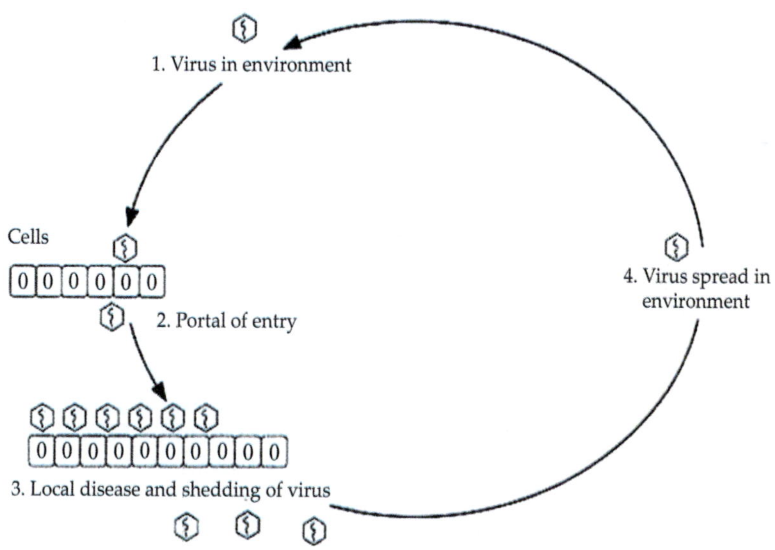

Figure 2.11. *Virus spread during localized infection. Numbers indicate sequence of events.*

Source: https://www.ncbi.nlm.nih.gov/books/NBK8149/bin/ch45f1.jpg

Extracellular spread involves the release of virus particles into the extracellular fluid from the infected cell. These virus particles can then infect adjacent cells susceptible to the virus. This mode of spread is typical for most viruses.

Intracellular spread occurs through different mechanisms. One method is through the fusion of infected cells with neighboring uninfected cells. This allows the virus to directly transfer from one cell to another, bypassing the extracellular environment. Another method involves cytoplasmic bridges or connections between infected and uninfected cells, facilitating the transfer of virus particles between them.

Certain viruses, such as herpesviruses, paramyxoviruses, and poxviruses, are capable of spreading through both intracellular and extracellular routes. This dual capability gives these viruses advantages, including protection from antibodies that cannot penetrate cell membranes effectively (Grieder et al., 1995). After establishing infection at the portal of entry, viruses may spread to cells beyond adjacent ones through various pathways within the local site. This can occur through liquid spaces such as lymphatics or by diffusion through surface fluids like the mucous layer in the respiratory tract. Additionally, infected migratory cells such as lymphocytes and macrophages can carry and spread the virus within local tissues.

Once infection is established, local virus multiplication continues, leading to localized virus shedding and the development of localized disease. In this context, the sites of initial implantation also serve as target organs and locations for shedding in many infections. A few examples of respiratory tract infections that fit this pattern include influenza, the common cold, and parainfluenza virus infections. Similarly, gastrointestinal tract infections caused by viruses such as rotaviruses and picornaviruses also fall into this category. Localized skin infections like warts, cowpox, and molluscum contagiosum demonstrate this localized spread as well.

In some cases, localized infections can spread over body surfaces to infect distant sites. For instance, in the absence of viremia (virus in the bloodstream), picornaviruses causing epidemic conjunctivitis can spread directly from the eye (the initial site of implantation) to the pharynx and intestine.

Other viruses have the ability to spread internally to distant target organs and sites of excretion, causing disseminated infections. Finally, some viruses can cause both localized and disseminated disease, as seen in infections like herpes simplex and measles (Figure 2.12).

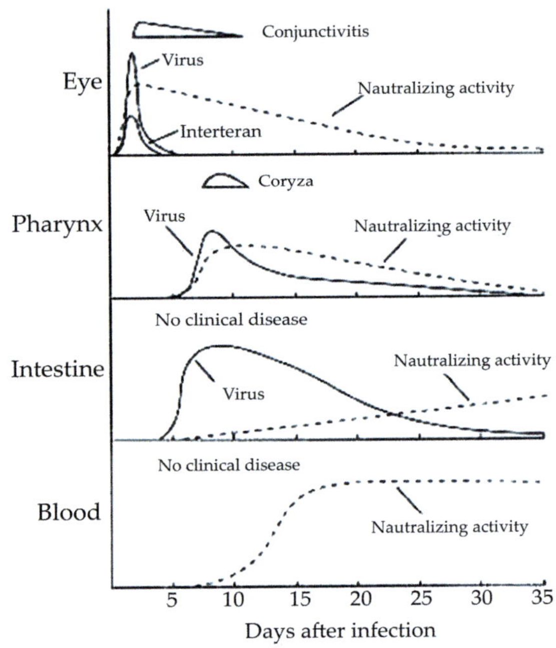

Figure 2.12. *Spread of picornavirus over body surfaces from eye to pharynx and intestine during natural infection. Local neutralizing antibody activity is shown.*

Source: https://www.ncbi.nlm.nih.gov/books/NBK8149/bin/ch45f2.jpg.

2.3.4. Dissemination from the Portal of Entry

2.3.4.1. Dissemination in the Bloodstream

The major routes of extensive virus dissemination across the body are the peripheral nerves and blood, which proliferating virus contacts connect to at the portal of entry. Circulation is the most typical method via which viruses move systemically (Figure 2.13). After an initial period of replication at the portal of entry (the alimentary and respiratory tracts), viruses like those that cause poliomyelitis, smallpox, and measles spread through the blood. Because the virus kills expendable and easily replaced cells, it frequently causes no significant symptoms or signs of illness there. Progeny of the virus spread to the lymphoid tissue via the afferent lymphatics, and then to the endothelial cells, particularly those of the lymphoreticular organs, in close proximity to the bloodstream, via the efferent lymphatics.

After the initial spread from the portal of entry, viruses may cause a brief primary viremia. Subsequently, the virus released directly into the bloodstream triggers a secondary viremia, which typically lasts several days. During secondary viremia, the virus comes into contact with the capillary system throughout all body tissues (Singh et al., 1995).

To enter a target organ from the bloodstream, viruses may replicate within endothelial cells lining the capillaries or within fixed macrophages positioned nearby. The virus can then exit on the side of the capillary facing the target organ. Alternatively, viruses may pass through small gaps in the capillary endothelium or penetrate the capillary wall by hitching a ride with an infected migrating leukocyte.

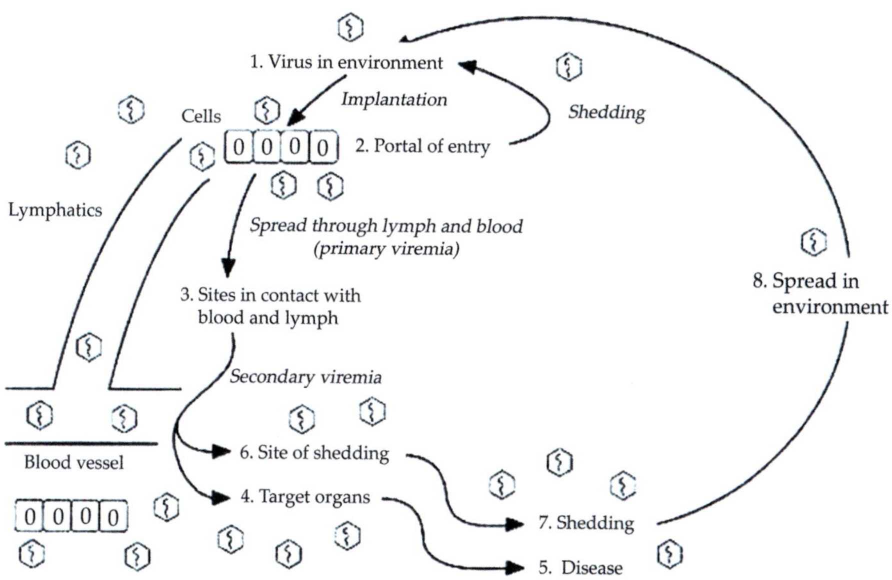

Figure 2.13. *Virus spread through bloodstream during a generalized infection. Numbers indicate sequence of events.*

Source: https://www.ncbi.nlm.nih.gov/books/NBK8149/bin/ch45f3.jpg.

Once inside the target organ or site of excretion, the virus replicates and spreads using mechanisms similar to those seen during local dissemination from the portal of entry. Disease manifests when the virus replicates in sufficient numbers to damage essential cells. For example, in poliomyelitis, the central nervous system serves as the target organ, while the alimentary tract serves as the portal of entry and the site of shedding. In other instances, such as with some viruses, the target organ and site of shedding are the same.

2.3.4.2. Dissemination in Nerves

Although less frequent than bloodstream dissemination, nerve dissemination is a mode of transmission for some significant illnesses (Figure 2.14). This pathway is present in infections with the poliomyelitis virus, herpes virus, and rabies virus. For instance, the rabies virus that enters the body through animal bites replicates subcutaneously and inside muscle tissue to infect nerve terminals. Evidence suggests that the virus spreads centrally in perineural cells and neurites (axons and dendrites), where it is protected from antibodies. The rabies virus travels down this nerve pathway to the central nervous system, the source of the disease. The salivary glands, the site of shedding, are reached by the centrifugal spread of the rabies virus through the nerves (Singh et al., 1995).

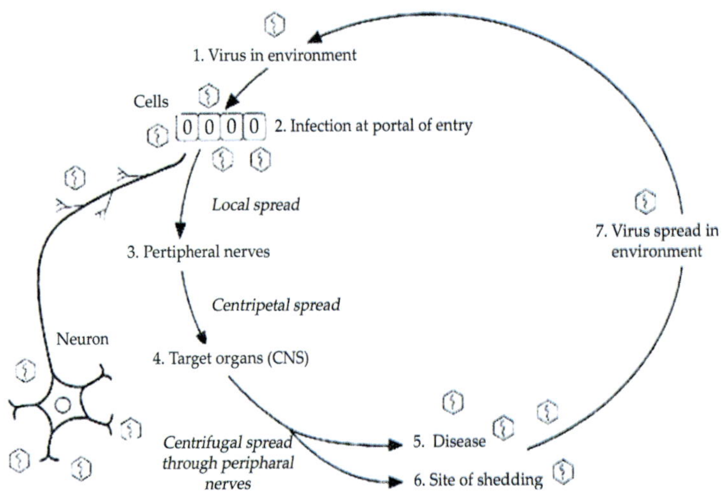

Figure 2.14. *Virus spreads through nerves during a generalized infection. Numbers indicate the sequence of events.*

Source: https://www.ncbi.nlm.nih.gov/books/NBK8149/bin/ch45f4.jpg.

2.3.5. Incubation Period

During most virus infections, individuals typically do not exhibit signs or symptoms of disease during the initial stages of virus dissemination. This period, known as the incubation period, begins from the moment of virus exposure and extends through the phase of virus spread, concluding when virus replication in the target organs induces observable disease. Occasionally, individuals may experience mild fever and malaise during viremia, but these symptoms are often fleeting and provide little diagnostic value. The duration of the incubation period varies depending on the virus and the route of infection. For infections where the virus travels a short distance to reach the target organ (e.g., infections where disease results from virus replication at the portal of entry), the incubation period tends to be brief, typically lasting 1 to 3 days. In contrast, infections that spread through the body stepwise have longer incubation periods. For instance, generalized infections caused by togaviruses may have a shorter-than-expected incubation period due to direct intravascular injection (e.g., through insect bites) of a rapidly multiplying virus (Strayer et al., 1990).

The mechanisms underlying the prolonged incubation periods seen in persistent infections (lasting months to years) are not well understood. In such infections, the infected cell often avoids lysis or experiences delayed lysis. Disease may also result from a delayed immune response to viral antigens (e.g., observed in arenaviruses infecting rodents), or from unknown mechanisms in slow viral infections where immune responses are absent (e.g., scrapie-kuru group). Additionally, mutations in host genetic material can lead to cellular transformation and cancer, contributing to disease progression in some persistent infections.

2.3.6. Multiplication in Target Organs

Virus replication in the target organ follows a pattern similar to replication at other body sites, with a few notable distinctions in systemic infections:

- **Timing of Target Organ Involvement:** In systemic infections, the virus typically reaches the target organ late in its stepwise progression through the body. This sequential movement means that by the time the virus infects the target organ, earlier infected sites may have already initiated local recovery mechanisms.
- **Origin of Clinical Disease:** Clinical disease originates from the target organ where virus replication reaches levels sufficient to cause observable symptoms and tissue damage.

At each stage of the virus's progression through the body, local recovery mechanisms are activated in response to infection. These mechanisms include local body defenses such as interferon production, inflammation, and immune responses specific to the infected tissues. Therefore, when the virus finally infects the target organ, the previously infected sites may have already undergone various stages of recovery.

The process is illustrated in Figure 2.12, depicting the sequential infection and recovery stages in different tissues during a spreading surface infection scenario. Although circulating interferons and immune responses play a role in terminating viremia (the presence of the virus in the bloodstream), they may act too late to prevent the seeding of the virus into the target organ and sites of shedding.

Nonetheless, these systemic defense mechanisms can diffuse to varying degrees into the target organs. This diffusion helps to slow down virus replication and limit disease severity by interfering with viral replication processes and supporting local immune responses within the target organ.

Depending on how well the virus evades or overwhelms host defenses, its replication in the target organ can reach levels that impair organ function, leading to disease or even death. Here are some additional details on the consequences:

- **Constitutional Symptoms:** Virus replication in the target organ can lead to constitutional symptoms such as fever and malaise. These symptoms arise from several factors:

- **Toxic Products:** Toxic substances produced during virus replication and cell necrosis can diffuse into the bloodstream and affect the entire body.
- **Inflammatory Mediators:** The release of inflammatory mediators like lymphokines can contribute to systemic symptoms.
- **Bronchospasm:** In respiratory infections, the release of leukotriene C4 can cause bronchospasm, exacerbating symptoms such as difficulty breathing.

- **Immune Reactions:** Viral antigens can trigger immune responses, leading to immune-mediated disease manifestations. This can include autoimmune-like reactions where the immune system mistakenly targets host tissues due to molecular mimicry or other mechanisms.
- **Secondary Infections:** Some viruses impair leukocyte function or cause immunosuppression, making the host more susceptible to secondary bacterial infections. This can complicate the clinical course of viral infections, especially in immunocompromised individuals.

Overall, the severity and nature of disease caused by viruses in the target organ depend on the interplay between viral replication dynamics, host immune responses, and the ability of the virus to manipulate host cell functions. Understanding these mechanisms is crucial for developing effective strategies for treating and preventing viral diseases (Strayer et al., 1990).

2.3.7. Shedding of Virus

Due to the vast diversity among viruses, they utilize nearly every conceivable site for shedding. The most common sites include:

- **Respiratory Tract:** Many viruses, such as influenza viruses and rhinoviruses, are shed through respiratory secretions, facilitating transmission through coughing, sneezing, and respiratory droplets.
- **Alimentary Tract:** Viruses causing gastroenteritis, such as rotaviruses and noroviruses, are shed through feces, leading to fecal-oral transmission.
- **Blood and Lymph:** Arboviruses, transmitted by biting insects like mosquitoes, are shed in blood, which is a mechanism for their transmission.
- **HIV:** Shedding occurs in blood and semen, which are significant routes of transmission for this virus.
- **Milk:** Some viruses, including certain retroviruses and cytomegalovirus, can be shed in breast milk, potentially transmitting infection from mother to child during breastfeeding.
- **Urinary Tract:** Cytomegalovirus, among others, is shed in urine, contributing to its spread.
- **Genital Tract:** Herpes simplex virus type 2 sheds in genital secretions, which is a major route of transmission through sexual contact or vertical transmission to the fetus during childbirth.
- **Saliva:** Rabies virus is primarily

shed in saliva, which is how it can be transmitted through bites.

- **Germ Cells:** Viruses that integrate into the DNA of host cells, such as some tumor viruses, can be shed through germ cells, potentially affecting offspring.

2.3.8. Congenital Infections

Infection of the fetus presents unique challenges and consequences compared to infections in adults or older children. Several factors influence the impact of virus infection on the fetus:

- **Immature Immune System:** The immune and interferon systems of very young fetuses are immature, which reduces their ability to mount an effective immune response against viruses.
- **Partial Placental Barrier:** While the placenta provides some protection, it is not fully effective in preventing viruses from crossing into the fetal circulation. Maternal immunity and interferon transmission to the fetus through the placenta may also be limited, further compromising fetal defenses.
- **Vulnerability of Fetal Organs:** During the first trimester of pregnancy, fetal organs are rapidly developing and highly vulnerable to disruption by virus replication. This vulnerability increases the risk of congenital anomalies if virus replication interferes with normal organ development.
- **Cellular Susceptibility and Hormonal Changes:** Fetal cells may be more susceptible to virus replication due to their undifferentiated state, particularly in early development. Hormonal changes during pregnancy can also affect susceptibility to viral infections.
- **Pathogenesis Leading to Congenital Anomalies:** To cause congenital anomalies, viruses must reach the fetus during maternal viremia. Once in the fetus, viruses can multiply and disrupt the development of organs, potentially leading to malformations.
- **Specific Viruses Involved:** Certain viruses are notorious for causing congenital anomalies. Cytomegalovirus (CMV) and rubella virus are among the most common viruses that can lead to developmental abnormalities when infecting the fetus.
- **Postnatal Shedding:** Infants born with congenital infections may shed the virus from sites where it persists in the body. This shedding can contribute to the transmission of the virus to others (Wold, 1994).

ROLE MODEL

DAVID BALTIMORE

He has contributed to immunology, virology, cancer research, biotechnology, and recombinant DNA research.

Biography

David Baltimore (born March 7, 1938, New York, U.S.) is an American virologist who shared the Nobel Prize for Physiology or Medicine in 1975 with Howard M. Temin and Renato Dulbecco. Working independently, Baltimore and Temin discovered reverse transcriptase, an enzyme that synthesizes DNA from RNA. Baltimore also conducted research that led to an understanding of the interaction between viruses and the genetic material of the cell. The research of all three men contributed to an understanding of the role of viruses in cancer development.

Baltimore and Temin both studied the process by which certain tumor-causing RNA viruses (those whose genetic material is composed of RNA) replicate after infecting a cell. They simultaneously demonstrated that these RNA viruses, now called retroviruses, contain the blueprint for an unusual enzyme—a polymerase called reverse transcriptase—that copies DNA from an RNA template. The newly formed viral DNA then integrates into the infected host cell, an event that can transform the infected cell into a cancer cell.

Baltimore received an undergraduate degree in chemistry from Swarthmore College, Pennsylvania (B.A., 1960), and went on to study animal virology at the Rockefeller Institute (now Rockefeller University) in New York City, where he obtained a doctorate in 1964, and at the Massachusetts Institute of Technology (MIT) in Boston. He worked with Dulbecco at the Salk Institute in La Jolla, California (1965–68), studying poliovirus replication mechanism.

Baltimore joined the faculty of MIT in 1968, accompanied by Alice Huang, a postdoctoral fellow who had worked on vesicular stomatitis virus (VSV) at the Salk Institute. In Boston, Baltimore and Huang, who had married, showed that VSV, an RNA virus, reproduced itself by means of an unusual enzyme (an RNA-dependent RNA polymerase) that copies RNA by a process not involving DNA.

Baltimore then turned his attention to two RNA tumor viruses—Rauscher murine leukemia virus and Rous sarcoma virus—to discover whether a similar enzyme was at work in their replication. It was through these experiments that he discovered reverse transcriptase. This discovery proved an exception to the "central dogma" of genetic theory, which states that the information encoded in genes always flows unidirectionally from DNA to RNA (and thence to proteins) and cannot be reversed. Since its discovery, reverse transcriptase has become an invaluable tool in recombinant DNA technology.

Baltimore became director of the Whitehead Institute for Biomedical Research in Cambridge, Massachusetts, in 1983 and in 1990, left to become president of Rockefeller University. In 1989, he figured prominently in a public dispute over a 1986 paper published in the Cell journal that he had coauthored while still at MIT. The co-author of the article, Thereza Imanishi-Kari, was accused of falsifying data published in the paper. Baltimore, who was not included in charges of misconduct, stood behind Imanishi-Kari, although he did retract the article. Because of his involvement in the case, however, he was asked to resign as president of Rockefeller University, and in 1994 he returned to MIT. In 1996 a U.S. government panel cleared Imanishi-Kari of the charges of scientific misconduct. The case was analyzed in The Baltimore Case (1998) by Daniel Kevles.

Baltimore was president of the California Institute of Technology from 1997 to 2006 when he was elected to a three-year term as president of the American Association for the Advancement of Science (AAAS). Among his other appointments, he served as a member of the Encyclopedia Britannica Editorial Board of Advisors.

CLASS ACTIVITY

VIRUS REPLICATION PATHWAY ANALYSIS AND PRESENTATION

Objective: To enhance graduate-level students' understanding of viral replication mechanisms by engaging them in a detailed analysis and presentation of the replication pathways of different types of viruses.

Assessment: Facilitate a class discussion to compare and contrast the replication mechanisms of different viruses. Highlight common themes and unique strategies used by different types of viruses. Discuss the implications of these mechanisms for antiviral drug development and therapeutic interventions.

SUMMARY

- Virus replication refers to the process by which viruses produce new infectious particles, or virions, within a host cell. This process begins when a virus attaches to a specific receptor on the surface of a susceptible host cell
- Viral pathogenesis, conversely, refers to the mechanisms by which viruses cause disease in the host organism. This process is multifaceted, involving direct damage to infected cells, immune responses triggered by the infection, and the broader physiological effects of the virus on the host
- The biological "purpose" of any replication cycle is the generation of new viral genomes and proteins in sufficient quantities to ensure the propagation of the viral genome (a clear example of the "selfish gene"); this requires that the extracellular viral genome is protected from enzymatic degradation and can be introduced into further target cells for further rounds of replication
- The development of *in vitro* cell culture systems was a watershed development in virology: not only did it become possible to dissect the intracellular events accompanying virus replication in a manner similar to that of the study of bacteriophages in bacterial cells, it also provided a means of quantifying the amount of infectious virus in samples and virus stocks
- Satellite viruses are subviral particles that contain a DNA or RNA genome coding for a capsid protein but that are dependent upon the presence of another virus for replication. The vast majority are found among plants, but a few impact medical virology and public health
- Hepatitis delta virus (HDV) is a unique example of a defective virus that requires co-infection with hepatitis B virus to provide its outer HBsAg-containing envelope. The HDV genome is a branched or rod-like, single-stranded RNA molecule similar to plant viroids in conformation. However, unlike other viroids, it codes for a single protein, the delta antigen
- Virus released into the medium can be titrated separately from a virus that remains cell-associated. Shortly after infection, the inoculated virus "disappears"—infectious particles cannot be detected, even if the cells are disrupted
- The cellular receptor for most orthomyxoviruses is the terminal sialic acid on an oligosaccharide side-chain of a glycoprotein (or glycolipid) exposed at the cell surface: the viral ligand is in a cleft at the distal tip of each monomer of the trimeric viral hemagglutinin glycoprotein

REVIEW QUESTIONS

1. What are satellite viruses and viroids?
2. How is genetic diversity generated in viruses?
3. What processes are involved in the penetration and uncoating of viruses?
4. How do viruses replicate their nucleic acids?

5. What are the strategies for viral gene expression?
6. How do regulatory genes and post-transcriptional processing affect viral replication?

REFERENCES

1. Albrecht, T., Boldogh, I., & Fons, M. (1990). Cell activation signals and the pathogenesis of human cytomegalovirus. *Intervirology, 31*, 68.
2. Bose, S., Jardetsky, T. S., & Lamb, R. A. (2015). Timing is everything: Fine-tuned molecular machines orchestrate paramyxovirus entry. *Virology, 479–480*, 518–530.
3. Coen, D. M. (1994). Acyclovir-resistant, pathogenic herpesviruses. *Trends in Microbiology, 2*, 481.
4. Fields, B. N. (1983). How do viruses cause different diseases? *Journal of the American Medical Association, 250*, 1754.
5. Flint, S. J., Enquist, L. W., Racaniello, V. R., Rall, G. F., & Skalka, A.-M. (2016). *Principles of virology* (4th ed.). ASM Press.
6. Freed, E. O. (2015). HIV-1 assembly, release and maturation. *Nature Reviews Microbiology, 13*, 484–496.
7. Grieder, F. B., Davis, N. L., & Aronson, J. F. (1995). Specific restrictions in the progression of Venezuelan equine encephalitis virus-induced disease resulting from single amino acid changes in glycoproteins. *Virology, 206*, 994.
8. Harrison, S. C. (2015). Viral membrane fusion. *Virology, 479–480*, 498–507.
9. Hogle, J. M. (2002). Poliovirus cell entry: Common themes in viral cell entry pathways. *Annual Review of Microbiology, 56*, 677–702.
10. Knipe, D. M., & Howley, P. M. (Eds.). (2013). *Fields virology* (6th ed.). Lippincott Williams and Wilkins.
11. Noack, J., Bernasconi, R., & Molinari, M. (2014). How viruses hijack the ERAD tuning machinery. *Journal of Virology, 88*, 10272–10275.
12. Ortin, J., & Martin-Benito, J. (2015). The RNA synthesis machinery of negative-stranded viruses. *Virology, 479–480*, 532–544.
13. Singh, I. P., Chopra, A. K., & Coppenhaver, D. H. (1995). Vertebrate brains contain a broadly active antiviral substance. *Antiviral Research, 27*, 375.
14. Strayer, D. S., Laybourne, K. A., & Heard, H. K. (1990). Determinants of the ability of malignant fibroma virus to induce immune dysfunction and tumor dissemination in vivo. *Microbial Pathogenesis, 9*, 173.
15. Wold, W. E., Hermiston, T. W., & Tollefson, A. E. (1994). Adenovirus proteins that subvert host defenses. *Trends in Microbiology, 2*, 437.

CHAPTER 3

Patterns of Some Viral Diseases of Humans

LEARNING OBJECTIVES

After studying this chapter, you will be able to:

- Understand the dynamics of human–virus interactions;
- Define patterns of specific viral diseases of humans;
- Explain why some viral infections targeting specific organ systems.

KEY TERMS FROM THIS CHAPTER

Chickenpox
Dengue
Encephalopathies
Hepatitis
Herpes
HIV/AIDS
Influenza
Measles
Mumps
Norovirus
Polio
Rabies
SARS
Zika

INTRODUCTION

Viral diseases represent a significant portion of the global health burden, affecting millions of people each year. These diseases are caused by viruses, which are microscopic infectious agents that require a host cell to replicate. The patterns of viral diseases in humans are influenced by a variety of factors, including the nature of the virus, the mode of transmission, the population's immunity, and environmental conditions. Understanding these patterns is crucial for developing effective public health strategies, including vaccination programs, antiviral treatments, and preventive measures (Adams and Carstens, 2012).

Epidemiologically, viral diseases can be categorized based on their transmission routes. Respiratory viruses, such as influenza and SARS-CoV-2, are spread primarily through airborne droplets and can lead to widespread outbreaks, especially in densely populated areas. These viruses often exhibit seasonal patterns, with a higher incidence in colder months when people are more likely to gather indoors. Enteric viruses, like norovirus and rotavirus, are transmitted through the fecal-oral route, often via contaminated food or water. These infections can cause significant morbidity, particularly in children and in areas with poor sanitation.

The patterns of some viral diseases are also shaped by the virus's ability to mutate and evolve. For instance, HIV and hepatitis B and C viruses have high mutation rates, allowing them to evade the host's immune system and establish chronic infections. These chronic viral infections can lead to long-term health complications, including liver cirrhosis and cancer. Additionally, certain viral diseases, such as those caused by the herpesviruses (e.g., herpes simplex virus and varicella-zoster virus), can remain latent in the body and reactivate later, leading to recurrent episodes of illness.

Emerging and re-emerging viral diseases pose ongoing challenges to public health. The emergence of new viruses, such as the novel coronaviruses and the re-emergence of diseases like dengue and Zika virus infections, highlights the dynamic nature of viral pathogens and the importance of surveillance and rapid response systems. Climate change, global travel, and urbanization are key factors driving the spread and emergence of these diseases. By understanding the epidemiological patterns of viral diseases, health authorities can better predict outbreaks, implement control measures, and ultimately reduce the impact of these infections on human populations.

3.1. THE DYNAMICS OF HUMAN–VIRUS INTERACTIONS

We've seen that a variety of factors, from the influence of particular genes controlling aspects of pathogenesis to more arbitrary factors that can be categorized as significant in the overall virulence of the disease, control the process of infection and the ensuing disease. From a purely medical perspective, it is important to understand the nature of the viral disease, or more precisely, the viral infection, its severity, the host's fate, and the fate of the virus that causes the disease. However, just as significant, the characteristics of the dynamic interaction between the virus and host offer crucial hints regarding the duration of a specific virus's association with its host. Moreover, the type of interaction offers hints about the host's evolutionary background (Taylor et al., 2012).

3.1.1. Particular Restrictions on the Character of Viral Disease and Mode of Persistence Are Placed by the Stable Association of Viruses with Their Natural Host.

Active rounds of infection somewhere in their reservoir are what keep viruses alive. It has been observed that when the population of susceptible individuals is depleted, a virus infection that results in immunity to reinfection will cause the virus to go extinct in a small population. Furthermore, even in a sizable reservoir, the host population will plummet and, in the worst circumstances, may go extinct if the virus infection directly or indirectly causes the deaths of a sufficient number of people. A virus that can only spread within that population will likewise go extinct. These restrictions result in several evolutionary constraints on the dynamics of the virus–host interaction, which can be precisely characterized by means of the mathematics of population biology and epidemiology. Large host populations are necessary for viruses whose infection results in an acute illness that is cleared up and immunity, as the disease cannot be too fatal or the virus cannot persist. Even so, effective dissemination of the infection is still required, even if the symptoms are minimal or nonexistent. This latter pattern of virus infection is typical of viruses that have animal reservoirs with large populations, like herds of ungulates or flocks of migratory birds. Urbanized human populations also fulfil this requirement because the agricultural/urban revolution began around 10,000 years ago, which led to sharp increases in our population.

On the other hand, there were not a lot of sedentary people in early human history. Instead, our ancestors were nomadic, small-scale groups of people arranged in families. In general, this organization resembles that of wolves and big cats, two predator species.

Only a virus that can create a persistent infection in its host with little to no reduction in the host's ability to survive and proliferate can persist in such a population, as no virus that is cleared and produces immunity can. Additionally, this persistence needs to permit the infectious virus to exist when it is most advantageous to infect fresh, vulnerable people (i.e., infants and adults from different groups that are occasionally encountered). Many viruses possess the genetic resources and replication mechanisms necessary to cause these kinds of infections, and, remarkably, genetic examination of these viruses reveals long-standing connections to humans.

The diagram in Figure 3.1 illustrates two primary strategies of virus replication. While some viruses are limited by a narrow host range, others, particularly those with RNA as their genetic material, have a wide host range and can easily infect different species. As a result, diseases caused by these viruses in new or additional host populations may have high mortality rates due to the lack of constraints on the virus's ability to spread.

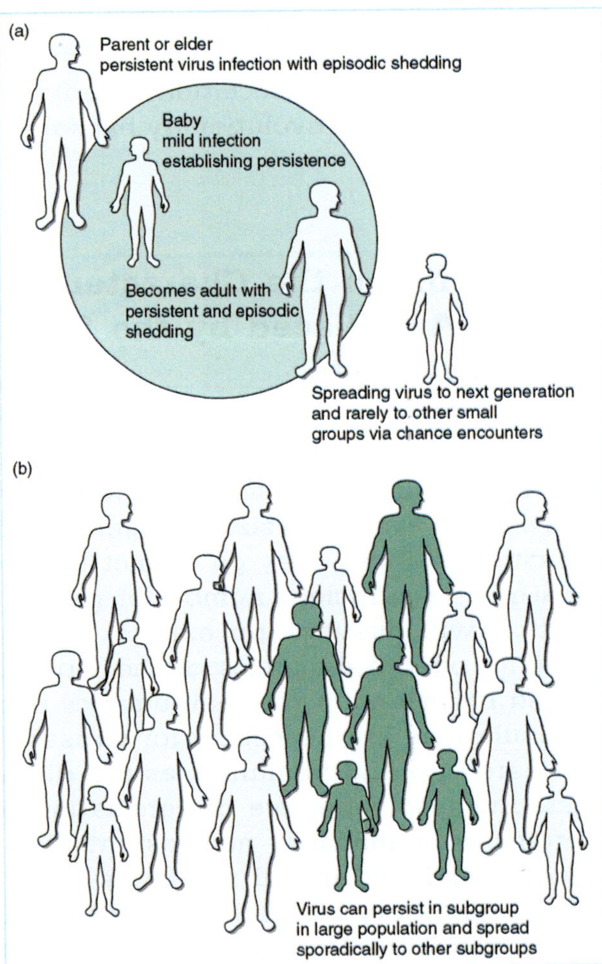

Figure 3.1. *Virus maintenance in small and large populations.*

Source: https://www.blackwellpublishing.com/content/BPL_Images/Content_store/Middle_Sized/Wagner9781405103466/036to048.pdf.

The above-described simple patterns of virus infection and persistence are further complicated by additional factors Notably, limits on mortality are eliminated if the incubation period—the interval between the initial infection and the onset of symptoms—is longer than the host's generation time. This is the situation with rabies, which essentially has a 100% death rate in infected carnivores, which are its natural host, but can reproduce even after infection due to its protracted incubation period. Given the length of human life, this may not seem to play a significant role in the persistence of high-mortality viral infections. However, a prolonged incubation period between the initial infection and final pathology is what accounts for the link between some persistent viral infections in humans and the appearance of tumors and immunopathologies much later in life (Adams and Carstens, 2012).

3.1.2. Classification of Human Disease-Causing Viruses According To Virus–Host Dynamics

Particularly for viruses that infect humans, we can create a basic classification of viruses by utilizing the characteristics of the virus-host interaction. These standards are helpful in arranging specific data about virally-induced diseases in relation to potential progressions and consequences.

Some of the basic criteria that can be incorporated into such a scheme include:

- Do viral disease symptoms take a short or long time to develop after first encountering the virus?
- Are the symptoms of initial infection relatively mild or severe?
- Can the infected individual be expected to recover completely, or partially?
- Does the virus stay associated with the victim following apparent recovery?
- If the association is lasting, is the virus maintained in an infectious form either sporadically or constantly?

3.1.2.1. Viruses with a Long History of Association with Human Populations are Typically Linked to Viral Diseases That Result in the Persistence of the Virus in the Host

Humans and other animals are susceptible to various viral infections that are caused by viruses that are either exclusive to the host population or also exist in another animal population that humans come into contact with. Two basic patterns have been observed in the course of human virus replication: acute infections with effective virus clearing after recovery from the acute infection, and persistent infections with incomplete virus clearing. One could argue that chronic infections are examples of stable viral-host relationships, meaning that the viruses are able to persist in host populations without causing significant harm. Acute infections, on the other hand, might be caused by viruses that have recently spread from another host (Adams and Carstens, 2012). These viruses may have originated from zoonotic infections in humans. Human rhinoviruses, among others, would not support this model because there are no known animal reservoirs, although influenza A viruses and hantaviruses are examples that do.

Table 3.1. *Some Viruses that Infect Humans*

Virus Group	Genome	Primary Reservoir	How Long Associated with Humans	Virus Type	Acute Disease	Primary Infection	Mortality Rate	Persistent/ Latent?	Reactivation	Chronic Disease/ Complications
Herpesvirus	DNA	Humans	Ancient	HSV-1	Facial lesion	Epidermis	Nil	Yes	Frequent at site	Encephalitis (rare)
				HSV-2	Genital lesion	Epidermis	Nil	Yes	Frequent at site	Encephalitis (rare)
				VZV	Chickenpox	Epidermis	Nil	Yes	Once	Shingles/ disseminated infection upon immune suppression
				HCMV	Monon-ucleosis	Hemato-poietic tissue	Nil	Yes	Asymptomatic/ infrequent?	Disseminated infection upon immune suppression/ retinitis
				EBV	Monon-ucleosis	Lymphoid tissue	Nil	Yes	Asymptomatic/ infrequent?	Lymphoma/ carcinoma
				HHV-6	Roseola	Lymphoid tissue	Nil	Yes	Asymptomatic/ infrequent?	?
				HHV-7	?	Lymphoid tissue	Nil	Yes	Asymptomatic/ infrequent?	?
				HHV-8	?	Lymphoid tissue	Nil	Yes	Asymptomatic/ infrequent?	Kaposi's sarcoma
Polyom-avirus	DNA	Humans	Ancient	JC	None	Kidney/ bladder	Nil	Yes	Infrequent shedding?	Encephalitis upon immune suppression
				BK	None	Kidney/ bladder	Nil	Yes	Infrequent shedding?	Kidney infection
Papillom-avirus	DNA	Humans	Ancient	>60 types	Warts	Epidermis	Nil	Yes	Constant shedding at the site	Cervical carcinoma (types 6, 11, 16, and 18)
Adenovirus	DNA	Humans	Ancient?	>12 types	Mild resp-iratory	Respiratory tract	Nil	Yes	Infrequent shedding?	?
Poxvirus	DNA	Humans	Recent	Variola	Smallpox	Epidermis	Moderate to high	No	N/A	None

Orthomy-xovirus	RNA	Birds, pigs	Sporadic/ current	Influenza A	Influenza A	Respiratory tract	Usually low/ rare high	No	N/A	None
		Humans	Sporadic/ current	Influenza B	Influenza A	Respiratory tract	Nil	No?	N/A	None
Coronavirus	RNA	Humans	?	Human coronavirus	Cold	Nasopharynx	Nil	No	N/A	None
		Palm civets	Current	SARS	Acute respiratory failure	Respiratory tract	Moderate to high	No	N/A	None
Picornavirus	RNA	Humans	Recent?	Poliovirus	None to mild digestive upset	GI tract	Nil	No	N/A	Paralysis
		Humans	Recent?	Hepatitis A virus	Hepatitis	Liver	Low	No?	N/A	Rare
		Humans	?	Rhino-viruses	Cold	Respiratory tract	Nil	No	N/A	None
Herpesvirus	RNA	Humans	?	Hepatitis E	Hepatitis; severe in newborns	Liver	Nil, except in pregnant women; moderate in newb-orns	No	N/A	Rare
Flavivirus	RNA	Birds	Sporadic/ current	West Nile	Encephalitis	Brain	Low	No	N/A	Neurological
		Primates	Sporadic/ current	Yellow fever	Encephalitis	Brain	Moderate	No	N/A	Neurological
		Humans	?	Hepatitis C virus	Hepatitis	Liver	Low	Occasional	Chronic hepatitis with no virus shedding	Liver failure/ carcinoma
Rhabdovirus	RNA	Carn-ivores	Sporadic/ current	Rabies	Encephalitis	Brain	100%	No	N/A	N/A
Togavirus	RNA	Horses	Sporadic/ current	Equine encephalitis virus	Encephalitis	Brain	Low	No	N/A	Neurological

		Humans?	?	Rubella virus (German measles)	Rash	Skin/ developing nerve tissue	Nil in adults; severe neur- ological symp- toms in deve- loping fetus	No	N/A	Fetal infection
Paramyx- ovirus	RNA	Carn- ivore?	Recent	Measles	Rash	Respiratory tract	Low/ moderate	Yes/no	Chronic virus antigen present/no infectious virus	SSPE
		Human	?	Mumps	Glandular inflam- mation	Respiratory tract	Nil	No	N/A	Infertility
		?	?	Respiratory syncytial virus	Mild respiratory in adults	Nasopharynx	Nil in adults	Yes	Virus shed from nasopharynx	Infections of newborns
Hepatitis delta	RNA	Humans	?	Hepatitis D	Hepatitis	Liver	Usually low but infection with hepatitis B leads to acute liver failure	Yes	Virus antigen present and infectious in blood	Liver failure
Bunyavirus	RNA	Mosq- uitoes	?	LaCrosse encephalitis virus	Encephalitis	CNS	Low; more severe for children	No	N/A	None
		Rodents	?	Hantavirus	Severe respiratory failure	Respiratory tract	Moderate especially in young adults	No	N/A	Respiratory failure
Hepadnavirus	RNA/ DNA	Humans	?	Hepatitis B	Hepatitis	Liver	Low	Yes	Virus antigen present and infectious in blood	Liver failure/ carcinoma

Retrovirus/ RNA tumor virus	RNA/ DNA	Humans	Ancient	Human T-cell leukemia	None	Lymphoid tissue	Nil?	Yes	Infectious virus shed?	Lymphoma/ paraparesis
Retrovirus/ lentivirus	RNA/ DNA	Chimpanzees/ mangabeys	1930 +/−20 years?	HIV-1/HIV-2	None/ influenza or mononucleosis-like symptoms	Lymphoid tissue	100%	Yes	Infectious virus shed	Immunodeficiency

Source: Edward K. Wagner, Martinez J. Hewlett, David C. Bloom, David Camerini; Basic Virology Third Edition; ISBN-13: 978-1-4051-4715-6.

Table 3.1 lists these virus groups along with some of the symptoms that humans who are infected with their experience.

Many significant human viruses are either asymptomatic or only cause mild symptoms during the initial infection phase, after which the virus and host form a stable relationship that lasts for the duration of the host's life. Some viruses are continuously shed during this relatively stable association, while persistent infections with other viruses result in the loss of any detectable virus. In the latter scenario, different kinds of stress on the host can cause viral recrudescence, which is the reemergence of an infectious virus accompanied by minimal or absent symptoms and the possibility of viral transmission to other people.

Human herpesviruses, human papilloma and related polyoma viruses, human retroviruses, and two entirely unrelated viruses—the hepatitis B and hepatitis D viruses—that cause chronic infection and severe liver damage are notable members of this virus group. While virus recrudescence ensures that there will be opportunities for virus spread at favorable times, virus persistence ensures that the virus can survive without infecting new people for the duration of the human host. (Adams and Carstens, 2012)

3.1.2.2. Viral Diseases Associated with Acute, Severe Infection are Suggestive of Zoonoses

A virus that exhibits acute, severe infection with variable or high mortality rates suggests that it either has a primary reservoir other than humans (a zoonosis) or it has recently "jumped" from such a reservoir to establish and maintain itself in the human population if a pattern of virus infection leading to persistent infectious virus correlates with the ancient association between humans and specific groups of viruses. One common example of such an illness is the influenza A virus. This virus primarily infects birds, but because of its wide host range, it can also create secondary reservoirs in pigs and humans. While not all virus strains introduced into the human population have high death rates, on rare occasions a high-mortality strain adapted for swine or birds can become established in the human population with serious, even catastrophic, consequences. A direct virus transmission from birds to humans, followed by a mutation that allowed the virus to spread from person to person, caused the massive global influenza epidemic (H1N1) in 1918–19, also known as a pandemic. The recently identified H5N1 avian flu, which originated in eastern China, raises serious concerns that it might have the same effect.

There are countless more instances of zoonotic viruses linked to serious human illness. The recent outbreak of severe acute respiratory syndrome (SARS) in Hong Kong was caused by a corona virus that originated in small carnivores and then spread to humans, most likely the palm civet. Other examples include the early 1990s outbreak of hantavirus respiratory disease in the southwest of the United States, which originated from a natural rodent reservoir. Other examples include the intermittent occurrence of West Nile virus, which is carried by avian populations, and different insect-borne encephalitis viruses.

The virus that causes AIDS, the human immunodeficiency virus (HIV), is the deadliest virus that has recently been introduced into the human population from a zoonosis. HIV comes in two varieties, with a 25% sequence difference between them. Both virus types have been demonstrated to be closely related to retroviruses that are carried asymptomatically in nonhuman primates in West Africa. It is evident that contact with these animals, most likely through their use as delicacies in regional cuisine, is how the viruses entered the human population. The virus is transmitted by intravenous drug use and sexual contact.

It can be deduced that other viruses that propagate as irregular epidemics with elevated fatality rates among humans have relatively recent pasts in human populations. One prominent example is smallpox, which is entirely adapted to humans but, because of its lack of persistence and ability to develop a robust and long-lasting immunity, depends on a large population for maintenance and spread. It can be assumed that related poxviruses were introduced into humans after agriculture and urbanization developed because they are found in numerous animals that sustain large populations. The same is true for viruses that spread through acute infection, such as the measles, polio, and others, which clear the immune system of any viruses that get in their way.

3.2. PATTERNS OF SPECIFIC VIRAL DISEASES OF HUMANS

Classifying human viruses based on their historical association with humans is beneficial from both an epidemiological and evolutionary standpoint. However, from a medical standpoint, it is equally useful to classify these viruses based on the type of disease they cause and its aftereffects. The subsequent sections include a classification of some significant viral diseases.

3.2.1. Acute Infections Followed by Virus Clearing

3.2.1.1. Colds and Respiratory Infections

Colds and respiratory infections are ubiquitous and affect individuals of all ages, races, and socioeconomic backgrounds. While they are often perceived as minor inconveniences, they can have significant implications for both individual well-being and public health. Understanding the causes and dynamics of these infections is crucial for developing effective prevention and treatment strategies. (Thapa et al., 2017)

3.2.1.1.1. Types of Respiratory Infections

Respiratory infections encompass a broad spectrum of illnesses, ranging from mild to severe. The common cold, caused primarily by rhinoviruses, adenoviruses, and coronaviruses, is the most prevalent respiratory infection globally. Influenza, caused by influenza viruses, poses a greater risk of complications and can lead to severe illness and death, particularly among vulnerable populations. Pneumonia, an infection of the lungs, can be caused by bacteria, viruses, or fungi and is a leading cause of morbidity and mortality worldwide.

3.2.1.1.2. Causes and Transmission

Respiratory infections are typically caused by viral or bacterial pathogens that are transmitted through respiratory secretions. The most common mode of transmission is through respiratory droplets generated when an infected individual coughs, sneezes, or talks. These droplets can be inhaled by nearby individuals or deposited on surfaces, where they can remain infectious for varying periods.

3.2.1.1.3. Symptoms

The symptoms of respiratory infections can vary depending on the causative agent and the individual's immune response. Common symptoms include cough, congestion, runny nose, sore throat, fever, fatigue, and body aches. In more severe cases, respiratory infections can lead to difficulty breathing, chest pain, and complications such as pneumonia.

3.2.1.1.4. Risk Factors

Certain factors increase the risk of contracting respiratory infections and experiencing severe complications. These include advanced age, young age, underlying medical conditions such as asthma or chronic obstructive pulmonary disease (COPD), weakened immune system, smoking, overcrowded living conditions, and exposure to environmental pollutants.

3.2.1.1.5. Prevention

Preventing respiratory infections involves both individual and community-level measures. Personal hygiene practices such as frequent hand washing, covering coughs and sneezes, and avoiding close contact with sick individuals can reduce the risk of transmission. Vaccination against influenza and certain types of pneumonia can also help prevent infection and reduce the severity of illness. In addition, strategies such as proper ventilation, maintaining physical distance, and wearing masks during outbreaks can limit the spread of respiratory pathogens in community settings.

3.2.1.1.6. Treatment

Treatment for respiratory infections depends on the causative agent and the severity of symptoms. For viral infections such as the common cold and influenza, supportive care aimed at relieving symptoms is typically recommended. This may include rest, hydration, over-the-counter medications for fever and pain, and in some cases, antiviral medications. Antibiotics are only effective against bacterial infections, so they are not routinely prescribed for viral respiratory infections unless complications such as bacterial pneumonia arise.

3.2.1.1.7. Impact on Public Health and Society

Respiratory infections have significant implications for public health, healthcare systems, and society as a whole. They contribute to absenteeism from work and school, decreased productivity, increased healthcare utilization, and economic burden. Severe outbreaks of respiratory infections, such as influenza pandemics or epidemics of emerging viruses like SARS-CoV-2, can strain healthcare resources, disrupt supply chains, and lead to social and economic upheaval.

3.2.1.2. Influenza

Influenza, commonly known as the flu, is a highly contagious viral infection that affects the respiratory system. It is caused by influenza viruses, primarily influenza A and influenza B, which belong to the Orthomyxoviridae family. Influenza viruses are known for their ability to mutate rapidly, leading to the emergence of new strains and seasonal outbreaks (Bravo et al., 2010).

3.2.1.2.1. Types of Influenza Viruses

There are three main types of influenza viruses: influenza A, influenza B, and influenza C. Influenza A viruses are further classified based on their surface proteins, hemagglutinin (H) and neuraminidase (N). The different combinations of these proteins result in various subtypes, such as H1N1 and H3N2, which circulate in humans and animals.

3.2.1.2.2. Transmission

Influenza viruses are primarily transmitted through respiratory droplets when an infected person coughs, sneezes, or talks. These droplets can be inhaled by nearby individuals, leading to infection. Influenza can also spread by touching contaminated surfaces and then touching the mouth, nose, or eyes.

3.2.1.2.3. Symptoms

The symptoms of influenza typically include fever, chills, cough, sore throat, runny or stuffy nose, muscle or body aches, fatigue, and headache. Some individuals, particularly children, may experience gastrointestinal symptoms such as nausea, vomiting, and diarrhea. Influenza symptoms usually develop one to four days after infection and can range from mild to severe.

3.2.1.2.4. Complications

While most people recover from influenza without complications, certain groups are at higher risk of developing severe illness. These include young children, older adults, pregnant women, individuals with underlying medical conditions such as asthma or diabetes, and those with weakened immune systems. Complications of influenza can include pneumonia, bronchitis, sinus infections, ear infections, and worsening of underlying health conditions.

3.2.1.2.5. Seasonal Influenza

Influenza is a seasonal illness, with peak activity typically occurring during the winter months in temperate regions. Seasonal influenza epidemics are caused by the circulation of different influenza virus strains, which vary from year to year. Vaccination is the most effective way to prevent seasonal influenza and its complications (Bravo et al., 2010).

3.2.1.2.6. Pandemic Influenza

In addition to seasonal influenza, there is the potential for influenza pandemics, which occur when a novel influenza virus emerges and spreads rapidly worldwide. Pandemics are caused by influenza viruses with significant antigenic changes that humans have little to no immunity against. Past influenza pandemics, such as the 1918 Spanish flu and the 2009 H1N1 pandemic, have had devastating effects on global health.

3.2.1.2.7. Prevention and Treatment

Preventing influenza involves vaccination, good hand hygiene, avoiding close contact with sick individuals, and staying home when experiencing flu-like symptoms. Antiviral medications such as oseltamivir (Tamiflu) and zanamivir (Relenza) can help reduce the severity and duration of influenza symptoms if taken early in the course of illness. However, antiviral drugs are most effective when used as soon as possible after symptoms appear.

3.2.1.3. Variola

Variola, commonly known as smallpox, is a highly contagious and often deadly viral infection caused by the variola virus. Smallpox is unique among infectious diseases in that it was the first human disease to be eradicated through a global vaccination campaign led by the World Health Organization (WHO). Here's an in-depth exploration of variola:

3.2.1.3.1. History

Smallpox has plagued humanity for centuries, with evidence of the disease dating back thousands of years. Smallpox outbreaks were responsible for significant morbidity and mortality throughout history, affecting populations worldwide. The disease is believed to have originated in Africa or Asia and spread through trade and conquest to other regions of the world.

3.2.1.3.2. Transmission

Smallpox is primarily spread through respiratory droplets from infected individuals or through direct contact with infectious bodily fluids or contaminated objects. The virus enters the body through the respiratory tract or breaks in the skin and mucous membranes, leading to infection and subsequent illness.

3.2.1.3.3. Symptoms

The symptoms of smallpox typically appear 10 to 14 days after exposure to the virus. The initial symptoms resemble those of a flu-like illness, including fever, fatigue, headache, and body aches. After a few days, a characteristic rash develops, consisting of raised bumps that progress from macules to papules, vesicles, pustules, and scabs over the course of about two weeks. Smallpox lesions are most concentrated on the face, arms, and legs but can appear anywhere on the body.

3.2.1.3.4. Variola Major and Variola Minor

There are two forms of smallpox: variola major and variola minor. Variola major is the more severe form of the disease, with a higher mortality rate and more extensive rash and systemic symptoms. Variola minor, also known as alastrim, is a milder form of smallpox with a lower mortality rate and less severe symptoms.

3.2.1.3.5. Complications

Smallpox can lead to various complications, including secondary bacterial infections of the skin, pneumonia, encephalitis (inflammation of the brain), and blindness if the eyes are affected by the rash. The mortality rate for variola major is estimated to be around 30%, while variola minor has a mortality rate of less than 1%.

3.2.1.3.6. Eradication Efforts

The eradication of smallpox is one of the greatest achievements in public health history. The WHO launched a global vaccination campaign in the 1960s, employing a strategy of mass vaccination, surveillance, and containment of outbreaks. Through coordinated efforts involving governments, healthcare workers, and international organizations, the last naturally occurring case of smallpox was reported in 1977 in Somalia. In 1980, the WHO declared smallpox eradicated, marking the first time a human disease had been eliminated through deliberate efforts.

3.2.1.3.7. Biological Warfare Concerns

Despite the eradication of smallpox, concerns remain about the potential use of the variola virus as a biological weapon. Smallpox is highly contagious, with a high mortality rate, making it a significant threat if deliberately released. Efforts to prevent the reintroduction of smallpox include the maintenance of vaccine stocks, surveillance for any sign of smallpox-like illness, and research into antiviral drugs and vaccines.

3.2.2. Infection of an "Accidental" Target Tissue Leading To Permanent Damage despite Efficient Clearing

As shown in Figure 3.2, certain viruses have the ability to target and harm an organ or organ system in a way that prevents the infected individual from fully recovering despite developing good immunity. An example of this is paralytic poliomyelitis, which is caused by the poliovirus. This small enteric virus with an RNA genome (a picornavirus) is typically spread through ingestion of fecal matter from an infected person. While most infections are localized to the small intestine and may result in mild symptoms like enteritis and diarrhea, some individuals may experience paralysis. The virus is encountered by the immune system through interactions with lymphatic tissue in the gut, leading to the development of an effective immune response that offers protection against future infections (Cotran et al., 2005).

Polio that is paralyzing can result from poliovirus infection. Only small intestine cells and motor neurons have the cellular surface protein CD155, which the virus must bind to in order to enter the cell. Rarely, a neurovirulent strain of an infection with a particular genotype that exhibits strong tropism for (propensity to infect) neurons can cause the virus to infect and kill motor neurons. In this case, paralysis results from the neurons' destruction.

It should be highlighted that paralysis brought on by a neuronal infection is a "dead end" and does not help the virus spread among people. It may be ironic to say that the paralytic side effects of poliovirus infections have had negative selection advantages because there would not have been a need to create a vaccine to prevent poliovirus infection if such a dramatic outcome had not occurred!

Rubella represents a variation on the theme of neuronal targets accidentally destroyed by an otherwise relatively benign course of acute viral infection. This illness, also known as German measles, is brought on by an RNA virus and manifests as a mild infection that frequently has no symptoms other than a slight rash. An immunocompetent person may have a mild infection, but the virus has a strong affinity for neural tissue differentiation and replication. Consequently, there is a very high chance that a child born to a rubella-infected mother in the first trimester of pregnancy will suffer from severe neurological damage. During a localized rubella epidemic, immunizing women who intend to become pregnant is an effective way to prevent such harm (Cotran et al., 2005).

3.2.3. Persistent Viral Infections

The absence of significant consequences for most infected individuals does not rule out the possibility of debilitating or fatal consequences, even though persistent viral infections frequently point to a lengthy history of coevolution between virus and host. In instances where the afflicted person's immune system is weak or nonexistent, this is particularly true (Figure 3:2), providing some instances of chronic infections and the complications that may result from them.

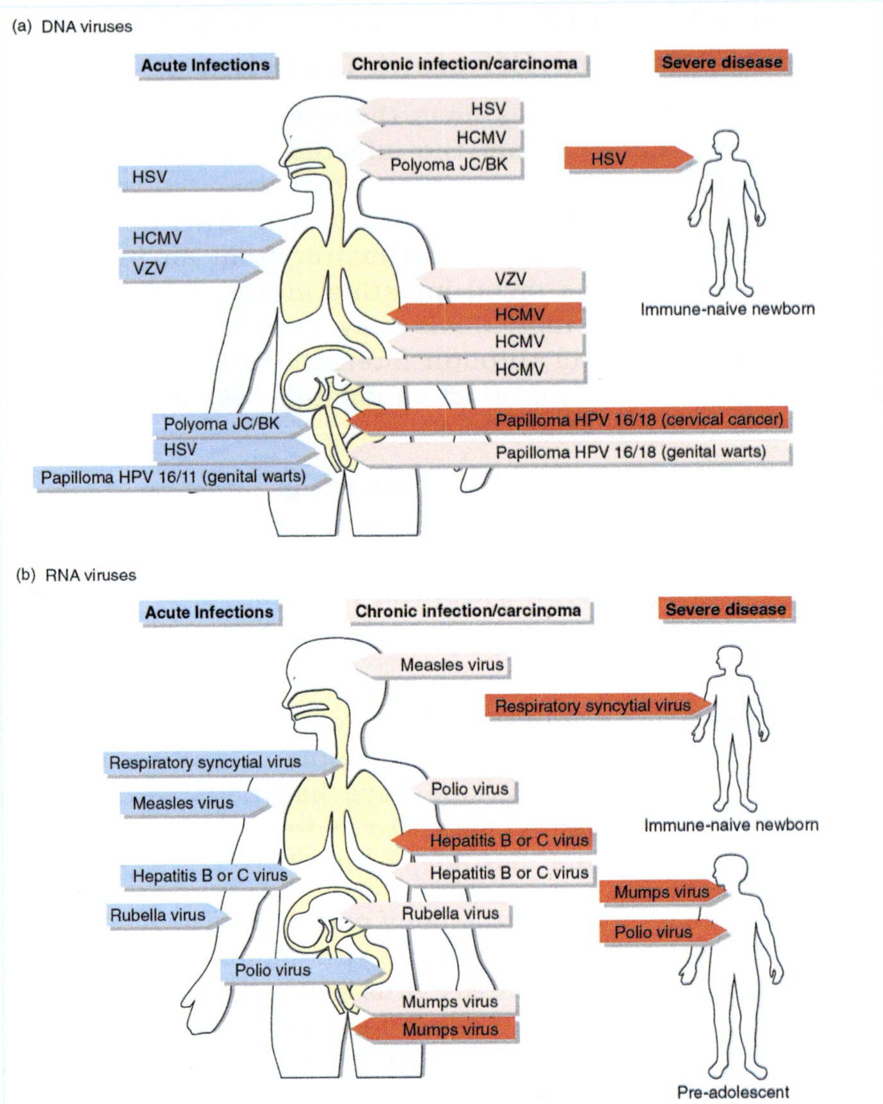

Figure 3.2. *Examples of virus infection of specific organs or organ systems. (a) DNA genome viruses. (b) RNA genome viruses. Blue labels indicate acute infections, while pink labels indicate infections that result in either chronic disease states or carcinomas. Red labels indicate acute infections that can result in severe disease.*

Source: Edward K. Wagner, Martinez J. Hewlett, David C. Bloom, David Camerini; Basic Virology Third Edition; ISBN-13: 978-1-4051-4715-6.

3.2.3.1. Papilloma and Polyomavirus Infections

Viral infections are a significant public health concern worldwide, causing a wide array of diseases ranging from mild to severe. Among the many viruses that infect humans, papillomaviruses (PVs) and polyomaviruses (PyVs) stand out due to their association with various clinical conditions, including benign and malignant tumors. In this chapter, we delve into the characteristics, epidemiology, pathogenesis, clinical manifestations, diagnosis, treatment, and prevention strategies for papillomavirus and polyomavirus infections (Bravo et al., 2010).

3.2.3.2. Papillomaviruses (PVs)

3.2.3.2.1. Characteristics

- PVs belong to the Papillomaviridae family, comprising a large group of DNA viruses with a circular double-stranded genome.
- They exhibit a high degree of species specificity and tissue tropism, infecting epithelial tissues in a variety of organisms, including humans.
- PVs are classified based on their genomic sequences and association with specific clinical conditions.

3.2.3.2.2. Epidemiology

- Human papillomavirus (HPV) is the most well-known member of the PV family and is highly prevalent globally.
- HPV infections are primarily transmitted through sexual contact, with certain high-risk types being implicated in the development of cervical, anal, and oropharyngeal cancers.
- The prevalence of HPV varies geographically and among different populations, influenced by factors such as sexual behavior, vaccination rates, and socioeconomic status (Bravo et al., 2010).

3.2.3.2.3. Pathogenesis

- HPV infects basal epithelial cells, where it establishes a persistent infection, often remaining asymptomatic.
- High-risk HPV types can cause dysplastic changes in infected epithelial cells, leading to the development of precancerous lesions and, eventually, invasive carcinomas.
- The viral oncoproteins E6 and E7 play crucial roles in HPV-mediated carcinogenesis by disrupting cell cycle regulation and promoting genomic instability.

3.2.3.2.4. Clinical Manifestations

- HPV infections can manifest as benign warts, including common warts, plantar warts, and genital warts.
- Persistent infection with high-risk HPV types can result in the development of various cancers, most notably cervical cancer, but also cancers of the anus, vulva, vagina, penis, and oropharynx.
- Clinical presentation varies depending on the site of infection and the HPV type involved.

3.2.3.2.5. Diagnosis

- HPV infections are diagnosed through molecular techniques such as polymerase chain reaction (PCR) assays and nucleic acid hybridization.
- Cervical cancer screening programs often utilize HPV testing alongside cytology (PAP smear) to identify women at risk of developing cervical neoplasia.

3.2.3.2.6. Treatment and Prevention

- Treatment modalities for HPV-related conditions include topical agents, surgical excision, cryotherapy, and laser therapy.

- Vaccination against HPV has proven highly effective in preventing HPV infection and associated diseases, with the introduction of HPV vaccines significantly reducing the incidence of cervical precancers and genital warts.

3.2.3.3. Polyomaviruses (PyVs)

3.2.3.3.1. Characteristics

- PyVs are small, non-enveloped viruses with a circular double-stranded DNA genome.
- They are named for their ability to induce tumors (Greek: poly, "many" + oma, "tumor") in experimental animals.
- PyVs are widely distributed in nature and infect various vertebrate species, including humans.

3.2.3.3.2. Epidemiology

- Human polyomaviruses, including BK virus (BKV) and JC virus (JCV), are prevalent in the general population, with primary infection typically occurring during childhood or adolescence.
- BKV and JCV establish latent infections in the host, with reactivation occurring under conditions of immunosuppression.

3.2.3.3.3. Pathogenesis

- BKV and JCV primarily infect renal epithelial cells and cells of the central nervous system, respectively.
- In immunocompromised individuals, BKV can cause hemorrhagic cystitis and nephropathy, whereas JCV is associated with progressive multifocal leukoencephalopathy (PML), a demyelinating disease of the brain.

3.2.3.3.4. Clinical Manifestations

- BKV-associated diseases include hemorrhagic cystitis and nephropathy, which commonly occur in transplant recipients and individuals undergoing immunosuppressive therapy.
- JCV-induced PML typically affects immunocompromised individuals, such as those with HIV/AIDS or recipients of immunosuppressive medications, leading to progressive neurological deficits and often fatal outcomes.

3.2.3.3.5. Diagnosis

- Laboratory diagnosis of BKV and JCV infections involves the detection of viral DNA in clinical specimens, such as urine, blood, or cerebrospinal fluid, using PCR-based assays.
- Histopathological examination of biopsy specimens may reveal characteristic viral inclusions in affected tissues.

3.2.3.3.6. Treatment and Prevention

- Management of BKV-associated diseases focuses on reducing immunosuppression and supportive care, although antiviral agents such as cidofovir and leflunomide may be used in severe cases.
- No specific antiviral therapy exists for PML, and treatment primarily

involves immune reconstitution and supportive measures (Cotran et al., 2005).

- Prevention of PyV infections relies on strategies to minimize immunosuppression in high-risk individuals and to limit exposure to contaminated bodily fluids.

3.2.3.4. Herpesvirus Infections and Latency

Herpesvirus infections are characterized by an acute infection that first appears, followed by an apparent recovery in which viral genomes are preserved in the absence of the production of infectious viruses in a particular tissue. The hallmark of latency is periodic reactivation, or "recrudescence," followed by (typically) milder acute infection symptoms. Viruses like HSV, EBV, and VZV are examples.

The viral genome is kept in a particular cell type during a latent infection and is not actively replicated. While EBV sustains itself in B lymphocytes, HSV maintains latent infections in sensory neurons. Certain virus genes essential for the viral genome's survival or for mediating the reactivation process are frequently expressed in latent infections.

The host must take an active role in reactivation. The body's natural defense against reinfection and immunity must momentarily wane. The host's response to physical or psychological stress can cause such a decline. Reactivation of HSV is frequently associated with a host under stress from weariness or anxiety. Shingles, an excruciating recurrence throughout the sensory nerve net that supplies the latent virus site, are caused by VZV reactivation. In contrast to HSV reactivation, VZV reactivation causes the nerve ganglion to be destroyed and is linked to an aging-related generalized decrease in immunity (Cotran et al., 2005). Controlling and preserving herpesvirus latency and identifying its replication sites depend on effective immunity. If a newborn comes into contact with the virus—for example, through infection from a mother who has a primary acute infection—they run the risk of developing profound disseminated HSV infections of their central nervous system (CNS). One of the leading causes of death for patients receiving organ transplants and immune suppression is disseminated human CMV infections. Furthermore, in certain patients with advanced immunodeficiencies brought on by HIV infection, CMV infections of the eye are a major cause of blindness. Additionally, one of the main causes of neurological abnormalities in developing fetuses is a primary CMV infection in pregnant women.

3.2.3.5. Other Complications Arising from Persistent Infections

Certain viruses can cause persistent infections that, in rare cases, result in a neoplasm or cancerous growth by causing ongoing tissue damage and cellular genes that control cell division (called tumor suppressor genes or oncogenes) to mutate. Examples include infections with slow-transforming retroviruses like the human T-cell leukemia virus (HTLV), certain genital papillomavirus infections, EBV infections, and chronic liver infections caused by hepatitis B and hepatitis C. The latter necessitates the additional involvement of co-carcinogens or additional cancer-causing factors.

Numerous researchers believe that abnormal immune responses to viral protein antigens that are continuously present in the body as a result of a persistent infection are the cause of autoimmune diseases like multiple sclerosis (MS). The infectious virus need not reappear as a result of such persistent infections. For instance, measles infection typically results in rash and healing even though some viral genomes and antigens remain in some tissues, such as brain tissue. Although the exact mechanism underlying this persistence is unknown, it is known that reduced levels of viral antigens are present on the surface of these cells carrying viral genomes and that virus maturation is inhibited. Antigens confer lifetime immunity against measles, but they can also cause immune complications if the host's immune system attacks and destroys measles-bearing neuronal tissue that would otherwise be healthy.

Such an autoimmune response results in the rare but deadly disease known as subacute sclerosing panencephalitis (SSPE), which strikes children a few years after contracting measles. While other serious measles aftereffects are frequent, SSPE is an uncommon consequence of measles infection. Vision impairment is among the most common. The virus multiplies within the host and infects the surface epithelium, causing a recognizable rash and lesions on the tongue, lips, and conjunctiva of the eyes. Conjunctival virus infection can clear up, but moving the eye muscles in reaction to light or while reading can cause additional infection of the eye musculature, which can cause irreversible damage. For this reason, people with measles should be kept out of the light and not use their eyes as much as possible.

3.2.4. Viral and Subviral Diseases with Long Incubation Periods

The majority of diseases caused by viruses have minimal to moderate fatality rates. It goes without saying that a virus with an excessively high death rate will infect all of its hosts so quickly that there will no longer be a viable population of susceptible people. However, there are times when this rule is not followed. High mortality rates can result from the introduction of viral disease into a virgin population (possibly as a result of intrusion into a novel ecosystem). Two notable instances are the introduction of measles during European expansion, which decimated indigenous populations in the Western Hemisphere, and the spread of smallpox throughout Europe in the Middle Ages. Another manifestation of infection with a virus that has an exceptionally long incubation period between the time of infection and the onset of disease symptoms results in an exception to the low mortality rule.

3.2.4.1. Rabies

Rabies is a deadly viral disease that primarily affects mammals, including humans. It is caused by the rabies virus, a member of the Lyssavirus genus. This zoonotic disease has a profound impact on public health worldwide, especially in regions where canine rabies is endemic. The virus is typically transmitted through the saliva of an infected animal, usually through bites or scratches. While rabies is preventable through prompt medical intervention, once symptoms appear, it is almost always fatal. (Cotran RS, Kumar V, Fausto N (2005)).

3.2.4.1.1. History of Rabies

Rabies has a long history, with references dating back to ancient civilizations. The first recorded case of rabies is believed to have occurred around 2300 BC. In ancient Mesopotamia, a legal code prescribed penalties for owners of rabid dogs that caused injury. The disease was also described by Aristotle in ancient Greece, who noted that it was incurable once symptoms appeared. In the 19th century, Louis Pasteur developed the first effective rabies vaccine, a milestone that significantly reduced the incidence of the disease.

3.2.4.1.2. Epidemiology

Rabies is found on every continent except Antarctica, with the highest burden in Asia and Africa. More than 95% of human rabies deaths occur in these regions, where dog bites are the primary mode of transmission. The World Health Organization (WHO) estimates that rabies causes approximately 59,000 human deaths annually, although this figure may be underestimated due to underreporting and misdiagnosis. Rabies in wildlife, such as bats, raccoons, skunks, and foxes, also pose a significant risk, particularly in the Americas and Europe.

3.2.4.1.3. Pathophysiology

The rabies virus is neurotropic, meaning it specifically targets the nervous system. After entering the body, usually through a bite, the virus binds to nicotinic acetylcholine receptors at the neuromuscular junction. It then travels retrogradely along peripheral nerves to the central nervous system (CNS). This process can take weeks to months, depending on the entry site and the viral load. Once the virus reaches the CNS, it causes encephalitis, leading to severe neurological symptoms and, ultimately, death. The virus also spreads centrifugally to other tissues, including the salivary glands, facilitating further transmission.

3.2.4.1.4. Clinical Presentation

The clinical course of rabies can be divided into several stages: incubation, prodromal, acute neurologic, coma, and death. The incubation period typically ranges from one to three months but can vary from a few days to several years. The prodromal phase is characterized by nonspecific symptoms such as fever, headache, and malaise. Patients may also experience pain or paresthesia at the site of the bite. The acute neurologic phase includes the classic symptoms of rabies: hydrophobia (fear of water), aerophobia (fear of air), hyperactivity, and agitation. Paralytic rabies, a less common form, manifests as flaccid paralysis. Without intervention, the disease progresses to coma and death within days to weeks (Domingo et al., 1999).

3.2.4.1.5. Diagnosis

Diagnosing rabies can be challenging, especially before the onset of specific symptoms. Laboratory confirmation is typically achieved through direct fluorescent antibody (DFA) testing of brain tissue, the gold standard for postmortem diagnosis. Rabies can be diagnosed in live patients by detecting viral RNA in saliva, cerebrospinal fluid, or skin biopsies using reverse transcription-polymerase chain reaction (RT-PCR). Serologic tests to detect rabies-specific antibodies can also support the diagnosis, especially in vaccinated individuals.

3.2.4.1.6. Prevention

Preventing rabies involves a combination of strategies, including vaccination, public health education, and animal control. Human rabies vaccines are highly effective and can be administered as pre-exposure prophylaxis (PrEP) for individuals at high risk, such as veterinarians and laboratory workers. Post-exposure prophylaxis (PEP), consisting of wound cleansing, rabies immunoglobulin (RIG), and a series of rabies vaccinations, is crucial for anyone exposed to the virus. Immediate and thorough washing of the wound with soap and water is a critical first step in reducing the risk of infection (Domingo et al., 1999).

3.2.4.1.7. Control

Controlling rabies in the animal population, particularly dogs is essential to reducing the incidence of human rabies. Mass vaccination of dogs, responsible pet ownership, and controlling stray dog populations are key strategies. In wildlife, oral rabies vaccination programs have been successful in some regions. Public health campaigns to educate communities about rabies prevention, the importance of seeking medical care after an animal bite, and the role of vaccination are vital components of rabies control efforts.

3.2.4.2. HIV–AIDS

HIV (human immunodeficiency virus) and AIDS (acquired immunodeficiency syndrome) represent one of the most significant global public health issues in modern history. Since its identification in the early 1980s, HIV/AIDS has led to millions of deaths and continues to be a major health concern, particularly in sub-Saharan Africa. This essay will explore the history, epidemiology, pathophysiology, clinical presentation, diagnosis, treatment, prevention, and ongoing challenges in the fight against HIV/AIDS.

3.2.4.2.1. History of HIV/AIDS

The first recognized cases of AIDS were reported in the United States in 1981 among gay men in New York and California, presenting with rare opportunistic infections and cancers. By 1983, researchers had identified HIV as the causative agent of AIDS. Early responses to the epidemic were hampered by stigma and lack of understanding, but the subsequent development of antiretroviral therapy (ART) in the mid-1990s transformed HIV from a fatal disease to a manageable chronic condition.

3.2.4.2.2. Epidemiology

HIV/AIDS affects millions globally, with approximately 38 million people living with HIV as of 2020. Sub-Saharan Africa bears the highest burden, accounting for about 67% of the global total. Other regions, including Eastern Europe, Central Asia, and parts of the Americas, also have significant epidemics. Despite progress in reducing new infections and deaths, disparities in access to prevention, testing, and treatment services persist, contributing to ongoing transmission and morbidity.

3.2.4.2.3. Pathophysiology

HIV is a retrovirus that targets the immune system, specifically CD4+ T cells, which are crucial for immune response. The virus binds to and enters these cells, integrating its genetic material into the host's DNA. This process involves several steps, including reverse transcription, integration, and replication. Over time, HIV replication leads to the depletion of CD4+ T cells, weakening the immune system and rendering the body vulnerable to opportunistic infections and certain cancers. Without treatment, this progressive immune deficiency leads to AIDS.

3.2.4.2.4. Clinical Presentation

The clinical course of HIV infection can be divided into three stages: acute infection, clinical latency, and AIDS.

- **Acute Infection:** This stage occurs 2-4 weeks after exposure and is characterized by flu-like symptoms such as fever, sore throat, rash, and lymphadenopathy. This phase, also known as acute retroviral syndrome, is highly infectious due to high viral loads.

- **Clinical Latency:** Also known as the asymptomatic or chronic phase, this stage can last several years. While the virus continues to replicate, individuals may not exhibit symptoms, but they remain infectious.
- **AIDS:** This final stage is defined by a CD4+ T cell count below 200 cells/µL or the presence of AIDS-defining conditions, such as Pneumocystis pneumonia, Kaposi's sarcoma, and other opportunistic infections and cancers. Without treatment, individuals with AIDS typically survive about three years.

3.2.4.2.5. Diagnosis

HIV diagnosis involves serological tests to detect HIV antibodies and antigens. The most common tests include:

- **ELISA (Enzyme-Linked Immunosorbent Assay):** Initially used to screen for HIV antibodies.
- **Western Blot:** Used to confirm positive ELISA results, though less common now.
- **Rapid Tests:** Provide results within 20-30 minutes and are suitable for point-of-care testing.
- **Nucleic Acid Tests (NATs):** Detect viral RNA, which are used for early diagnosis and to monitor viral load.

Early diagnosis is crucial for timely intervention and reducing transmission.

3.2.4.2.6. Treatment

The introduction of antiretroviral therapy (ART) has revolutionized HIV treatment. ART involves a combination of drugs from different classes, including:

- **Nucleoside Reverse Transcriptase Inhibitors (NRTIs):** Block reverse transcription, a key step in viral replication.
- **Non-Nucleoside Reverse Transcriptase Inhibitors (NNRTIs):** Also inhibit reverse transcriptase but through a different mechanism.
- **Protease Inhibitors (PIs):** Prevent viral protease enzymes from processing viral proteins.
- **Integrase Strand Transfer Inhibitors (INSTIs):** Block the integration of viral DNA into the host genome.
- **Entry Inhibitors:** Prevent the virus from entering CD4+ cells.

Effective ART suppresses viral replication, maintaining viral loads at undetectable levels, preserving immune function, and preventing progression to AIDS (Domingo et al., 1999).

3.2.4.2.7. Prevention

Preventing HIV transmission involves a combination of strategies:

- **Condom Use:** Consistent and correct use of condoms reduces the risk of sexual transmission.
- **Pre-Exposure Prophylaxis (PrEP):** A daily oral medication for individuals at high risk, which significantly reduces the risk of acquiring HIV.
- **Post-Exposure Prophylaxis (PEP):** A course of antiretroviral drugs taken within 72 hours of potential exposure to prevent infection.
- **Harm Reduction:** Programs for people who inject drugs, such as needle exchange services and opioid substitution therapy.

- **Mother-To-Child Transmission Prevention:** ART for HIV-positive pregnant women significantly reduces the risk of transmission to the baby.
- **Voluntary Medical Male Circumcision (VMMC):** Reduces the risk of heterosexual men acquiring HIV.

3.2.4.3. Prion Diseases

Prion diseases, also known as transmissible spongiform encephalopathies (TSEs), are a group of rare, fatal neurodegenerative disorders affecting humans and animals. These diseases are caused by the misfolding of prion proteins, which leads to brain damage characterized by sponge-like holes in neural tissue. In humans, prion diseases manifest in various forms, including Creutzfeldt-Jakob disease (CJD), variant Creutzfeldt-Jakob disease (vCJD), Gerstmann-Sträussler-Scheinker syndrome (GSS), fatal familial insomnia (FFI), and kuru.

Creutzfeldt-Jakob disease is the most common human prion disease, with sporadic, familial, and acquired forms. Sporadic CJD (sCJD), accounting for about 85% of cases, typically occurs in individuals aged 60-65 without any known risk factors. Familial CJD (fCJD) is caused by inherited mutations in the prion protein gene (PRNP), while acquired CJD can result from exposure to contaminated medical equipment or tissues. Variant CJD, which emerged in the 1990s in the United Kingdom, is linked to the consumption of beef contaminated with bovine spongiform encephalopathy (BSE), or "mad cow disease."

GSS and fatal familial insomnia are both inherited prion diseases, each associated with specific mutations in the PRNP gene. GSS typically presents with a combination of dementia and motor symptoms, such as ataxia, while FFI is characterized by progressively worsening insomnia, leading to severe autonomic dysfunction and, ultimately, death. Kuru, historically found among the Fore people of Papua New Guinea, was transmitted through ritualistic cannibalism. The elimination of this practice has led to a decline in new kuru cases.

The diagnosis of prion diseases relies on clinical assessment, neuroimaging, and the detection of abnormal prion proteins in cerebrospinal fluid or brain tissue. There is currently no cure for prion diseases, and treatment is primarily supportive, focusing on alleviating symptoms and improving the quality of life for patients. Research efforts continue to explore potential therapeutic approaches and to better understand the mechanisms underlying prion propagation and neurodegeneration.

3.3. SOME VIRAL INFECTIONS TARGETING SPECIFIC ORGAN SYSTEMS

While each vertebrate host's organ system serves a crucial or important purpose in the organism's life, a few are so important that they can cause major problems or even death if they are disrupted. Among these are the circulatory system, the immune system, the liver, and the central nervous system (CNS), which influences all facets of behavior, both innate and learned. These systems are frequently infected with viruses that pose a serious risk to the person's life and can cause irreversible tissue damage. For instance, the primary sign of AIDS is the destruction of immune system cells targeted by HIV, which results in neoplasms and opportunistic infections that cause death. Not all viral infections result in death, but other viruses can cause diseases just as devastating as HIV. Examining a few liver and central nervous system viruses offers some intriguing instances of both severe and constrained disease progressions (Domingo et al., 1999).

Important illustrations of various aspects of viral infection and pathogenesis can also be seen in the various patterns of aftereffects that follow infection of a common target organ. First, very specific interactions between a particular virus and the cell type it infects lead to specific tissue or cell tropism. Different outcomes of infection are evident depending on the type of cell infected, the severity of the symptoms, and the type of damage caused by the infection. Persistent infection is a multifaceted phenomenon. It results, in part, from the virus interacting with the immune system of the host and changing it. Persistence frequently refers to the virus's ability to adjust to its ongoing relationship with the target cell. Third, trying to understand the relationships among viruses does not really benefit from the classification of viruses according to the diseases they cause. Finally, fourth, viruses can target the same organ and spread through a variety of different pathways. Just as crucial as the virus's initial point of entry is how it moves throughout the host.

3.3.1. Viral Infections of Nerve Tissue

It is easy to distinguish between the central and peripheral segments of the vertebrate nerve net. The peripheral part connects spinal cord circuits to transfer impulses from the brain to the brain. Viral infections of nerve tissue are classified as infections of particular neuronal populations, including neurons of the brain and brain stem itself (encephalitis), the covering of the brain (meningitis), and the neurons of the spinal cord (myelitis).

The brain and CNS occupy a special place in the body and are shielded from potentially dangerous circulating pathogens as well as the rest of the body by a physical and physiological barrier. This barrier, also known as the blood-brain barrier, acts as a partial but effective defense against infections. In addition to physically destroying tissue as a result of an active infection, viruses that migrate through neurons can also enter the nervous system directly through olfactory neurons (which are not isolated from the central nervous system) or through other, less well-defined mechanisms. Since a particular subset of CNS cells called microglial cells perform functions similar to or identical to those of macrophages in other tissues, it is undeniable that pathogen invasion of the CNS is not all that uncommon (Porterfield and Htraavik,1994).

Numerous viruses can infect nerve tissue; some of these infections are harmless, but others specifically target nerve tissue. When viruses do infect nerve tissue, they typically target one or more areas. Although the distinction is not perfect, many viruses, like enteroviruses and genital herpes simplex virus (HSV-2), are known to cause meningitis, while others, like rabies and facial herpes simplex virus (HSV-1), are nearly invariably linked to encephalitis. Aseptic meningitis, also known as viral meningitis, is generally less fatal than most viral infections linked to encephalitis; however, all viral infections are dangerous and can result in severe illnesses.

Although some viral infections of the brain can be serious, this is not always the case. With proper symptomatic care, some CNS viral infections have relatively mild outcomes. Viruses that affect the brain can be categorized based on the extent of brain involvement and whether it is the primary or secondary target.

3.3.2. Viral Encephalitis with Favorable Prognosis for Recovery

Many of the viruses that result in encephalitis have RNA genomes and are spread by arthropod vectors from zoonoses, with human infection often being accidental. These viruses are commonly referred to as arboviruses, although this classification is not precise and includes two groups of viruses that are not closely related by other criteria. It can be challenging to assess the symptoms of encephalitis in wild animals, but certain equine encephalitis viruses are known to cause severe illness in horses. In humans, symptoms of viral encephalitis typically include drowsiness, mild malaise, and occasionally coma. Mosquito-borne encephalitis viruses typically do not directly invade neural tissue, instead infecting supporting tissue. The host's response to this infection and resulting inflammation are what lead to the neurological symptoms observed (Prusiner et al., 1996).

Since the primary target of such encephalitis virus infections is tissue at the periphery of neural tissue, the infection can be treated, and full recovery can occur as long as the host's immune system functions as intended. Lethargy and malaise during the disease's symptomatic phase leave infected people open to other environmental dangers, such as infection with other pathogens. However, the illness usually goes away if these risks are managed with appropriate care. Although encephalitis viruses frequently strike humans by accident, it is unclear whether human encephalitis symptoms play a significant part in the virus's ability to spread. Like all arthropod-borne illnesses, the blood-associated virus is ingested by the arthropod during the viremic stage of infection, and any behavioral effects are unintentional. However, it's possible that the sluggishness that accompanies an active illness makes afflicted animals more susceptible to arthropod bites, which could play a role in the disease's natural transmission.

3.3.3. Viral Infections of the Liver (Viral Hepatitis)

Liver diseases are particularly significant in many fields of medicine due to the liver's vital physiological role as well as the liver's frequent passage through all circulating blood and lymph. The term "hepatitis viruses" refers to a group of diverse, unrelated viruses that cause damage to the liver. Any hepatitis virus can cause severe liver damage to the host that is infected. Hepatitis virus-induced liver failure is a primary cause of liver transplantation. Furthermore, several of these viruses create persistent carrier states, in which the virus persists for several months or even years after infection. At the moment, A, B, C, delta (D), and E are the five human hepatitis viruses that are comparatively well characterized. Each has a different disease severity and set of aftereffects (Rotbart, 1991).

3.3.3.1. Hepatitis A

The poliovirus and this virus are related. It is transmitted by tainted food or water, and it results in a general feeling of unwellness and a potentially serious but manageable loss of liver function. With the right medical attention, a patient can usually fully recover their liver function, completely eradicate the virus from their body, and develop effective immunity against future infections. For those who are susceptible to infection, there is a reasonably effective hepatitis A vaccine available.

3.3.3.2. Hepatitis B

Retroviruses and the hepatitis B virus are related but clearly different. In contrast to hepatitis A, the B virus is primarily transmitted through blood, either during intercourse or other blood-contamination incidents (such as sharing needles). Persistent viremia and liver damage are the consequences of the primary infection. Hepatitis B infection is endemic among intravenous drug users, commercial sex partners, and their clients. It poses a particular risk to medical personnel due to the potential for needle stick transmission from contaminated blood. In Southeast Asia, where birth trauma can transmit the virus from mother to child, the disease is endemic.

Acute illness with accompanying liver failure can result from a hepatitis B virus infection, or it can be asymptomatic. The virus is frequently totally eradicated, resulting in either a full or partial recovery of liver function. Unfortunately, a significant portion of those who contract the virus go on to become chronic, asymptomatic carriers of it. Indeed, a major contributing factor to some human liver cancers (carcinomas) common in Southeast Asia is chronic hepatitis B infection. Fulminant infection, a third type of hepatitis B virus infection, is characterized by a quick onset of severe liver damage and frequently results in death.

3.3.3.3. Hepatitis C

The virus that causes hepatitis C initially appeared to share some general similarities with a wide range of bacterial, plant, and animal viruses, such as hepatitis A and poliovirus. That being said, it has been established that this virus belongs to the family Flaviviridae. The virus, which can spread through tainted blood and blood products, is estimated to be responsible for up to 25% of cases of acute viral hepatitis globally. Although there isn't any proof right now that arthropod vectors are effectively spreading it, it's still possible. In contrast to individuals infected with the hepatitis A virus, a considerable percentage of victims fail to develop a viable immune response to the infection, resulting in a chronic infection that can persist for years and cause cumulative liver damage and carcinoma.

3.3.3.4. Hepatitis D

The hepatitis B virus is necessary for the hepatitis delta (D) virus to replicate, making it a defective virus. In Southeast Asia, a major hub for hepatitis B infection, it is not particularly common despite this requirement. When hepatitis D and B are co-infected in the same person, the incidence of acute or chronic liver disease is not significantly higher than when hepatitis B is the only virus involved. On the other hand, cirrhosis of the liver and acute illness followed by persistent viral secretion are frequently associated with hepatitis D virus infection in a person who has previously contracted hepatitis B.

3.3.3.5. Hepatitis E

Hepatitis E is also transmitted by tainted water and maybe food, just like hepatitis (a) It is widespread worldwide and has been linked to major epidemics in Russia and India due to contaminated drinking water. Although this virus usually causes a mild illness, pregnant women may experience a high mortality rate from the disease. After the acute phase of infection, recovery from the acute phase is usually complete and no evidence of chronic infection is present (Shope,1994).

CASE STUDY

OUTBREAK MANAGEMENT AND CONTROL OF A VIRAL DISEASE: THE COVID-19 PANDEMIC

Background

The COVID-19 pandemic, caused by the novel coronavirus SARS-CoV-2, emerged in late 2019 in Wuhan, China. It rapidly spread across the globe, causing widespread illness, significant mortality, and severe disruption to societies and economies. This case study examines the public health response, challenges, and strategies implemented to manage and control the viral outbreak.

Challenges

- **Rapid Transmission:** COVID-19 exhibited a high transmission rate, spreading through respiratory droplets, close contact, and potentially airborne routes in certain settings.
- **Healthcare Strain:** The influx of patients overwhelmed healthcare systems, leading to shortages of medical supplies, hospital beds, and healthcare workers.
- **Vaccine Development:** Developing, testing, and distributing vaccines on an unprecedented scale was a significant scientific and logistical challenge.
- **Public Compliance:** Ensuring public adherence to preventive measures such as social distancing, mask-wearing, and vaccination faced varying levels of compliance due to misinformation and resistance.
- **Economic Impact:** Lockdowns and other restrictions severely impacted economies, leading to job losses, business closures, and financial instability.
-

Solution

A multi-faceted public health approach was implemented globally, involving containment, mitigation, and vaccination strategies.

Implementation Steps:

- Early Detection and Surveillance:
  - **Testing and Contact Tracing:** Rapid expansion of testing capabilities to identify and isolate infected individuals. Contact tracing efforts were intensified to track and contain the spread of the virus.

- **Genome Sequencing:** Monitoring viral mutations through genome sequencing to identify new variants and adjust public health responses accordingly.

- Containment Measures:
  - **Lockdowns and Quarantines:** Implementing lockdowns, travel restrictions, and quarantine protocols to limit movement and reduce transmission.
  - **Social Distancing and Mask Mandates:** Mandating social distancing and mask-wearing in public spaces to reduce the spread of the virus.
- Healthcare System Support:
  - **Resource Allocation:** Ensuring adequate supply of personal protective equipment (PPE), ventilators, and medical supplies. Establishing temporary hospitals and care facilities to handle patient overflow.
  - **Healthcare Workforce Support:** Providing mental health support, training, and incentives to healthcare workers to sustain their efforts during the crisis.
- Public Health Communication:
  - **Information Campaigns:** Launching campaigns to educate the public about preventive measures, the importance of vaccination, and combating misinformation.
  - **Transparency and Trust:** Maintaining transparency about the evolving situation, vaccine development, and safety to build public trust.
- Vaccination Campaign:
  - **Accelerated Vaccine Development:** Supporting global collaboration for the rapid development, testing, and approval of COVID-19 vaccines.
  - **Mass Vaccination Programs:** Rolling out mass vaccination campaigns prioritizing vulnerable populations and frontline workers. Establishing distribution channels and vaccination centers.
- Economic Support:
  - **Financial Aid Packages:** Implementing financial aid programs, unemployment benefits, and business relief packages to mitigate economic impacts.
  - **Stimulus Measures:** Introducing stimulus measures to support economic recovery and sustain businesses and individuals affected by the pandemic.

Results

- **Reduction in Transmission:** The combined efforts of lockdowns, social distancing, and mask mandates helped to slow the transmission of the virus, particularly in the early stages of the pandemic.

- **Healthcare System Stabilization:** Resource allocation and support measures helped stabilize healthcare systems, allowing them to manage the surge in COVID-19 cases more effectively.
- **Vaccination Success:** The rapid development and deployment of vaccines significantly reduced severe illness, hospitalizations, and deaths associated with COVID-19. Vaccination efforts contributed to achieving herd immunity in many regions.
- Economic Recovery: Financial aid and stimulus measures provided crucial support to individuals and businesses, aiding in economic recovery as restrictions were gradually lifted.

CLASS ACTIVITY

Engage students with a quick brainstorming session, listing common viral diseases such as influenza, HIV/AIDS, and COVID-19, and discuss their modes of transmission, symptoms, and preventive measures.

SUMMARY

- Viral diseases of humans are illnesses caused by viruses, which are microscopic infectious agents that require a host cell to replicate. These diseases encompass various health conditions, affecting various organ systems and leading to acute and chronic illnesses.
- Viruses are maintained by active rounds of infection somewhere in their reservoir. We have seen that a virus infection leading to immunity to reinfection will lead to virus extinction in a small population once the pool of susceptible individuals is exhausted.
- A number of important human viruses are either asymptomatic or cause relatively mild symptoms of primary infection, which is followed by a stable association between virus and host that lasts as long as the latter lives. During this more or less stable association, some viruses are constantly shed, while persistent infections with others lead to the loss of any detectable virus.
- Colds and respiratory infections are ubiquitous and affect individuals of all ages, races, and socioeconomic backgrounds. While they are often perceived as minor inconveniences, they can have significant implications for both individual well-being and public health.
- Respiratory infections encompass a broad spectrum of illnesses affecting the respiratory system, including the upper respiratory tract (e.g., nose, throat) and lower respiratory tract (e.g., lungs). These infections can be caused by viruses, bacteria, fungi, or other pathogens, but viral respiratory infections are among the most common.
- The disease caused by infection with smallpox (variola) virus is an example of a much more severe disease than flu, with correspondingly higher mortality rates. There are (or were) two forms of the disease: variola major and variola minor. These differed in the severity of symptoms and death rates.
- Viruses that persist in the individual as chronic or latent infections are common. Often, the course of the initial infection is similar to that seen in an acute infection, followed by efficient clearance, as described previously. In a persistent infection, however, the virus cannot be cleared either because of virus-induced deficiencies in the immune response or because the virus maintains infection in localized areas where immunity is not complete.
- AIDS, which is characterized by a latent period in which HIV can be transmitted, followed by severe disease, is an example of a "new" viral disease. In humans, virus spread is often the result of behavioral patterns of infected individuals during HIV's long latent period. This spread pattern makes it unlikely that there is any selective pressure over time toward ameliorating the late severe symptoms.
- The vertebrae nerve net can be readily divided into peripheral and central portions. The peripheral portion moves impulses to and from the brain through connecting circuits in the spinal cord. Viral infections of nerve tissue can be divided into infections of specific groups of neurons: neurons of the spinal cord

(myelitis), the covering of the brain (meningitis), and neurons of the brain and brain stem itself (encephalitis).

- Diseases of the liver hold a special place in many types of medicine, both because of the important physiological role of this organ and because all circulating blood and lymph pass through the liver frequently. Several different and unrelated viruses target the liver; these are collectively known as hepatitis viruses.

REVIEW QUESTIONS

1. What features of pathogenesis are shared by measles virus, varicella-zoster virus, and variola virus?
2. What features distinguish an acute from a persistent infection?
3. Provide examples of viral diseases that predominantly affect the respiratory system.
4. What are some of the unique features of infection by rabies virus?
5. Distinguish encephalitis produced by herpesvirus from infection with an arbovirus such as La Crosse encephalitis virus.
6. How do viral infections like hepatitis primarily target the liver?
7. The disease subacute sclerosing panencephalitis (SSPE) is a complication that may follow infection with measles virus. Discuss the possible mechanisms occurring during the development of this rare disease.

REFERENCES

1. Adams, M. J., & Carstens, E. B. (2012). Ratification vote on taxonomic proposals to the International Committee on Taxonomy of Viruses (2012). *Archives of Virology*, *157*(7), 1411–1422. https://doi.org/10.1007/s00705-012-1299-6 (accessed on 4th October 2024).
2. Bravo, I. G., de Sanjose, S., & Gottschling, M. (2010). The clinical importance of understanding the evolution of papillomaviruses. *Trends in Microbiology*, *18*, 432–438.
3. Cotran, R. S., Kumar, V., & Fausto, N. (2005). *Robbins and Cotran pathologic basis of disease* (7th ed.). Elsevier/Saunders.
4. Domingo, E., Webster, R. G., & Holland, J. J. (Eds.). (1999). *Origin and evolution of viruses*. Academic Press.
5. Porterfield, J. S., & Htraavik, T. (1994). Encephalitis viruses. In R. G. Webster & A. Granoff (Eds.), *Encyclopedia of virology* (pp. 1–42). Academic Press.
6. Prusiner, S. B., Telling, G., Cohen, G., & DeArmond, S. (1996). Prion diseases of humans and animals. *Seminars in Virology*, *7*, 159–174.
7. Rotbart, H. A. (1991). Viral meningitis and the aseptic meningitis syndrome. In W. M. Scheld, R. J. Whitley, & D. T. Durack (Eds.), *Infections of the central nervous system* (pp. 237–246). Raven Press.

8. Shope, R. E. (1994). Rabies-like viruses. In R. G. Webster & A. Granoff (Eds.), *Encyclopedia of virology* (pp. 789–794). Academic Press.

9. Taylor, M., Kobiler, O., & Enquist, L. W. (2012). Alphaherpesvirus axon-to-cell spread involves limited virion transmission. *Proceedings of the National Academy of Sciences of the United States of America*, *109*(42), 17046–17051. https://doi.org/10.1073/pnas.1212926109 (accessed on 4th October 2024).

10. Thapa, S., Gokhale, S., Sharma, A. L., Sapkota, L. B., Ansari, S., & Gautam, R. (2017). Burden of bacterial upper respiratory tract pathogens in school children of Nepal. *BMJ Open Respiratory Research*, *4*(1), e000203. https://doi.org/10.1136/bmjresp-2017-000203 (accessed on 4th October 2024).

CHAPTER 4

Virus Structure and Classification

LEARNING OBJECTIVES

After studying this chapter, you will be able to:

- Define virus;
- Illustrate the structure of viruses;
- Explain the virus classification and taxonomy.

KEY TERMS FROM THIS CHAPTER

Bacteriophage	Capsid	Capsomers
Complex viruses	Cytopathic effects	Envelope
Icosahedral	Nucleocapsid	Phage
Plaque	Positive strand	Protomers
Spike	Virion	Virus

INTRODUCTION

Viruses are among the simplest and most intriguing forms of life, characterized by their unique structure and mechanisms of action. Understanding the structure and classification of viruses is fundamental to virology, medicine, and molecular biology, as it provides critical insights into how viruses infect hosts, propagate, and evolve.

Viruses are acellular entities primarily consisting of genetic material enclosed in a protein coat, called a capsid. The genetic material can be either DNA or RNA, and it may be single-stranded or double-stranded. This nucleic acid encodes the information necessary for the virus to replicate and infect host cells. The capsid, composed of protein subunits called capsomeres, protects the viral genome and aids in its delivery into host cells (Baker et al., 1999).

Many viruses also possess an outer lipid envelope derived from the host cell membrane, embedded with viral glycoproteins. These glycoproteins are crucial for recognizing and attaching to host cells, determining the virus's host range and specificity. The envelope is more common in animal viruses and is typically acquired as the virus exits the host cell by budding.

The size and shape of viruses can vary widely. Common shapes include icosahedral, helical, and complex structures. The icosahedral structure, seen in viruses like adenoviruses, provides stability and efficient packing of the viral genome. Helical viruses, such as tobacco mosaic virus, have a rod-like appearance with the genome spiraled within the capsid. Complex viruses, such as bacteriophages, possess intricate structures with specialized components for infecting bacterial cells.

Viruses are classified based on their genetic material, replication strategy, morphology, and host range. The Baltimore classification system is widely used, categorizing viruses into seven groups based on their genome type and replication method.

4.1. VIRUS

An infectious agent that can only reproduce inside of its host is called a virus. Numerous living things, including bacteria, plants, and animals, are susceptible to virus infection. Viruses have a very simple structure and are so small that they must be seen under a microscope. A virus particle is made up of a viral genome, or genetic material, enclosed in a capsid, a protein shell, when it is not dependent on its host. A membrane known as an envelope encloses the protein shell of certain viruses. Viral genomes are incredibly diverse because they can be either DNA or RNA, single- or double-stranded, linear or circular, and differ in length and number of DNA or RNA molecules.

The process of viral replication commences when a virus attaches itself to its host cell and penetrates the cell wall or membrane. After removing its coat from the protein, the virus injects its genome into the host cell. The host cell's machinery is then forced to replicate the viral genome and produce viral proteins in order to create new capsids, as a result of the viral genome seizing control of the cell. After that, new viruses are created by assembling the viral particles. Lysis is the process by which the new viruses burst out of the host cell, killing it. In order to create an envelope around the capsid, certain viruses seize a piece of the host membrane during the lysis process (Baker et al., 1999).

The newly replicated viruses may then infect additional hosts. Human diseases like influenza, chicken pox, AIDS, the common cold, and rabies are caused by a variety of viruses. Vaccination is the main method of preventing viral infections. An unaffected person is given a vaccine consisting of inactive viral particles to boost their immunity against the illness.

4.1.1. General Properties of Viruses

Viruses are a special class of infectious agents that are distinguished by their basic acellular structure and mode of reproduction. A complete virus particle, also known as a virion, is made up of one or more DNA or RNA molecules that are sometimes encased in additional layers of protein coating. These extra layers, which can include more proteins, lipids, and carbohydrates, can be extremely complex. There are two stages in which viruses can live: extracellular and intracellular. The extracellular phase, or virions, have few or no enzymes and are unable to proliferate on their own without living cells. During the intracellular stage, viruses are essentially replicating nucleic acids that stimulate the host's metabolism to produce virion components; in due course, the entire virus particle or virions are released.

At least three characteristics set viruses apart from living cells: (i) their basic, acellular structure; (ii) the presence of either RNA or DNA, but not both, in nearly all virions (the human cytomegalovirus contains four mRNAs and a DNA genome); and (iii) their incapacity to divide into procaryotes and eucaryotes and reproduce independently of them. Though obligatorily intracellular parasites like viruses, bacteria like chlamydia and rickettsia do not fit the first two requirements.

4.1.2. The Cultivation of Viruses

In contrast to bacteria and eucaryotic microorganisms, viruses cannot be cultivated in the same manner because they are unable to reproduce independently of living cells. Researchers have been cultivating animal viruses for a long time by injecting the viruses into appropriate host animals or fertilized chicken eggs that are incubated for six to eight days after laying. The egg's shell surface is first cleaned with iodine and punctured with a tiny, sterile drill to prepare it for virus culture. Following inoculation, gelatin is used to seal the drill hole, and the egg is then incubated. Viruses must be injected into the appropriate area because they may only be able to reproduce in specific areas of the embryo. For instance, the mumps virus prefers the allantoic cavity, but the myxoma virus grows well on the chorioallantoic membrane. A localized tissue lesion called a "pock" may result from the infection; this lesion's appearance is frequently indicative of the virus.

Animal viruses have recently been cultured in tissue (cell) culture on animal cell monolayers. The invention of growth media for animal cells and the development of antibiotics that can stop bacterial and fungal contamination have made this technique possible. A virus inoculum is applied to a layer of animal cells in a specially prepared petri dish, and the viruses are given time to settle and attach to the cells. After that, a thin layer of agar is applied to the cells to stop the spread of the virions, allowing only neighboring cells to become infected by newly formed virions. Because of this, there are frequently localized regions of cellular lysis and destruction known as plaques, which can be identified if stained with dyes that can differentiate between living and dead cells, like trypan blue or neutral red. It's not always the case that viral growth causes cells to lyse and form plaques. Cytopathic effects are the microscopic or macroscopic degenerative changes or abnormalities that animal viruses, in particular, can cause in host cells and tissues. Although plaque formation from cell lysis is not always the result of cytotoxic effects, it can still be fatal.

Young, actively growing bacterial cells are cultured in broth or agar cultures, which are home to bacteriophages, also known as phages for short. Cell lysis can cause so many host cells to be destroyed that turbid bacterial cultures can clear quickly. The bacteriophage sample is combined with cool, liquid agar and an appropriate bacterial culture to create agar cultures. The mixture is rapidly transferred into a petri dish with sterile agar on the bottom. Bacteria in the top agar layer multiply and grow once it hardens, creating a continuous, opaque layer known as "lawn." A virion infects and multiplies in a nearby cell whenever it settles in the top agar. Bacterial lysis eventually causes a clearing or plaque in the grass. As is evident, the phage under cultivation is frequently identified by its plaque appearance (Baker et al., 1999).

4.1.3. Common Characteristics of Viruses

Originally described in "The World of Viruses," viruses were thought to be filterable agents that could spread illness. Since then, scientific methods and microscopy improvements have allowed for more accurate categorization of viruses and their characteristics. While molecular and cellular assays have expanded our understanding of how viruses function and are related, electron microscopy has allowed us to see viruses in great detail. Taken together, we have learned that although they can be quite diverse, viruses share several common characteristics:

4.1.3.1. Viruses are Small in Size

The human immunodeficiency virus and influenza have a more typical size of approximately 100 nm in diameter, but the smallest viruses are only about 20 nm in diameter. Since human cells typically have a diameter of 10–30 μm (microns), the viruses that infect them are typically 100–1000 times smaller than the average human cell.

Some viruses, though, are noticeably bigger than 100 nm. Poxviruses, like the smallpox-causing variola virus, can grow to a length of almost 400 nm. In contrast, filoviruses, like the deadly Ebola and Marburg viruses, have a diameter of only 80 nm but can spread into long threads that can reach over 1000 nm. Recently, a number of very large viruses that infect amoebas have been identified. The megavirus has a diameter of 400 nm, while the pandoravirus has an ovoid or elliptical structure that approaches 1000 nm in length. It is a common misconception that all viruses are larger than bacteria. While most bacteria have a size range of 2000–3000 nm, some strains of bacteria known as mycobacteria can have sizes as small as ten times smaller, falling within the range of these large viruses. Thus, while being small is a common feature of viruses, their sizes can vary from a few nanometers to larger than some bacteria (Figure 4.1).

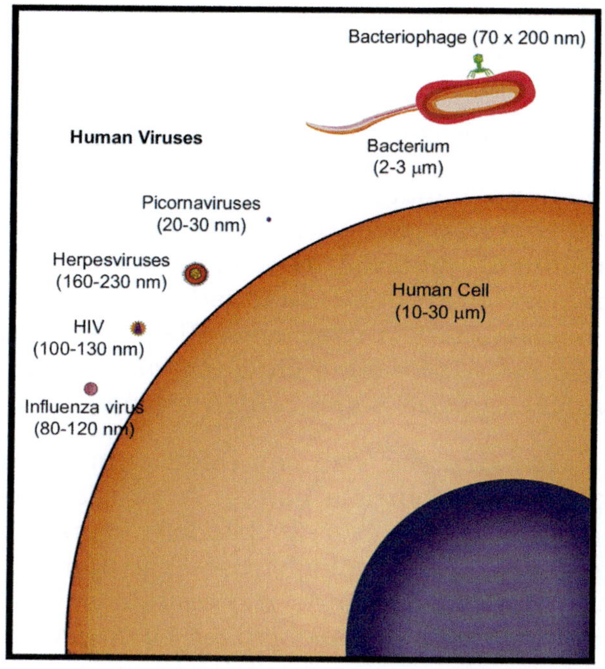

Figure 4.1. *Virus and cell size comparison.*

Source: https://www.ncbi.nlm.nih.gov/pmc/articles/PMC7150055/bin/f02-01-9780128009475.jpg.

Human viruses range in diameter from 20 to 200 nm, though they can be larger or smaller. By contrast, the average length of a human cell is 10–30 μM, while bacteria are typically 2-3 μM.

4.1.3.2. Viruses are Obligate Intracellular Parasites, Meaning That They Completely Depend Upon the Internal Environment of the Cell to Create New Infectious Virus Particles or Virions.

All viruses attach themselves to a cell's surface to enter the cell. The virus breaks down into smaller pieces, and the nucleic acid that makes up its genetic material contains instructions for the proteins that will self-assemble to form new virions. The term "de novo replication" comes from the Latin "from new." Viral particles use the energy and machinery of the host cell to build new virions entirely from scratch, in contrast to cells, which divide into two equal pieces and grow in size.

4.1.3.3. The Genetic Material of Viruses can be Composed of DNA or RNA.

The genetic material found in all living cells, including those of humans, animals, plants, and bacteria, is double-stranded DNA, or dsDNA. In contrast, viruses possess genomes, or genetic material, that may consist of either RNA or DNA (but not both). Not all viruses have double-stranded genomes; some can have single-stranded DNA (ssDNA) genomes, and viruses with RNA genomes can have either single- or double-stranded genomes. However, a virus will only have one kind of nucleic acid genome; thus, viruses with both ssDNA and ssRNA genomes, for example, are not found (Bourne et al., 2006).

The genome size can differ significantly between viruses, much like how the size of the virus particle varies. A virus's genome typically has between 7000 and 20,000 base pairs (bp) or 7–20 kilobase pairs (kb). Larger viruses do not always have large genomes, but smaller viruses can naturally hold less nucleic acid than larger virions. While most viruses do not contain much nucleic acid, some dsDNA viruses have very large genomes: herpesviruses have genomes that are 120-200 kb in total, and the very large pandoraviruses mentioned previously have the largest genomes: up to 2.5 million bases, rivaling the genome size of many bacteria! In comparison, eukaryotic cells have much larger genomes: a red alga has the smallest known eukaryotic genome, at 8 million base pairs; a human cell contains over 3 billion nucleotides in its hereditary material; the largest genome yet sequenced, at over 22 billion base pairs, is that of the loblolly pine tree.

4.1.3.3.1. Virion Size: Getting Smaller

1000 millimeters (mm) in a meter (m)	1 mm = 10^{-3} m
1000 micrometers (µm, or microns) in a millimeter	1 µm = 10^{-6} m
1000 nanometers (nm) in a micrometer	1 nm = 10^{-9} m

4.1.3.3.2. Virus Genome Size: Getting Bigger

1000 base pairs (nucleotide pairs, bp) in a kilobase pair (kb)	1 kb = 10^{3} bp
1000 kb in a megabase pair (mb)	1 mb = 10^{6} bp
1000 mb in a gigabase pair (gb)	1 gb = 10^{9} bp

4.2. STRUCTURE OF VIRUSES

Since viruses are so important and it was discovered that their structures were sufficiently simple to understand, there has been a significant amount of research done on virus morphology in recent decades. Several methods have been used to advance science, including immunology, X-ray diffraction, biochemical analysis, and electron microscopy. The broad characteristics of virus structure are becoming evident, even though the vast variety of viruses means that our knowledge is still incomplete.(Figure 4.2) (Bourne et al., 2006).

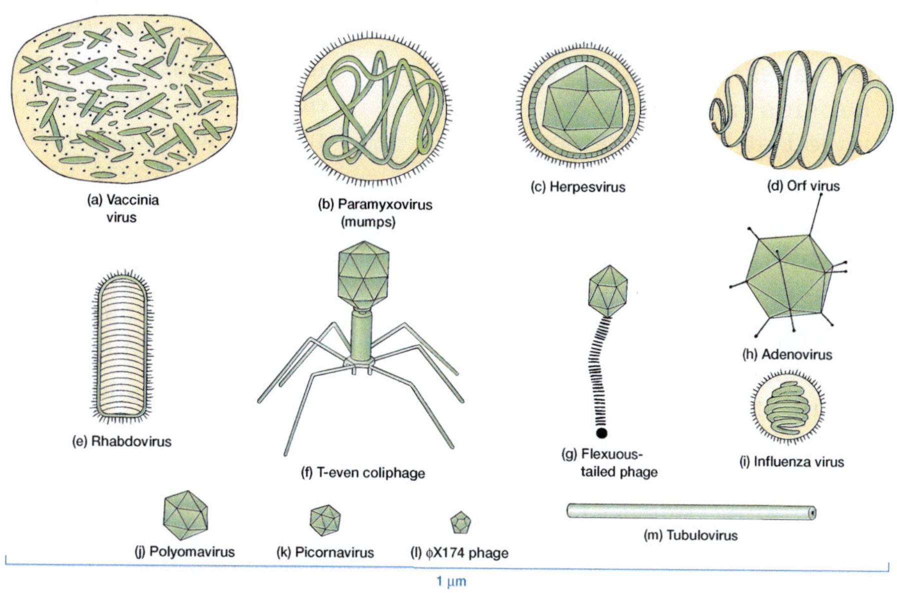

Figure 4.2. *The size and morphology of selected viruses. The viruses are drawn to scale. A 1 m line is provided at the bottom of the figure*

Source: https://asutoshcollege.in/new-web/Study_Material/Virus_07042020.pdf.

4.2.1. Virion Size

The diameter of virions varies from approximately 10 to 300 or 400 nm. The poxviruses, such as vaccinia, are about the same size as the smallest bacteria and are visible under a light microscope, while the smallest viruses are slightly larger than ribosomes. However, the majority of viruses must be examined using scanning and transmission electron microscopes because they are too tiny to be seen under a light microscope.

4.2.2. General Structural Properties

A nucleocapsid core serves as the building block for all virions, regardless of other components they may have. In fact, certain viruses are made entirely of nucleocapsids. The viral genetic material is protected and assisted in its transfer between host cells by the capsid, a protein coat that surrounds the nucleic acid, either DNA or RNA, that makes up the nucleocapsid.

There are four general morphological types of capsids and virion structures.

- Some capsids are icosahedral in shape. An icosahedron is a regular polyhedron with 20 equilateral triangular faces and 12 vertices. These capsids appear spherical when viewed at low power in the electron microscope.
- Other capsids are helical and shaped like hollow protein cylinders, which may be either rigid or flexible
- Many viruses have an envelope, an outer membranous layer surrounding the nucleocapsid. Enveloped viruses have a roughly spherical but somewhat variable shape even though their nucleocapsid can be either icosahedral or helical
- Complex viruses have capsid symmetry that is neither purely icosahedral nor helical. They may possess tails and other structures (e.g., many bacteriophages) or have complex, multilayered walls surrounding the nucleic acid (e.g., poxviruses such as vaccinia).
- Large macromolecular structures called icosahedral and helical capsids comprise numerous copies of one or more different kinds of protomers or subunits of proteins. The fact that the information contained in viral genetic material is utilized as efficiently as possible is likely the most significant benefit of this design strategy. For instance, the capsid of the tobacco mosaic virus (TMV) has a single class of small subunit with 158 amino acids. Out of the 6,000 nucleotides in the virus RNA, only roughly 474 are needed to code for the amino acids found in coat proteins. A large nucleic acid, like the TMV RNA, cannot be encased in a protein coat without utilizing most or all of the available genetic material to code for capsid proteins, unless the same protein is used repeatedly in the construction of capsids. About 2,900 of the 6,000 nucleotides needed to construct the TMV capsid would be needed if it were made up of six distinct protomers the same size as the TMV subunit, freeing up much less genetic material for other uses (Bresnahan and Shenk, 2000).

Protozomers typically interact specifically with one another and spontaneously associate to form the capsid once they have formed and are exposed to the right conditions. The process is known as self-assembly because the capsid is put together without the assistance of any outside party. Certain more intricate viruses have genes encoding unique components needed for the virion's assembly but not included in it.

For the infectious virus to infect other cells and people, the host cell must release the infectious virus particle. Whether it is dsDNA, ssDNA, dsRNA, or ssRNA, the virus's nucleic acid genome needs to be safeguarded throughout the process. The virus will come into contact with enzymes in the extracellular environment that can degrade or break down nucleic acid. Shearing the nucleic acid strands into fragments could also occur from physical stresses like air or liquid flow. Furthermore, just like human DNA, the genomes of viruses can be harmed by radioactivity or UV light. Should the virus's nucleic acid genome sustain damage, it will be incapable of generating new viral particles.

The virus envelops its nucleic acid in a protein shell called the capsid, derived from the Latin capsa, which means "box," to shield the delicate nucleic acid from this hostile environment. The capsid is made up of one or more distinct protein types that, like bricks fitting together to form a wall, repeat repeatedly to form the entire capsid. This pattern of repetition creates a robust but somewhat flexible capsid. In addition to being small in size, the capsid adequately shields the nucleic acid inside of it and is physically very difficult to break open. The nucleocapsid of the virion comprises the nucleic acid and the capsid together (Figure 4.3).

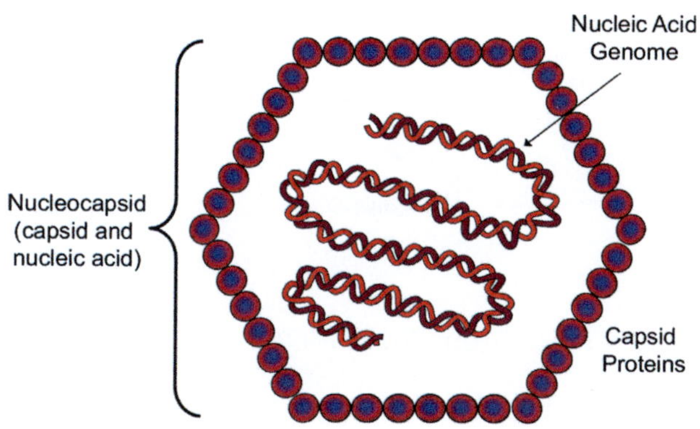

Figure 4.3. *Basic virus architecture.*

Source: https://www.ncbi.nlm.nih.gov/pmc/articles/PMC7150055/bin/f02-02-9780128009475.jpg.

Viral capsid proteins shield the delicate nucleic acid genome from environmental damage. The nucleocapsid is the combined name for the capsid and nucleic acid.

Keep in mind that most viruses have significantly small genomes. Genes contain the information needed to produce proteins, so smaller genomes have limited capacity to encode multiple proteins. This is why the capsid of the virus is made up of a single or only a few proteins that repeat to create the structure. If the nucleic acid of the virus consisted of more proteins, it would be too large to fit inside the capsid.

Similar to how a roll of magnets comes together on its own, capsid proteins are also able to self-assemble. H. was the first to demonstrate this. Robley Williams and Fraenkel-Conrat in 1955. When the protein subunits and RNA genome of the tobacco mosaic virus were reunited in a test tube, infectious virions spontaneously generated. This suggested that all the information needed to put together a virus already exists: hydrophobic and electrostatic forces will mainly hold the physical parts together while the virus assembles itself.

An envelope usually envelops the capsid of the virus. The cell's endoplasmic reticulum, Golgi complex, or even nuclear membrane can provide the lipid membrane that makes up the envelope. The plasma membrane is the source of the envelope most frequently (Cáceres et al., 2013).

Virus capsids are held together by bonds similar to those in living organisms. Covalent bonds are rare in capsids, as they involve atoms sharing electrons and are the strongest type of bond. Electrostatic interactions, which are weaker than covalent bonds, occur when oppositely charged atoms are attracted to each other. Ionic bonds are a type of electrostatic interaction that happens between oppositely charged atoms or ions. Hydrogen bonds are weak electrostatic forces between slightly charged atoms, typically involving hydrogen (slightly positively charged) and another partially negatively charged atom like oxygen. Van der Waals forces are weak interactions that occur when an atom becomes slightly charged due to random asymmetry in its electrons.

The way viruses assemble and adhere to cells is also influenced by the characteristics of water. Water is a polar molecule, meaning it has two distinct ends, similar to the positive and negative ends of a battery or magnet. Nonpolar molecules are defined as having no distinct ends. Since water is polar, it attracts other polar molecules as well. The term "water loving" describes these molecules. Hydrophobic, or "water avoiding," molecules are nonpolar and do not attract water. As a result, hydrophobic molecules will group together in an aqueous solution. This explains the phenomenon where water (polar) does not mix with oil (nonpolar).

These viruses frequently contain proteins known as matrix proteins, which serve to join the envelope and the capsid within. A virus is called nonenveloped or naked if it has no envelope. Additionally, each virus has an embedded virus attachment protein in its outermost layer. If the virus is enveloped, it will be found in the envelope; otherwise, it will be found in the capsid. The viral protein known as the virus attachment protein helps the virus dock to the host cell's plasma membrane, which is the initial stage of infection (Figure 4.4).

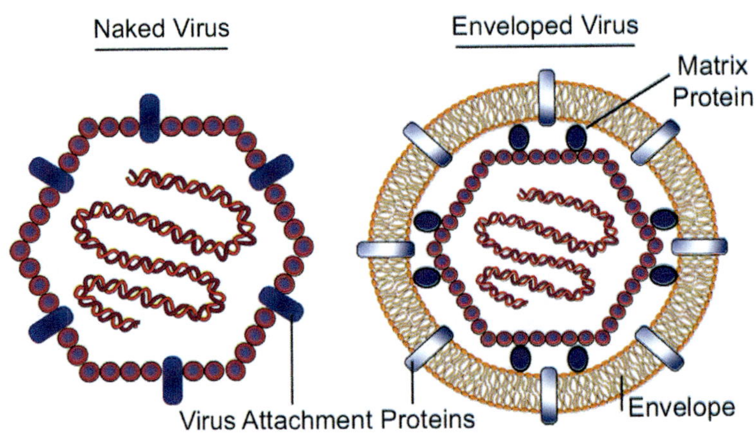

Figure 4.4. *Comparison between a naked and enveloped virion.*

Source: https://www.ncbi.nlm.nih.gov/pmc/articles/PMC7150055/bin/f02-03-9780128009475.jpg.

The capsid of an enveloped virion is wrapped with a lipid membrane derived from the cell. Virus attachment proteins located in the capsid or envelope facilitate the binding of the virus to its host cell.

4.2.3. Helical Capsid Structure

Every virus has a capsid made of proteins to shield its nucleic acid genome from the elements. Two shapes of virus capsids are commonly observed: icosahedral and helical. A spiral shape that curves cylindrically around an axis is called a helix (plural: helices). It is also a typical biological structure: DNA is a double helix of nucleotides, and many proteins contain helical-shaped sections. In a helical virus, the capsid proteins wind around the outside or inside of the nucleic acid to form a long tube or rod-like structure, while the viral nucleic acid coils into a helical shape (Clare and Orlova, 2010). The nucleocapsid is made up of the capsid and nucleic acid. The term "nucleocapsid protein" refers to the protein that encircles the nucleic acid. The helical nucleocapsid uncoils once inside the cell, allowing the nucleic acid to be accessible (Figure 4.5).

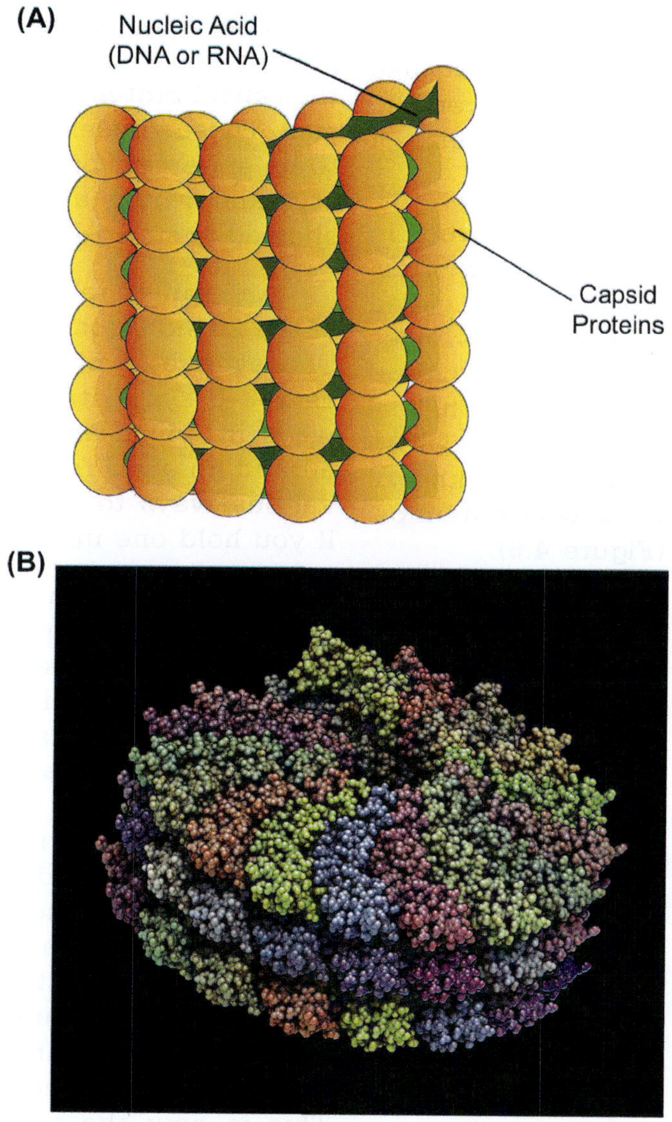

Figure 4.5. *Helical capsid structure. (A) Viral capsid proteins wind around the nucleic acid, forming a helical nucleocapsid. (B) Helical structure of tobacco mosaic virus. Rendering of 49 subunits was performed using QuteMol using a 2xea PDB assembly.*

Source: https://www.ncbi.nlm.nih.gov/pmc/articles/PMC7150055/bin/f02-04-9780128009475.jpg.

Forming a helical capsid is thought to have a few benefits. Initially, just one kind of capsid protein is needed. The capsid is formed by repeatedly repeating this protein subunit.

Comparing this structure to a capsid made of several proteins, it is simpler and requires less free energy to assemble (Dixon et al., 2012). Furthermore, the length of nucleic acid needed is decreased because only one nucleocapsid protein requires the expression of one gene rather than multiple. There are also no limits on the amount of nucleic acid that can be packed into the virion because the helical structure can go on forever; the length of the capsid will equal the size of the coiled nucleic acid.

Helical viruses can be covered in clothing or not. The tobacco mosaic virus, the first virus to be identified, is a naked helical virus. The majority of plant viruses are helical, and enveloped forms of helical plant viruses are extremely rare. All helical animal viruses, on the other hand, are enveloped. These comprise well-known viruses like Ebola, rabies, measles, mumps, and influenza viruses (Figure 4.6).

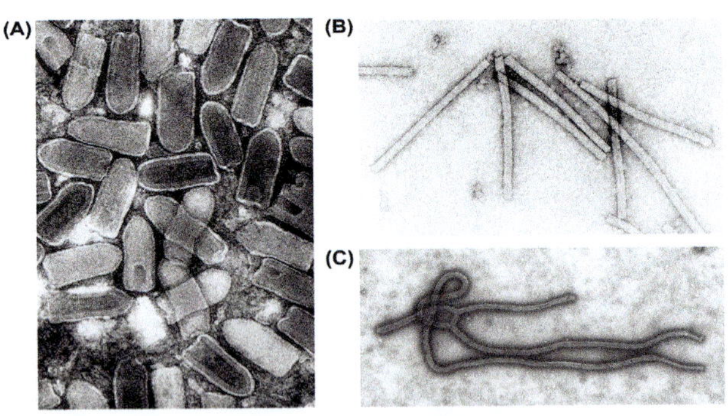

Figure 4.6. *Electron micrographs of helical viruses. (A) Vesicular stomatitis virus forms bullet-shaped helical nucleocapsids. (Image courtesy of CDC/Dr. Fred (a) Murphy.) (B) Tobacco mosaic virus forms long helical tubes. (Image courtesy of the USDA Beltsville Electron Microscopy Unit.) (C) The helical Ebola virus forms long threads that can extend over 1000 nm in length. (Image courtesy of CDC/Cynthia Goldsmith.)*

Source: https://www.ncbi.nlm.nih.gov/pmc/articles/PMC7150055/bin/f02-05-9780128009475.jpg.

4.2.4. Icosahedral Capsid Structure

The icosahedron is far more common than the helical architecture among the two main capsid structures. The genomes of icosahedral viruses are entirely enclosed in an icosahedral capsid that functions as a protein shell, unlike helical viruses where the capsid proteins wind around the nucleic acid. These viruses were originally believed to be spherical, but developments in X-ray crystallography and electron microscopy showed their structure was icosahedral.

An equilateral triangle makes up each of the 20 sides (or faces) of an icosahedron. The symmetry of an icosahedron, known as 2–3–5 symmetry, describes the various ways in which an icosahedron can rotate around an axis. You'll notice that there are various ways to rotate an icosahedral die if you hold one in your hand. Assume you.

Two mathematical parameters—amplitude and pitch—define a helix and are also applicable to helical capsid structures. The amplitude, which indicates the width of the capsid, is just the helix's diameter. The height or distance of one full turn of the helix is known as the pitch. Similar to how the height of the stairs on a one-story staircase can be added, we can calculate the pitch of the helix by measuring the rise, or distance, that each capsid subunit gains. A 10-foot staircase has 20 steps, each measuring 6 inches; a virus with 16.3 subunits per turn and a rise of 0.14 nm for each subunit produces a pitch of 2.28 nm. This is how the tobacco mosaic virus is constructed (Harrison,1984).

We put an imaginary pencil through the center of one of the icosahedron's edges while looking directly at it. The center of a triangle with its triangles facing up and down would be where your pencil would be.

The icosahedron has the same arrangement (symmetry) when rotated 180° clockwise: two triangles, one facing up and one facing down. You return to the starting point by keeping the icosahedron rotating. This axis of symmetry is called the "twofold axis" because, when you rotate the shape around it using your pencil, you will come across your starting structure twice in a single revolution: once at the beginning and once after you have rotated it 180°.

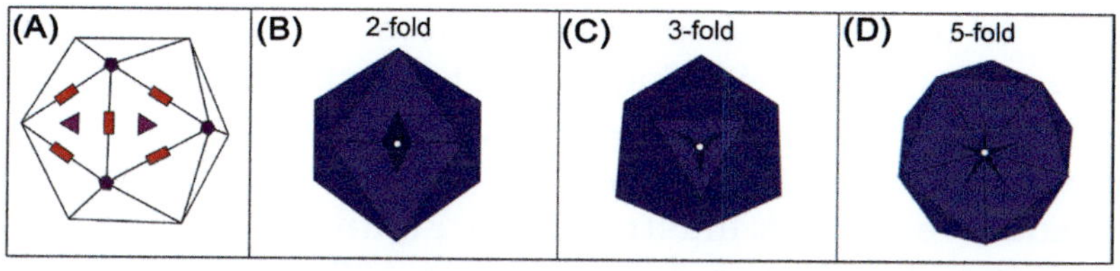

Figure 4.7. *Icosahedron terminology and axes of symmetry. (A) Icosahedron faces (fuchsia triangles), edges (red rectangles), and vertices (violet pentagons) are indicated on the white icosahedron. (B) The twofold axis of symmetry occurs when the axis is placed through the center of an edge. The threefold axis occurs when the axis is placed in the center of a face (C), and the fivefold axis passes through a vertex of the icosahedron (D).*

Source: https://www.ncbi.nlm.nih.gov/pmc/articles/PMC7150055/bin/f02-06-9780128009475.jpg.

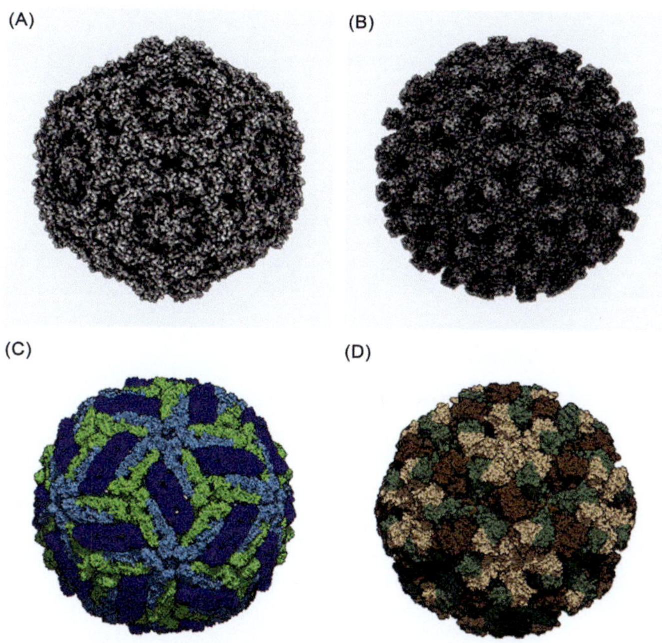

Figure 4.8. *Illustrations of viruses, as viewed on the twofold axis of rotation. 3D surface reconstructions of parvovirus B19 (A), human hepatitis B virus (B), dengue virus (C), and Norwalk virus (D). Illustrations created with QuteMol (IEEE. Trans. Vis. Comput. Graph 2006, 12, 1237–44) using 2G33 (J. Virol. 2006, 80, 11055–61), 1S58 (Proc. Natl. Acad. Sci. U. S. (a) 2004, 101, 11628–33), 1K4R (Cell 2002, 108, 717–25), and 1IHM (Science 1999, 286, 287–90) PDB assemblies.*

Source: https://www.ncbi.nlm.nih.gov/pmc/articles/PMC7150055/bin/f02-06-9780128009475.jpg.

Conversely, if you rotate the shape such that your pencil axis passes through the center of one of the icosahedron's small triangle faces, you will see the first view twice more, for a total of three views. The triple axis is this. Similarly, you will find symmetry five times in one rotation if your pencil axis passes through an icosahedron's tip or vertex. This is known as the fivefold axis. Because an icosahedron has twofold, threefold, and fivefold axes of symmetry, it is therefore known to possess 2–3–5 symmetry. When working with an icosahedral virus, this terminology is helpful because it can be used to identify particular spots on the virus or areas where the virion interacts with the cell surface. For example, if the interaction happens at the threefold axis, you can determine that a virus interacts with a cell surface receptor at one of the icosahedron's faces. One of the icosahedron's vertices (tips) will have a protein sticking out of the capsid at the fivefold axis. Each and every virus illustration in Figure 4.7. When looking at, consider the twofold axis of symmetry.(Figure 4.8) (Harrison,1984).

Every face (small triangle) of the icosahedral capsid comprises viral proteins. However, because viral proteins are not triangular, the entire face cannot be formed by a single protein subunit. As a result, at least three viral protein subunits are fitted together to form a face. These could be three distinct proteins, or they could all be the same protein. The structural unit is made up of all of the subunits. The virion's capsid is formed by repeating the structural unit.(Figure 4.9) (Kaufmann et al., 2004).

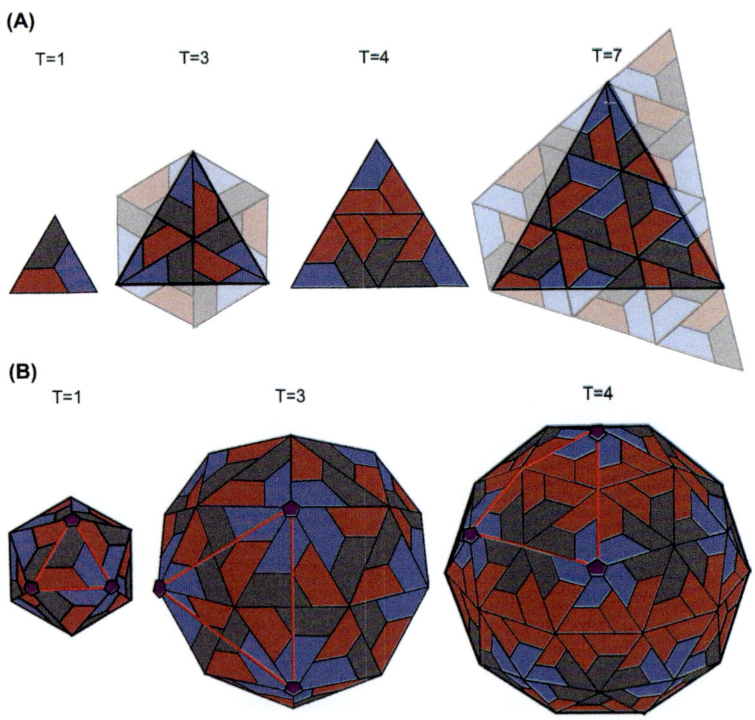

Figure 4.9. *Capsid architecture and triangulation number.*

Source: https://www.ncbi.nlm.nih.gov/pmc/articles/PMC7150055/bin/f02-08-9780128009475.jpg.

(i) Viral protein subunits make up the structural units of virus capsids. The quantity of structural units on each face of the icosahedron is indicated by the triangulation number (T). The icosahedron face of a T = 1 virus is made up of a single structural unit that is made up of three distinct protein subunits: gray, red, and blue. (ii) Capsids of virion with T = 1, T = 3, and T = 4. The triangular face of the icosahedron is delineated by red lines, and its vertices, or fivefold axes, are indicated by purple pentagons

Repetition is the key to understanding how some viruses are able to form very large icosahedral capsids. To create a larger icosahedron side, the structural unit can be repeated numerous times. Because the structural units form the icosahedron's triangle face, the number of structural units that create each side is known as the triangulation number (T). A single structural unit makes up each icosahedron face in a T = 1 virus (Lwoff and Tournier, 1971). The face of a T = 4 virus comprises four structural units. In a T = 3 virus, the face is formed by three structural units in total, but this happens as six half-units (half of each structural unit forms part of an adjacent face). This is an example of how the structural unit can overlap from one face to another. Comparing a T = 7 virus's structural units to the triangle face, they are similarly somewhat skewed. Only the most fundamental concepts are covered here due to the complexity of the geometry and math associated with the icosahedral capsid structure. In any case, the icosahedron can grow larger and larger without requiring the synthesis of new proteins by increasing the number of identical structural units on each face. A few viruses even have triangulation numbers greater than 25.

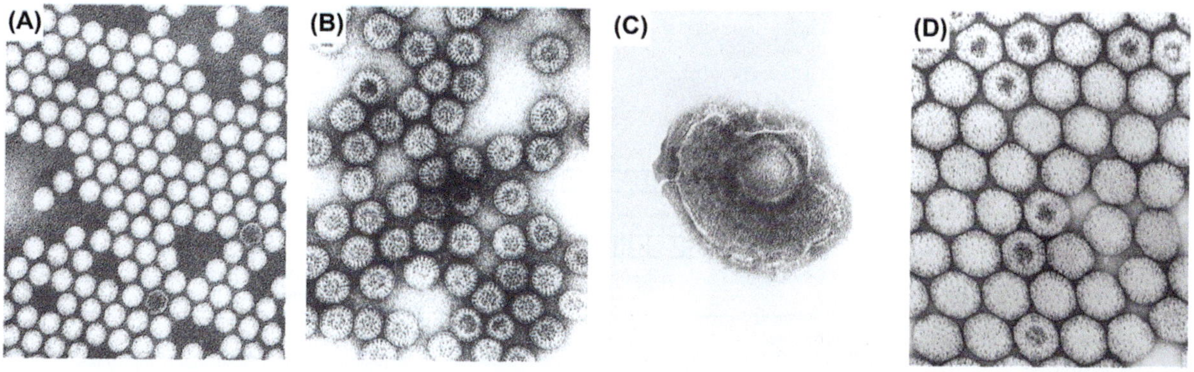

Figure 4.10. *Electron micrographs of icosahedral viruses. Poliovirus (A), rotavirus (B), varicella–zoster virus (C), the virus that causes chickenpox and shingles, and reovirus (D). Note that C is enveloped. Images courtesy of the CDC: Dr. Fred Murphy and Sylvia Whitfield (A), Dr. Erskine Palmer (B and D), and Dr. Erskine Palmer and B.G. Partin (C).*

Source: https://www.ncbi.nlm.nih.gov/pmc/articles/PMC7150055/bin/f02-09-9780128009475.jpg.

The proteins within the structural unit can be arranged into capsomeres, which are visible in electron micrographs and form three-dimensional structures. In icosahedral viruses, capsomeres typically appear as pentons or hexons on the surface of the icosahedron, forming a recognizable pattern. These capsomeres are morphological units that result from the interaction of proteins within the repeated structural units. Because icosahedral symmetry proteins require less energy to form than other structures, research indicates that the icosahedral virus structure is evolutionarily favored and explains why it appears so frequently.

Icosahedral viruses, such as the rhinovirus, hepatitis B virus, herpesvirus, and human papillomavirus, are commonly infecting animals. Icosahedral viruses have the same ability to be either naked or enveloped as their helical counterparts. The structure of the capsid is unrelated to the type of viral nucleic acid (dsDNA, ssDNA, dsRNA, and ssRNA); depending on the virus, icosahedral viral capsids can contain any of the nucleic acid types (Figure 4.10).

4.2.5. Nucleic Acids

In terms of the makeup of their genetic material, viruses are incredibly adaptable. They use single-stranded DNA, double-stranded DNA, single-stranded RNA, and double-stranded RNA, the four types of nucleic acids that can exist. Animal viruses contain all four varieties. The majority of plant viruses have genomes made of single-stranded RNA. Bacterial viruses typically contain double-stranded DNA, but phages can also contain single-stranded RNA or DNA (Matthews, 1985).

Table 4.1. *Types of Viral Nucleic Acids*

Nucleic Acid Type	Nucleic Acid Structure	Virus Examples
DNA		
Single-Stranded	Linear single strand Circular single strand	Parvoviruses ϕX174, M13, fd phages
Double-Stranded	Linear double strand	Herpesviruses (herpes simplex viruses, cytomegalovirus, Epstein-Barr virus), adenoviruses, T coliphages, lambda phage, and other bacteriophages
	Linear double strand with single chain breaks 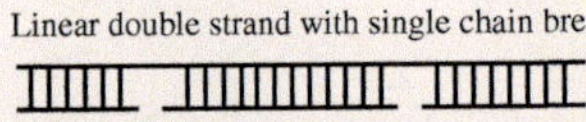	T5 coliphage
	Double strand with cross-linked ends 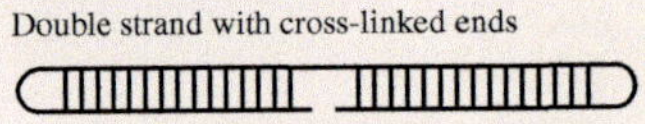	Vaccinia, smallpox
	Closed circular double strand	Polyomaviruses (SV-40), papillomaviruses, PM2 phage, cauliflower mosaic
RNA		
Single-Stranded	Linear, single stranded, positive strand	Picornaviruses (polio, rhinoviruses), togaviruses, RNA bacteriophages, TMV, and most plant viruses
	Linear, single stranded, negative strand	Rhabdoviruses (rabies), paramyxoviruses (mumps, measles)
	Linear, single stranded, segmented, positive strand	Brome mosaic virus (individual segments in separate virions)
	Linear, single stranded, segmented, diploid (two identical single strands), positive strand	Retroviruses (Rous sarcoma virus, human immunodeficiency virus)
	Linear, single stranded, segmented, negative strand	Paramyxoviruses, orthomyxoviruses (influenza)
Double-Stranded	Linear, double stranded, segmented	Reoviruses, wound-tumor virus of plants, cytoplasmic polyhedrosis virus of insects, phage ϕ6, many mycoviruses

Source: Modified from S. E. Luria et al., General Virology, 3d edition, 1983. John Wiley & Sons, Inc., New York, NY. https://asutoshcollege.in/new-web/Study_Material/Virus_07042020.pdf.

Table 4.1 summarizes many variations seen in viral nucleic acids. The size of viral genetic material also varies greatly. The smallest genomes (those of the MS2 and Q⊠

viruses) are around 1 × 10^6 daltons, just large enough to code for three to four proteins. MS2, Qβ, and other viruses save space using overlapping genes. At the other extreme, T-even bacteriophages, herpesvirus, and vaccinia virus have genomes of 1.0 to 1.6 × 10^8 daltons and may be able to direct the synthesis of over 100 proteins. In the following paragraphs the nature of each nucleic acid type is briefly summarized.

Tiny DNA viruses like ϕX174 and M13 bacteriophages or the parvoviruses possess single-stranded DNA (ssDNA) genomes. Some of these viruses have linear pieces of DNA, whereas others use a single, closed circle of DNA for their genome (Prasad et al., 1999).

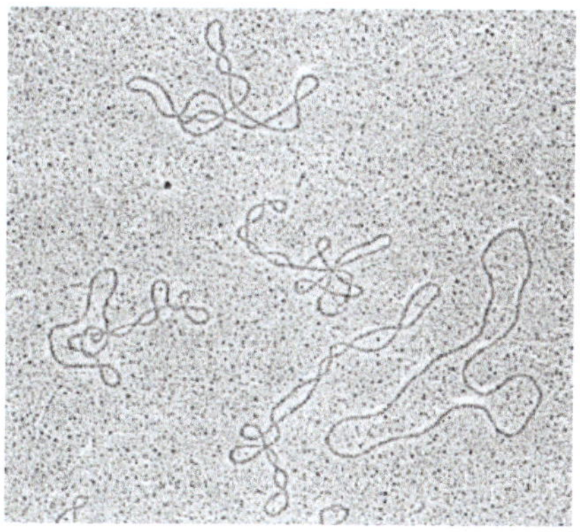

Figure 4.11. *Circular Phage DNA. The closed circular DNA of the phage PM2 (× 93,000). Note both the relaxed and highly twisted or supercoiled forms.*

Source: https://asutoshcollege.in/new-web/Study_Material/Virus_07042020.pdf.

Double-stranded DNA, or dsDNA, is the genetic material used by the majority of DNA viruses. Many viruses contain modified linear dsDNA, while others contain circular dsDNA. Because of its cohesive ends, which are single-stranded complementary segments 12 nucleotides long, the lambda phage can cycle when its linear dsDNA base pairs with another phage (Figures 4.11 and 4.12).

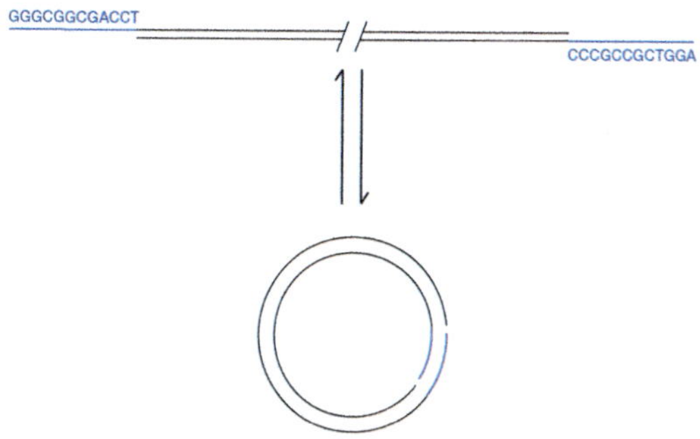

Figure 4.12. *Circularization of Lambda DNA. The linear DNA of the lambda phage can be reversibly circularized. This is made possible by cohesive ends (in color) that have complementary sequences and can base pair with each other.*

Source: https://asutoshcollege.in/new-web/Study_Material/Virus_07042020.pdf.

Numerous virus DNAs contain unique bases in addition to the typical nucleotides found in DNA. The T-even phages of E, for instance. Instead of cytosine, coli have 5-hydroxymethylcytosine. Typically, glucose is joined to the hydroxymethyl group (Prasad et al., 1999).

Single-stranded RNA (ssRNA) is the genetic material used by most RNA viruses. The RNA strand is referred to as the plus strand or positive strand if the base sequences of the RNA and viral mRNA are the same (viral mRNA is defined as plus or positive). On the other hand, the viral RNA genome is referred to as a minus or negative strand if it is complementary to the viral mRNA. Positive strand RNA viruses include those that cause polio, tobacco mosaic, brome mosaic, and Rous sarcoma; negative-strand RNA viruses include those that cause rabies, mumps, measles, and influenza. A large number of these RNA genomes are segmented, meaning they are split up into different parts. Every segment or fragment is thought to code for a single protein. Even though some virus genomes may consist of up to 10 or 12 segments, most segments are most likely all enclosed in the same capsid. For reproduction to be successful, all segments do not have to be found in the same virion. The genome of the brome mosaic virus consists of four segments split up among three distinct virus particles. For infectivity, all three of the largest segments must be present. The various brome mosaic virions are still able to successfully infect the same host in spite of this intricate and apparently ineffective arrangement.

Plus, strand viral RNA frequently shares similarities with mRNA beyond just nucleotide sequence equivalency. Similar to how eucaryotic mRNA typically has a 7-methylguanosine 5′ cap, the genomes of many plant and animal viruses are capped. Furthermore, the majority of all plus-strand RNA animal viruses share a poly-A stretch at the 3′ ends of their genome, making them structurally similar to eucaryotic mRNA on both ends. Plus, strand RNAs can actually control protein synthesis as soon as they enter the cell. The tobacco mosaic virus's genomes can actually take amino acids, and several single-stranded plant viral RNAs have 3′ ends that resemble eucaryotic transfer RNA. There is no evidence of capping in RNA bacteriophages. The genomes of a few viruses are double-stranded RNA (dsRNA). All seem to be divided into several segments; some, like reoviruses, have as many as ten or twelve. It is known that these dsRNA viruses can infect fungi, animals, plants, and even a single species of bacteria (Prasad and Schmid, 2012).

4.2.6. Viral Envelopes and Enzymes

An envelope-like outer membrane surrounds the majority of animal viruses, certain plant viruses, and at least one bacterial virus. Animal virus envelopes typically originate from nuclear or plasma membranes of the host cell; the lipids and carbohydrates found in these membranes are typical host components. On the other hand, virus genes code for envelope proteins, which can even protrude as peplomers or spikes from the envelope surface. There is a chance that these spikes help the virus attach to the surface of the host cell. They can also be used for virus identification because they vary amongst viruses. Enclosed viruses are known as pleomorphic because they usually have a somewhat variable shape due to the flexible, membranous structure of the envelope. Viruses, such as the bullet-shaped rabies virus, have envelopes that are firmly attached to the underlying nucleocapsid, giving the virion a consistent, recognizable shape. Solvents such as ether can cause the envelope of certain viruses to become so disrupted that lipid-mediated activities are blocked or envelope proteins become denatured and rendered inactive. It is then stated that the virus is "ether sensitive."

The influenza virus is a well-researched enveloped virus with spikes projecting about 10 nm from its surface at intervals of 7 to 8 nm. Some spikes contain neuraminidase enzymes, which may help the virus penetrate mucous layers to reach host cells. Other spikes have hemagglutinin proteins that bind virions to red blood cell membranes and cause hemagglutination. Hemagglutinins aid in virion attachment to host cells. Proteins, such as the exposed spike proteins on the outer envelope surface, are typically glycoproteins, meaning they have carbohydrates attached. The M or matrix protein, a nonglycosylated protein found on the inner envelope surface, helps stabilize it (Figures 4.13 and 4.14).

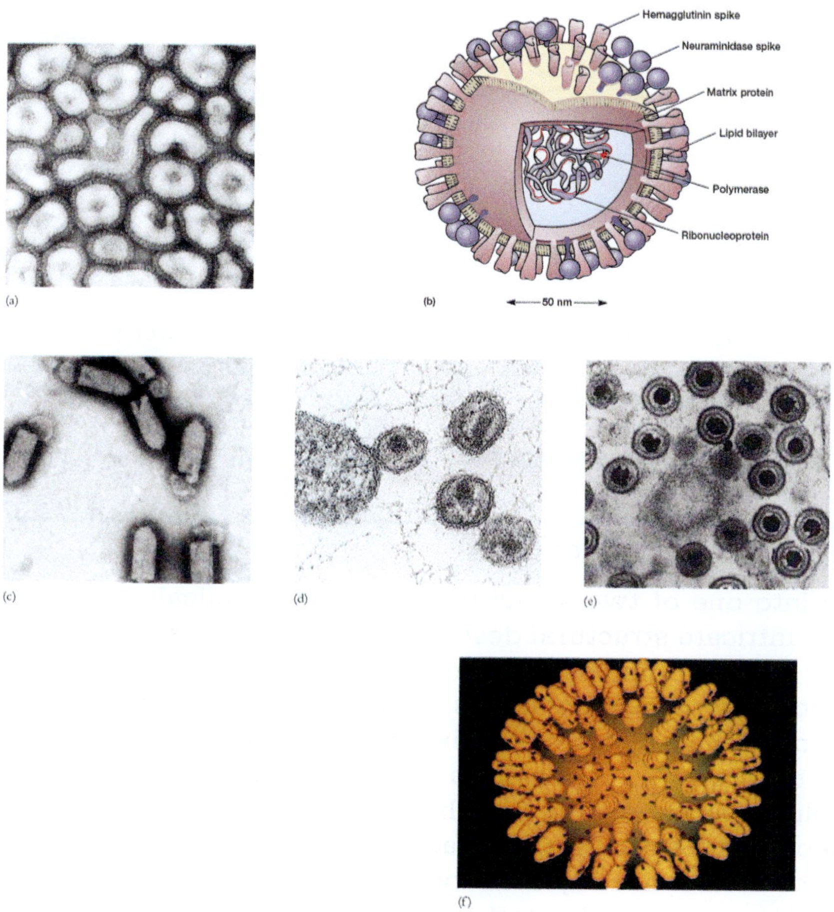

Figure 4.13. *Examples of enveloped viruses. (a) Human influenza virus. Note the flexibility of the envelope and the spikes projecting from its surface (×282,000). (b) Diagram of the influenza virion. (c) Rhabdovirus particles (×250,000). This is the vesicular stomatitis virus, a relative of the rabies virus, which is similar in appearance. (d) Human immunodeficiency viruses (×33,000). (e) Herpesviruses (×100,000). (f) Computer image of the Semliki Forest virus, which occasionally causes encephalitis in humans.*

Source: https://asutoshcollege.in/new-web/Study_Material/Virus_07042020.pdf.

It has come to light that virions do not lack enzymes, despite the initial theory that they only contained structural capsid proteins. Enzymes are sometimes linked to the envelope or capsid (e.g., neuraminidase of influenza). The majority of viral enzymes are most likely found inside the capsid. Numerous of these are connected to the replication of nucleic acids. As an illustration, the influenza virus uses RNA for its genetic material and is accompanied by an RNA-dependent RNA polymerase that functions as an RNA transcriptase, synthesizing mRNA in response to instructions from its RNA genome. Ribonucleoprotein is linked to the polymerase. Viruses may contain one or more enzymes necessary for completing their life cycles even though they do not have a true metabolism and cannot proliferate without the assistance of living cells.

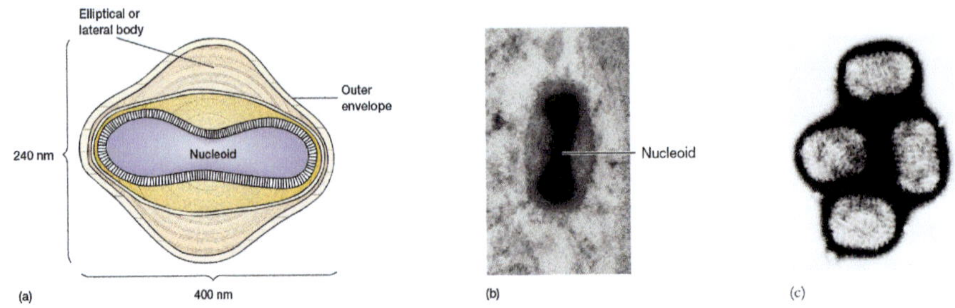

Figure 4.14. *Vaccinia virus morphology. (a) Diagram of vaccinia structure. (b) Micrograph of the virion clearly showing the nucleoid (×200,000). (c) Vaccinia surface structure. An electron micrograph of four virions showing the thick array of surface fibers (×150,000).*

Source: https://asutoshcollege.in/new-web/Study_Material/Virus_07042020.pdf.

4.2.7. Complex Viral Structures

Most viruses fall into one of two structural categories: helical or icosahedral. However, some viruses have intricate structural designs that defy strict adherence to an icosahedral or simple helical shape. Complexly structured viruses include many bacteriophages, geminiviruses, and pyxviruses. Poxviruses are big, oval- or brick-shaped viruses ranging from 200 to 400 nm. They include the viruses that cause cowpox and smallpox. Two "lateral bodies," the purpose of which is currently unknown, surround the dumbbell-shaped core of the complex virion, which contains the viral DNA. The geminiviruses also present complex structure (Tarini et al., 2006). These viruses that infect plants are made up of two icosahedral heads joined together, as their name implies. Bacteriophages, viruses that replicate and infect bacteria, are also referred to as prokaryotic or bacterial viruses. Numerous bacteriophages exhibit intricate structural features as well. For example, bacteriophage P2 comprises an icosahedral head that houses the nucleic acid and is connected to a cylindrical tail sheath that aids in the bacteriophage's attachment to the bacterial cell (Figure 4.15).

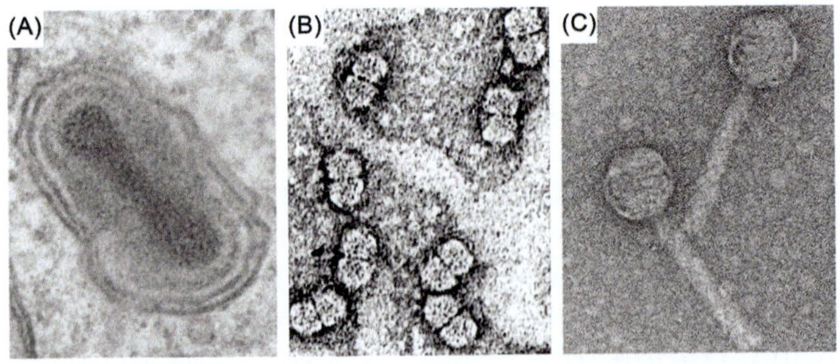

Figure 4.15. *Electron micrograph of viruses with complex architecture. Vaccinia virus (A), a virus belonging to the poxvirus family, has a complex capsid architecture with a dumbbell-shaped core. Geminiviruses (B) have a double-icosahedron capsid. Bacteriophages, such as P2 (C), often have complex capsid structures. Images courtesy of Ana Caceres et al. (A, PLoS Pathog. 2013, 9(11), e1003719), Kassie Kasdorf (B), and Mostafa Fatehi (C).*

Source: https://www.ncbi.nlm.nih.gov/pmc/articles/PMC7150055/bin/f02-10-9780128009475.jpg.

4.3. VIRUS CLASSIFICATION AND TAXONOMY

There are several reasons why the classification of viruses is helpful. By comparing newly discovered viruses to similar viruses, it enables scientists to compare and contrast viruses and reveal information about them. It also enables researchers to examine the genesis and evolutionary history of viruses. Virus classification is not easy, though, as there are currently over 2800 distinct viral species, each with a wide range of characteristics (Van Regenmortel, 1990).

A classification system was created in the 1970s by David Baltimore, a Nobel laureate. Viruses are categorized using the Baltimore classification system according to their replication strategy and type of nucleic acid genome. Additionally, the system distinguishes between positive strand (+) and negative strand (–) single-stranded RNA viruses. Messenger RNA (mRNA) in a cell is a positive strand because positive-strand (also known as positive-sense or plus-strand) RNA can be translated into proteins immediately, as will be covered in more detail in the upcoming chapter. It is necessary to first transcribe negative-strand (also known as negative-sense or minus-strand) RNA into positive-strand RNA before it can be translated into proteins. Baltimore also considered viruses that can produce DNA from an RNA template, a process known as reverse transcription, which is not possible for cells. Together, the seven classes are

- **Class I:** dsDNA viruses;
- **Class II:** ssDNA viruses;
- **Class III:** dsRNA viruses;
- **Class IV:** positive-sense ssRNA viruses;
- **Class V:** negative-sense ssRNA viruses;
- **Class VI:** RNA viruses that reverse transcribe;
- **Class VII:** DNA viruses that reverse transcribe.

Viruses can be classified in various ways, such as by size, capsid structure, nucleic acid type, physical properties, host species, or disease caused. The International Committee on Taxonomy of Viruses (ICTV) was established in 1966 to address the challenge of classifying viruses. Taxonomy is the science of categorizing organisms based on similar characteristics, and the ICTV uses a taxonomical hierarchy similar to that of living things. Viruses, not being alive, have their separate classification system that does not fit within the traditional tree of life. Unlike living organisms, viruses are classified only by order, family, genus, and species (Table 4.2).

Table 4.2. *Taxa Used to Classify Viruses*

Taxon	Notes	Example
Order	It ends in the -virales suffix; only about half of viruses are currently classified in order.	*Picornavirales*
Family	Ends in -viridae suffix; subfamilies are indicated with -virinae suffix.	*Picornaviridae*
Genus	Ends in -virus suffix.	*Enterovirus*
Species	Generally, the "common name" of the virus. Classifying and cataloging anything below the species classification (such as subtypes, serotypes, strains, isolates, or variants) is the responsibility of the specific field.	*Rhinovirus A* (Serotypes include Human rhinovirus 1, which includes strains human rhinovirus 1A and human rhinovirus 1B)

Source: https://www.ncbi.nlm.nih.gov/pmc/articles/PMC7150055/.

In order to group viruses that are most similar to one another, the ICTV classifies viruses according to a wide range of distinct features. The virus's chemical and physical characteristics are considered, including the kind of nucleic acid it encodes and the quantity of distinct proteins it can encode. Scientists can now quickly and easily sequence viral genomes thanks to advancements in DNA technologies. By comparing the nucleic acid sequences of two viruses, they can ascertain how closely related they are. Additional characteristics of virion are also considered, such as size, shape of the capsid, and presence or absence of an envelope. Vertebrate virus taxa: note that certain families are not yet categorized into orders. Take note of how different viruses from different families differ in size.(Figure 4.16) (Van Regenmortel, 1990).

Currently, the ICTV has classified seven orders of viruses, each of which has 103 families. However, there are still 77 virus families that have not been placed in any order. These include well-known viruses like papillomaviruses, poxviruses, and retroviruses. Proposals for new orders have been made, and as the taxonomy process moves forward, more will probably be made.

Newly discovered virus naming guidelines have been established by the ICTV. Viruses are not named using the Latin binomial names that are used for living things, where the genus and species are listed together (e.g., Yersinia pestis or Homo sapiens). Additionally, although in the past this was how some viruses were named, virus names should not contain any personal names. Instead, they should be meaningful and simple to use (Baker et al., 1999).

If a virus order, family, genus, or species is being directly mentioned, the virus name should be written in italics, with the initial letter capitalized. However, unless a proper name is encountered, capitalization and italics are not necessary when discussing anything other than viral classification. As an example, the Lassa virus—named for the Nigerian town where the first case of the virus occurred—and the lymphocytic choriomeningitis virus are two examples of the severe disease that can be caused by certain members of the Arenaviridae family. Some instances of the places where virus names originated are shown in Tables 4.3 and 4.4.

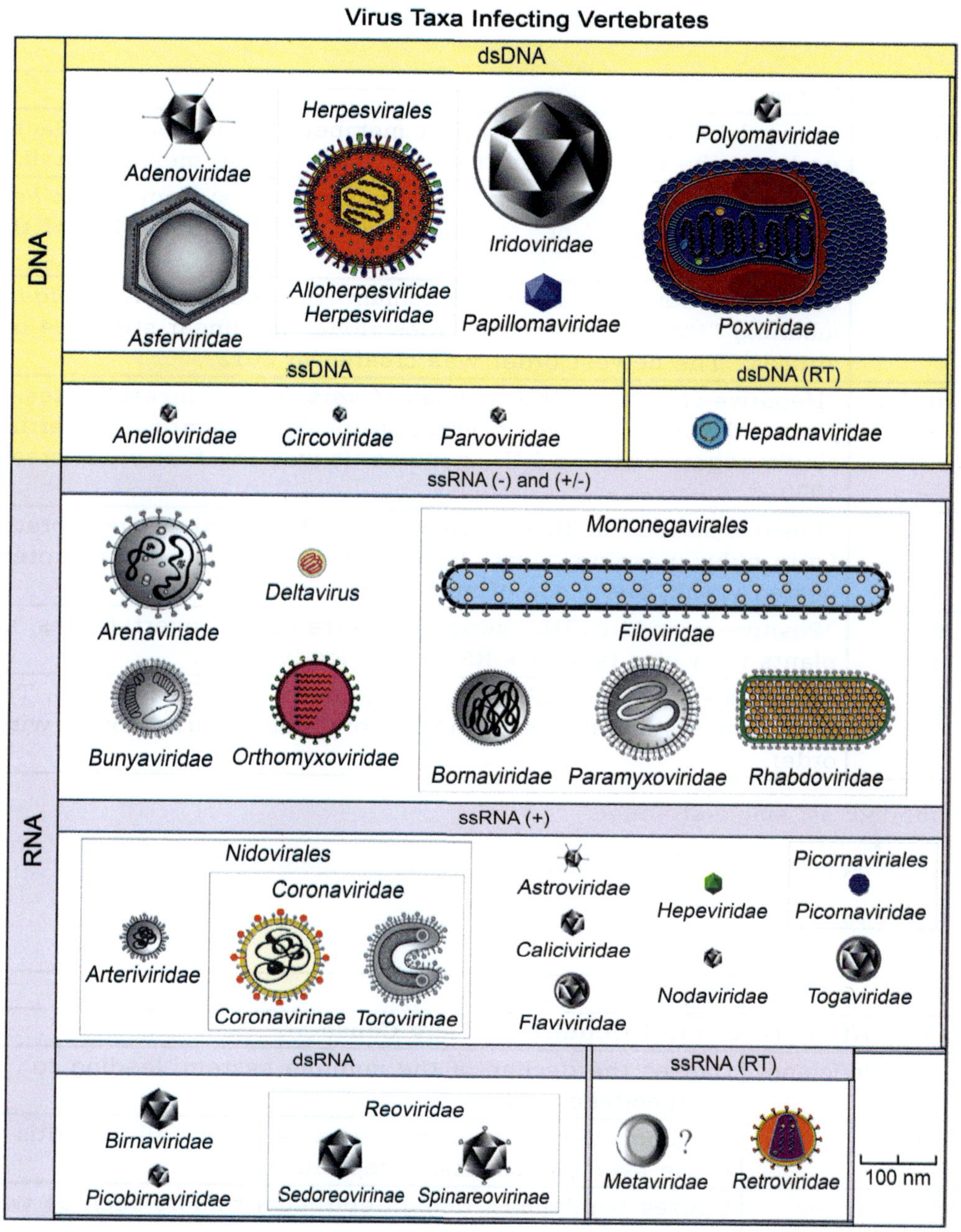

Figure 4.16. *Taxa of viruses that infect vertebrates. Viruses are categorized based on their type of nucleic acid (DNA viruses in yellow boxes and RNA viruses in blue boxes) and further classified based on distinguishing characteristics. Note the nucleic acid, size, and architectural differences between viruses of different families. Viruses in color will be discussed in later chapters. Modified with permission from Virus Taxonomy: Ninth Report of the International Committee on Taxonomy of Viruses, Elsevier, 2012.*

Source: https://www.ncbi.nlm.nih.gov/pmc/articles/PMC7150055/bin/f02-11-9780128009475.jpg.

Table 4.3. *Current Orders of the Virosphere*

Order	Notes
Caudovirales	Tailed dsDNA viruses that infect members of the domains Bacteria and Archaea; the name comes from the Latin *cauda*, meaning "tail."
Herpesvirales	dsDNA viruses of vertebrates and invertebrates; from Greek *herpes*, meaning "creeping" or "spreading" (describing the rashes of these viruses).
Ligamenvirales	dsDNA viruses that infect the domain Archaea; from Latin *ligamen*, meaning "thread" or "string" (describing the linear structure of the viruses). The newest order was created in 2012.
Mononegavirales	"Negative-strand" ssRNA viruses of vertebrates, invertebrates, and plants; the name derives from Latin for "one negative," referring to the single negative-strand RNA genome. It was the first order created in 1990.
Nidovirales	"Positive-strand" ssRNA viruses of vertebrates and invertebrates; from Latin *nidus* meaning "nest" because they encode several proteins nested within one piece of mRNA.
Picornavirales	"Positive-strand" ssRNA viruses of vertebrates, invertebrates, and plants from pico (small) + RNA + virales (viruses).
Tymovirales	"Positive-strand" ssRNA viruses of plants and invertebrates; Tymo is an acronym standing for Turnip Yellow Mosaic virus, found within this order.

ds, Double-stranded; ss, single-stranded.

Source: https://www.ncbi.nlm.nih.gov/pmc/articles/PMC7150055/.

Table 4.4. *Name Origins of Selected Viruses*

Name	Origin
Viruses Named After the Clinical Conditions They Cause	
Human immunodeficiency virus (HIV)	Causes the decline of the immune system, leading to immunodeficiency
Hepatitis virus	Although they are not in the same family, all hepatitis viruses cause liver inflammation (hepatitis)
Human papillomavirus (HPV)	Causes papillomas, benign epithelial tumors such as warts
Poxviruses	From pockests meaning "sac," referring to the blistery rash observed
Rabies virus	From Latin *rabies*, meaning "madness," describing the symptoms seen with disease progression
Viruses Named After Their Location of Discovery	
Coxsackievirus	Named after Coxsackie, New York, the location from where the first specimens were obtained

Ebola virus	Named after the Ebola River in northern Democratic Republic of the Congo (formerly Zaire), where the virus first emerged in 1976
Marburg virus	Named after Marburg, a town in Germany, where an outbreak occurred in 1967
Nipah virus	First identified in the Malaysian village of Kampung Sungai Nipah in 1998
Norwalk virus	Named after a 1968 outbreak in children at an elementary school in Norwalk, Ohio
West Nile virus	First isolated from a woman in the West Nile district of Uganda in 1937
Viruses Named After Their Properties	
Coronavirus	From Latin *corona*, meaning crown, referring to the crown-like appearance of the virions when viewed with an electron microscope
Herpesviruses	From Greek *herpein*, "to creep," referring to the lesions that slowly spread across the skin
Influenza virus	Originated in 15th century Italy, from an epidemic attributed to the "influence of the stars"
Picornaviruses	Pico meaning "small" + RNA viruses
Poliovirus	From Greek *polios*, meaning "gray," referring to the gray matter (cell bodies) in the spinal cord that it infects and damages
Viruses Named After People (Historically Assigned; Viruses can no Longer be Named After Individuals)	
Epstein–Barr virus	Named after Michael Anthony Epstein and Yvonne Barr, who discovered the virus
JC Virus	Named after a patient, John Cunningham, from which the virus was isolated
Rous sarcoma virus	Discovered by Peyton Rous in 1911

Source: https://www.ncbi.nlm.nih.gov/pmc/articles/PMC7150055/.

4.3.1. Principles of Virus Taxonomy

Compared with the classification of bacteria or eucaryotic microorganisms, viruses are in a far less satisfactory state. This is partly because their origin and evolutionary history are unknown to us. Generally, viruses are classified into multiple broad categories according to their host preferences, such as plant, animal, bacterial, bacteriophage, etc. In the past, virologists who worked with these groups couldn't come to a consensus on a common nomenclature and classification scheme. The International Committee for Taxonomy of Viruses created a standardized classification system in 1971, and since then, viruses have been divided into three orders, 56 families, 9 subfamilies, 233 genera, and 1,550 virus species. The nucleic acid type, nucleic acid strandedness, sense (positive or negative) of ssRNA genomes, presence or absence of an envelope, and host are the properties that the committee considers most important when defining families. Names of families of viruses end in viridae, names of subfamilies in virinae, and names of genus and species within viruses. For instance, the poxviruses belong to the family Poxviridae; the vertebrate poxviruses are found in the subfamily Chorodopoxvirinae. Numerous genera within the subfamily are differentiated based on host specificity and immunologic traits. Smallpox-causing variola major, vaccinia, and cowpox are among the species of the genus Orthopoxvirus (Van Regenmortel et al., 2000).

Table 4.5. *Some Common Virus Groups and Their Characteristics*

Nucleic Acid	Strandedness	Capsid Symmetry[a]	Presence of Envelope	Size of Capsid (nm)[b]	Number of Capsomers	Virus Group	Host Range[c]
RNA	Single	I	–	22–30	32	*Picornaviridae*	A
		I	+	40–70(e)	32	*Togaviridae*	A
		I?	+	100(e)		*Retroviridae*	A
		H	+	9(h), 80–120(e)		*Orthomyxoviridae*	A
		H	+	18(h), 125–250(e)		*Paramyxoviridae*	A
		H	+	14–16(h), 80–160(e)		*Coronaviridae*	A
		H	+	18(h), 70–80 × 130–240 (bullet shaped)		*Rhabdoviridae*	A
		I,B	–	26–35; 18–26 × 30–85		*Bromoviridae*	P
		H	–	18 × 300		*Tobamovirus*	P
		I	–	26–27	32	*Leviviridae* [Qβ]	B
RNA	Double	I	–	70–80	92	*Reoviridae*	A,P
		I	+	100(e)		*Cystoviridae*	B
DNA	Single	I	–	20–25	12	*Parvoviridae*	A
		I	–	18 × 30 (paired particles)		*Geminiviridae*	P
		I	–	25–35		*Microviridae*	B
		H	–	6 × 900–1,900		*Inoviridae*	B
DNA	Double	I	–	40	72	Polyomaviridae	A
		I	–	55	72	*Papillomaviridae*	A
		I	–	60–90	252	*Adenoviridae*	A
		I	+	130–180		*Iridoviridae*	A
		I	+	100, 180–200(e)	162	*Herpesviridae*	A
		C	+	200–260 × 250–290(e)		*Poxviridae*	A
		H	+	40 × 300(e)		*Baculoviridae*	A
		C	+	28 (core), 42(e)	42	*Hepadnaviridae*	A
		I,B	–	50; 30 × 60–900		*Caulimoviridae*	P
		I	–	60		*Corticoviridae*	B
		Bi	–	80 × 110, 110[d]		*Myoviridae*	B

[a]Types of symmetry: I, icosahedral; H, helical; C, complex; Bi, binal; B, bacilliform.
[b]Diameter of helical capsid (h); diameter of enveloped virion (e).
[c]Host range: A, animal; P, plant; B, bacterium.
[d]The first number is the head diameter; the second number, the tail length.

Source. https://asutoshcollege.in/new-web/Study_Material/Virus_07042020.pd.

Based on traits related to the type of host used, the structure and composition of the virion, the mode of reproduction, and the nature of any diseases caused, viruses are classified into various taxonomic groups. Some of the more important characteristics are:

- **Nature of the Host**—animal, plant, bacterial, insect, fungal
- **Nucleic Acid Characteristics**—DNA or RNA, single or double stranded, molecular weight, segmentation and number of pieces of nucleic acid (RNA viruses), the sense of the strand in ssRNA viruses;
- **Capsid Symmetry**—icosahedral, helical, binal;
- Presence of an envelope and ether sensitivity;
- Diameter of the virion or nucleocapsid;
- Number of capsomers in icosahedral viruses;
- Immunologic properties;

- Gene number and genomic map;
- Intracellular location of viral replication;
- The presence or absence of a DNA intermediate (ssRNA viruses), and the presence of reverse transcriptase;
- Type of virus release;
- Disease caused and/or special clinical features, method of transmission.

A few common virus groups are described using some of these properties, as shown in Table 4.5. When specific consideration is given to viruses that affect bacteria, animals, and plants, virus classification is further explored (Table 4.5).

CASE STUDY

HUMAN IMMUNODEFICIENCY VIRUS (HIV)

Structural Analysis

- Capsid and Genome:
  - HIV is an enveloped virus with a conical capsid that houses its RNA genome.
  - The genome is composed of two identical single-stranded RNA molecules, making it a retrovirus (Group VI in the Baltimore classification system).
- Enzyme Complex:
  - Inside the capsid, HIV contains three essential enzymes: reverse transcriptase, integrase, and protease, critical for its replication cycle.
- Envelope and Glycoproteins:
  - The virus is surrounded by a lipid bilayer envelope derived from the host cell membrane.
  - Embedded in the envelope are glycoproteins gp120 and gp41, which play crucial roles in binding and entry into host cells.
- Structural Proteins:
  - The matrix protein (p17) lines the inside of the envelope, providing structural integrity.
  - The capsid protein (p24) forms the conical core that encases the viral RNA and enzymes.

Classification

- Baltimore Classification:
  - HIV is classified as a Group VI virus, meaning it has an RNA genome that is reverse-transcribed into DNA within the host cell.
- ICTV Classification:
  - According to the International Committee on Taxonomy of Viruses, HIV belongs to the family Retroviridae, genus Lentivirus.
- Replication Mechanism:

- HIV uses reverse transcriptase to convert its RNA genome into DNA, which integrates into the host genome, enabling persistent infection.

Implications

HIV's ability to integrate into the host genome makes it particularly challenging to eradicate. Its classification as a retrovirus informs treatment strategies, leading to the development of antiretroviral therapies that target reverse transcriptase and other viral enzymes to inhibit replication and manage the disease.

Conclusion

Understanding the structure and classification of viruses like influenza is crucial for developing effective treatments and preventive measures. By comparing these two viruses, we can appreciate the diversity in viral mechanisms and the strategies required to combat viral diseases. This knowledge enhances our ability to respond to current viral threats and prepares us for future challenges in virology and infectious disease management.

CLASS ACTIVITY

Students will classify various viruses based on their genetic material, replication strategy, and host range. Provide diagrams of different virus structures for students to color, reinforcing their understanding of virus anatomy. Assign each student a different type of virus (e.g., bacteriophage, influenza, HIV). Provide them with materials to create a 3D model of their assigned virus.

SUMMARY

- Viruses are so small that a microscope is necessary to visualize them, and they have a very simple structure.
- Viral genomes are very diverse since they can be DNA or RNA, single- or double-stranded, linear or circular, and vary in length and in the number of DNA or RNA molecules.
- Viruses are a unique group of infectious agents whose distinctiveness resides in their simple, acellular organization and pattern of reproduction.
- Bacterial viruses or bacteriophages (phages for short) are cultivated in either broth or agar cultures of young, actively growing bacterial cells.
- The smallest of viruses are about 20 nm in diameter, although influenza and the human immunodeficiency virus have a more typical size, about 100 nm in diameter.
- The virus disassembles, and its genetic material (made of nucleic acid) encodes the instructions for the proteins that will spontaneously assemble into the new virions.
- Virus morphology has been intensely studied over the past decades because of the importance of viruses and the realization that virus structure was simple enough to be understood.
- Virions range in size from about 10 to 300 or 400 nm in diameter. The smallest viruses are a little larger than ribosomes, whereas the poxviruses, like vaccinia, are about the same size as the smallest bacteria and can be seen in the light microscope.
- Many viruses have an envelope, an outer membranous layer surrounding the nucleocapsid.
- Viral capsid proteins protect the fragile genome, composed of nucleic acid, from the harsh environment. Together, capsid and nucleic acid are known as the nucleocapsid.
- Helical viruses can be enveloped or naked. The first virus described, tobacco mosaic virus, is a naked helical virus.
- An icosahedron is a geometric shape with 20 sides (or faces), each composed of an equilateral triangle.
- The classification of viruses is useful for many reasons. It allows scientists to contrast viruses and to reveal information on newly discovered viruses by comparing them to similar viruses.
- Viruses are divided into different taxonomic groups based on characteristics that are related to the type of host used, virion structure and composition, mode of reproduction, and the nature of any diseases caused.

REVIEW QUESTIONS

1. What are the general properties of viruses?
2. Elaborate the structure of viruses.
3. What is helical capsid structure?
4. Write short notes on viral envelopes and enzymes.
5. Explain the principles of virus taxonomy.

REFERENCES

1. Baker, T. S., Olson, N. H., & Fuller, S. D. (1999). Adding the third dimension to virus life cycles: Three-dimensional reconstruction of icosahedral viruses from cryo-electron micrographs. *Microbiology and Molecular Biology Reviews*, *63*(4), 862–922.
2. Bourne, C. R., Finn, M. G., & Zlotnick, A. (2006). Global structural changes in hepatitis B virus capsids induced by the assembly effector HAP1. *Journal of Virology*, *80*, 11055–11061.
3. Bresnahan, W. A., & Shenk, T. (2000). A subset of viral transcripts packaged within human cytomegalovirus particles. *Science*, *288*, 2373–2376.
4. Cáceres, A., Perdiguero, B., & Gómez, C. E. (2013). Involvement of the cellular phosphatase DUSP1 in vaccinia virus infection. *PLoS Pathogens*, *9*.
5. Clare, D. K., & Orlova, E. V. (2010). 4.6 Å Cryo-EM reconstruction of tobacco mosaic virus from images recorded at 300 keV on a 4k × 4k CCD camera. *Journal of Structural Biology*, *171*, 303–308.
6. Dixon, L. K., Alonso, C., & Escribano, J. M. (2012). Virus taxonomy. In *Virus taxonomy: Ninth report of the international committee on taxonomy of viruses* (pp. 153–162). Academic Press.
7. Harrison, S. C. (1984). Structure of viruses. In *The microbe 1984: Part I, viruses* (pp. 1–42). Cambridge University Press.
8. Kaufmann, B., Simpson, A. A., & Rossmann, M. G. (2004). The structure of human parvovirus B19. *Proceedings of the National Academy of Sciences of the United States of America*, *101*, 11628–11633.
9. Kuhn, R. J., Zhang, W., & Rossmann, M. G. (2002). Structure of dengue virus: Implications for flavivirus organization, maturation, and fusion. *Cell*, *108*, 717–725.
10. Lwoff, A., & Tournier, P. (1971). Remarks on the classification of viruses. In K. Maramorosch & E. Kurstak (Eds.), *Comparative virology* (pp. 1–42). Academic Press.
11. Matthews, R. E. F. (1985). Viral taxonomy for the nonvirologist. *Annual Review of Microbiology*, *39*, 451–474.
12. Prasad, B. V. V., & Schmid, M. F. (2012). Principles of virus structural organization. In *Advances in experimental medicine and biology* (pp. 17–47). Springer.

13. Prasad, B. V., Hardy, M. E., Dokland, T., Bella, J., Rossmann, M. G., & Estes, M. K. (1999). X-ray crystallographic structure of the Norwalk virus capsid. *Science*, *286*, 287–290.

14. Tarini, M., Cignoni, P., & Montani, C. (2006). Ambient occlusion and edge cueing to enhance real-time molecular visualization. *IEEE Transactions on Visualization and Computer Graphics*, *12*, 1237–1244.

15. Van Regenmortel, M. H. (1990). Virus species, a much overlooked but essential concept in virus classification. *Intervirology*, *31*, 241–254.

16. Van Regenmortel, M. H. V., Fauquet, C. M., Bishop, D. H. L., Carstens, E. B., Estes, M. K., Lemon, S. M., Maniloff, J., Mayo, M. A., McGeoch, D. J., Pringle, C. R., & Wickner, R. B. (Eds.). (2000). *Virus taxonomy: The classification and nomenclature of viruses*. Academic Press.

CHAPTER 5

The Beginning and End of the Virus Replication Cycle

LEARNING OBJECTIVES

After studying this chapter, you will be able to:

- Describe the steps of the viral replication cycle;
- Focus on viral entry;
- Describe the animal virus entry into cells—the role of the cellular receptor;
- Describe the mechanisms of entry of nonenveloped viruses and entry of enveloped viruses;
- Know about entry of plant virus into cells;
- State the injection of bacteriophage DNA into Escherichia coli;
- Describe the nonspecific methods of introducing viral genomes into cells;
- State the three steps of exit;
- Learn about the generation of the virion envelope and egress of the enveloped virion.

KEY TERMS FROM THIS CHAPTER

Cell lysis	Cell-to-cell spread	Cytoplasmic trafficking
Eukaryotic cells	Fusion peptide	Genome replication
HIV infection	Immune system	Intracellular parasites
Late domain	Macropinocytosis	Microfilaments
Monoclonal antibodies	Multivesicular bodies	Packaging signal
Penetration	Picornavirus	Retrovirus

INTRODUCTION

Viruses are parasitic organisms that live inside cells. To create a progeny virus, viruses must enter their intended target cells and take control of the host's cellular machinery. The term "virus life cycle" refers to the events that occur during the multiple stages of virus propagation inside cells. The virus replicates its genome, assembles the viral components, sheds its capsid, transcribes its RNA, translates it to the viral proteins, enters the cell, localizes itself in the intracellular milieu, and then leaves the cell.

All viruses follow the same basic replication cycle, but the length of time it takes varies depending on the host cell's characteristics and the size and genetic complexity of the virus (Connolly et al., 2011).

5.1. STEPS OF THE VIRAL REPLICATION CYCLE

A virus must overcome several challenges in order to enter the host cells. The invaders encounter barriers in the form of cellular membranes. The plasma membrane is the first barrier that any animal virus must get past. For certain viruses, the second line of defense against genome replication in the nucleus is the nuclear membrane. Let's examine how viruses get past the obstacles.

Three stages comprise the life cycle of a virus: entry, genome replication, and exit (Figure 5.1). A businessman's life can be compared to the life cycle of a virus, with the entry representing his route to work, the genome replication representing his task at work, and the exit representing his route home. Entry is the first phase. The three stages of entry are attachment, penetration, and uncoating. During attachment, a virus particle enters the host cell and adheres to the cell surface. During uncoating, the virus sheds its capsid. The viral genome is used for gene expression and viral genome replication once it has been uncoated. Following the accumulation of viral proteins and genomes, a progeny virion particle is assembled and released extracellularly. The exit is made up of the assembly of virion and the release from the cell.

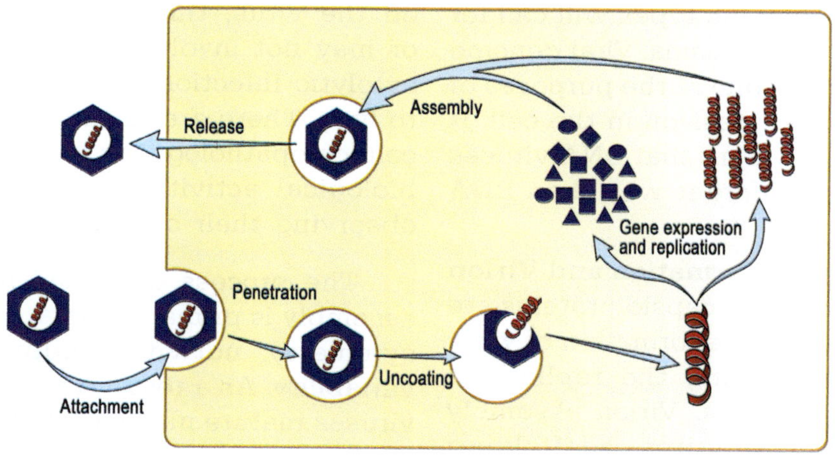

Figure 5.1. *The life cycle of the virus. The virus life cycle could be divided into six steps: attachment, penetration, uncoating, gene expression and replication, assembly, and release. The viral capsid (blue) and genome (brown) are schematically drawn for explanation. The nucleus is omitted for clarity.*

Source: Wang-Shick Ryu. Virus Life Cycle. Molecular Virology of Human Pathogenic Viruses. Elsevier. DOI: 10.1016/B978-0-12-800838-6.00003-5.

The basic replication process involves the following steps:

- **Virus Recognition, Attachment, and Entry into the Cell:** In order to introduce their genome into a host cell and ensure that it is transported by cellular functions to a location where the virus replication cycle can continue, viruses must be able to take advantage of specific features of the host cell in which they will replicate. This calls for either injecting the viral genome into the host cell, as is the case with many bacterial viruses, or forcing the cell to absorb the entire virus particle in a particular manner.
- **Viral Gene Expression and Genome Replication:** Nucleic acid-based viral genes must be decoded before being translated into viral proteins. Making mRNA is necessary for this. It goes without saying that different genome types will call for different mechanisms. Viral genome replication is one of the purposes of viral gene expression in the cell. It should be evident that DNA viruses work in a different way than RNA viruses.
- **Viral Capsid Formation and Virion Assembly:** Viral capsid proteins are necessary for the formation of virus structures during the replication of viral genomes. Virion assembly frequently requires scaffolding proteins, which are viral proteins necessary to form the capsid structure but not actually part of it. Another stage of viral gene expression is also frequently necessary. The virus must be released once capsids have formed. An enveloped virus would acquire a membrane envelope in order to release itself in this way.

This broad pattern has a great deal of variation and subtle difference. Think about how viruses enter the body. All animal and bacterial viruses use a specific cellular receptor for entry; however, there is no record of a plant virus doing so. Certain bacterial viruses have a very complicated viral entry process that involves biochemical reactions between the virus capsid protein components in order to inject the viral genome (Connolly et al., 2011).

The specifics of the virus release process also vary widely. Here, the majority of variations are observed in viruses that are discharged from eukaryotic cells. Certain infections cause cell death as a result of virus release (a cytocidal infection). Depending on the virus, this type of cell death may or may not involve cell lysis (cytolysis). A cytolytic infection is one that causes cells to lyse. There are further alterations to the cell (cytopathology) as well. Certain viruses' biological activity can be determined by observing their cytotoxic effects.

The process of capsid maturation and assembly is primarily dictated by the virus's symmetry, notwithstanding this kind of variability. As a result, icosahedral bacterial viruses mature in a manner that is strikingly similar to that of herpesviruses. Once more, the assembly of helical viruses is similar in plants, animals, and bacteria.

5.2. VIRAL ENTRY

A virus particle recognizes a viral receptor during entry, the initial stage of virus infection. The four stages of viral entry are uncoating, cytoplasmic trafficking, attachment, and penetration. Although the division into four steps is ambiguous, it serves the purpose of explanation because these steps are frequently connected to one another.

5.2.1. Attachment

The term "attachment" describes the initial interaction of virus particles with host cells, involving two types of host proteins on the plasma membrane: viral receptors and attachment factors.

Table 5.1. *Viral Receptors for Major Human Viruses*

Family: Prototype Virus	Receptor (Coreceptor)	Attachment Factor	Viral Antireceptor
DNA Virus			
Parvovirus: AAV	HSPG (FGFR, integrin)	HSPG	CAP
Polyoma: SV40	GM1 gangliosides	–	VP1
Adeonovirus: Ad5	Integrin	CAR	Fiber protein
Herpesvirus: HSV-1	Nectin-1/HVEM, PILRα	HSPG	gD, gB
Herpesvirus: CMV	EGFR	HSPG	
Herpesvirus: EBV	CD21	HSPG	gp350
RNA Virus			
Picornavirus: Poliovirus	PVR/CD155	–	VP1, VP2, VP3
Picornavirus: Coxsackie virus	CAR	–	
Picornavirus: Rhinovirus	ICAM-1 or LDL receptor	DC-SIGN, L-SIGN	VP1, VP2, VP3
Flavivirus: Hepatitis C virus	CD81, Claudin-1, Occludin	SR-B1, LDL receptor	E2
Coronavirus: SARS virus	ACE2	–	Spike
Orthomyxovirus: Influenza virus	Sialic acid	–	HA
Rhabdovirus: VSV	Phosphatidyl serine	–	G protein
Rhabdovirus: Rabies	NCAM-1/CD56	–	G protein
Reovirus: Reovirus	JAM-A	–	Spike protein s1
RT Virus			
Retrovirus: HIV-1	CD4 (CXCR4 or CCR5)	DC-SIGN, L-SIGN	gp120
Hepatitis B virus	NTCP	HSPG	Pre-S1

Source: Wang-Shick Ryu. Virus Life Cycle. Molecular Virology of Human Pathogenic Viruses. Elsevier. DOI: 10.1016/B978-0-12-800838-6.00003-5.

The viral particle interacts with the entry receptor more easily because the attachment factor on the cell surface attracts and retains the virus particles. As evidenced by the fact that different viruses bind to glycoaminoglycans like heparins, the attachment factors' wider specificity can be seen (Table 5.1). Viral receptors, in contrast to attachment factors, facilitate the entry of virus particles into cells by binding to them. Moreover, the viral receptors determine cell tropism and—more significantly—are virus-specific. For instance, HIV specifically recognizes CD4, and it infects T cells that express CD4 (Connolly et al., 2011). In the last thirty years, nearly all of the viral receptors for the main pathogenic viruses that affect humans have been found. Developments in molecular biology during the 1970s were essential to the discovery. Crucially, a few noteworthy details about the characteristics of viral receptors jump out. First, as can be seen in Table 5.1, the molecular makeup of the viral receptors is highly varied, ranging from glycoproteins to phospholipids. The viral receptor is even found in the carbohydrate moiety of membrane glycoproteins. For example, the influenza virus's entry receptor is found in the sialic acid residue of glycans. Secondly, the majority of receptors for viral entry serve specific cellular purposes. For example, the entry receptor of a certain picornavirus, LDL receptor, has a physiological function of taking up LDL particles into cells, and an entry receptor of a certain herpesvirus, EGFR, is an EGFR (Figure 5.2). Put another way, viruses use the proteins found in cells that serve physiological purposes as entry receptors by subverting them. Third, a large number of viral receptors, including CD4 and CAR, are members of the immunoglobulin superfamily (*refer to* Figure 5.2). Fourth, in addition to the primary receptors needed for entry, certain viruses also need coreceptors. For instance, chemokine receptors like CCR5 or CXCR4 are necessary for HIV to efficiently enter the body as a coreceptor.

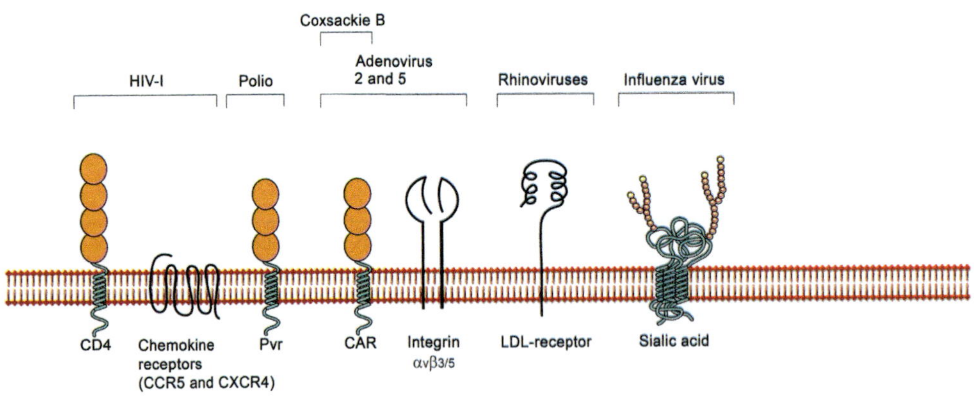

Figure 5.2. *The viral receptors of representative human pathogenic viruses. Virus particles recognize the ectodomain of the receptor molecules. Immunoglobulin (IgG) fold, which is shared by members of immunoglobulin superfamily (i.e., CD4, Pvr, and CAR), is drawn in a circle. The sialic acid moiety of glycan in membrane glycoproteins is marked by yellow. Pvr, poliovirus receptor; CAR, coxsackie virus and adenovirus receptor.*

Source: Wang-Shick Ryu. Virus Life Cycle. Molecular Virology of Human Pathogenic Viruses. Elsevier. DOI: 10.1016/B978-0-12-800838-6.00003-5.

Crucially, the susceptibility to a particular virus is determined by the receptor's presence in a particular cell. Cell tropism is therefore primarily dictated by the receptor. For example, T helper lymphocytes that express CD4 are more susceptible to HIV infection. Experimenting with CD4 expression makes a HeLa cell that is normally resistant to HIV infection susceptible to HIV infection. Conversely, the proteins on the surface of virus particles are the ones that the viruses use to identify their receptors. When a virus is enveloped, its glycoproteins bind to receptors on the viral envelope. For example, the target cell's CD4 molecule binds to the HIV virion particle's gp120. Capsid proteins can bind to receptors directly in the case of nude viruses. For example, the target cell's CAR molecule binds to the fiber protein of the adenovirus particle (Dimmock and Primrose, 1994).

While the presence or lack of a receptor is a major factor in determining a virus's susceptibility, host restriction factors that restrict viral infection can also influence cell tropism. For example, human TRIM5α inhibits simian immunodeficiency virus infection of human cells, while TRIM5α of monkey cells limits HIV infection of monkey cells. Therefore, the host range of primate lentivirus is determined by TRIM5α, a host restriction factor.

5.2.1.1. Strategies to Discovery Viral Receptor

Which three experimental techniques have been successfully applied to reveal viral receptors on the susceptible cells' cell surfaces? The initial method involves identifying the receptors through the biochemical purification of cell surface proteins that bind to the antireceptors of the virus (i.e., structural proteins of viruses). It is possible to achieve affinity purification of plasma membrane proteins by employing the viral structural proteins as a ligand. Alternately, many candidate proteins in SDS-PAGE gel may result from immunoprecipitation of plasma membrane proteins that bind to the viral structural protein. The next step involves exposing every protein band on the gel to mass spectroscopy (e.g., MALDI-TOF) in order to identify it. The second strategy is to use monoclonal antibodies (Mab) against plasma membrane proteins. It takes advantage of the possibility that an entry could be blocked by a receptor-specific antibody. First, a set of Mab is prepared to test against the susceptible cell's plasma membrane. Next, the capacity of each Mab to inhibit the viral infection is examined. The ligand for a particular Mab may be a viable candidate for the viral receptor if it inhibits the virus infection. Ultimately, immunoprecipitation and mass spectroscopy can be used to identify the ligand. The third method uses the cDNA expression library to identify the receptor through functional cloning. A non-susceptible cell is transfected with a single copy number from the library, and the cell's ability to contract infection is then monitored. High-throughput robotics is a must for this approach because the human cDNA library contains more than 30,000 genes (Dimmock and Primrose, 1994). The HCV receptor has recently been successfully cloned using this cutting-edge technology. Validation of the receptors' functionality is necessary after the possible molecules for the viral receptors are found using one of the three methods mentioned above. The purpose of this validation is to verify if transfecting a non-susceptible cell with cDNA is sufficient to turn it into a susceptible cell (Figure 5.3).

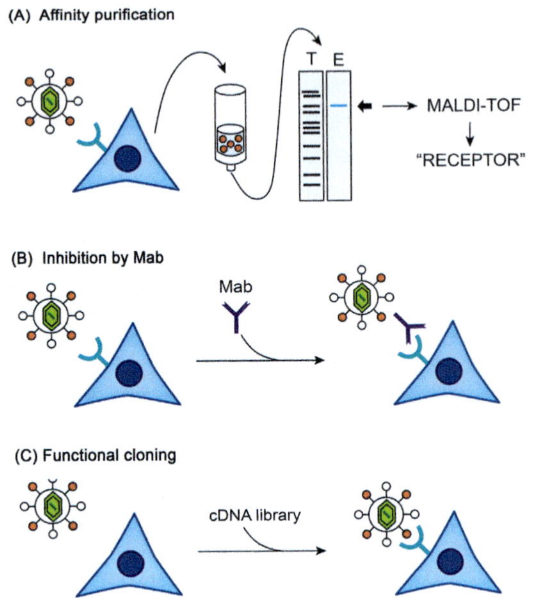

Figure 5.3. *(A) Affinity purification. (B) Inhibition by Monoclonal antibodies (Mab). (C) Functional cloning.*

Source: Wang-Shick Ryu. Virus Life Cycle. Molecular Virology of Human Pathogenic Viruses. Elsevier. DOI: 10.1016/B978-0-12-800838-6.00003-5.

5.2.2. Penetration

The virus particle attaches itself to the target cell, and the next stage is for it to enter the cytoplasm. Whether or not the penetration is enveloped, the method varies. Direct fusion or receptor-mediated endocytosis are the two mechanisms for dealing with enveloped viruses. The mechanism of receptor-mediated endocytosis is used to penetrate nonenveloped naked viruses.

5.2.2.1. Direct Fusion

As its name suggests, direct fusion combines two membranes (i.e., the cell membrane and the viral envelope) (Figure 5.4A). Here, the viral envelope remains on the plasma membrane while the viral nucleocapsid is transferred straight to the cytoplasm. One example of a virus that spreads by direct fusion is the retrovirus.

5.2.2.2. Receptor-Mediated Endocytosis

Most viruses rely on endocytic uptakes, or receptor-mediated endocytosis, which allows some viruses to enter the cytosol directly through the plasma membrane (Figure 5.4B). The virus particle-receptor complex initiates endocytosis by creating a coated pit on the plasma membrane, which results in the formation of endosomes after viral particles bind to the receptor. The virus particle consequently finds itself inside the endosome. The next step is to break down the endosome to get to the cytoplasm. Whether an endosome is enveloped or not, the process of endosome breakdown varies. Endosome breakdown occurs when an early endosome with an acidic pH triggers membrane fusion between the viral envelope and the endosomal membrane in viruses that are enveloped. More specifically, the envelope glycoprotein's fusion peptide is exposed (i.e., activated) as a result of a conformational shift brought on by a low pH; the fusion peptide then starts the fusion. One of the capsid proteins causes the endosome lysis of naked viruses that are not enveloped. Put differently, the mechanism of penetration for envelope viruses is membrane fusion (refer to Figure 5.4B), whereas the mechanism of penetration for nonenveloped viruses is membrane lysis (Figure 5.4C).

Most viruses enter cells through endocytosis mediated by receptors. What would be the benefits of receptor-mediated endocytosis over direct fusion? It is clear that, in contrast to direct fusion, receptor-mediated endocytosis avoids the actin cortex and the microfilament meshwork in the cortex, which act as a barrier to penetration (*see* Figure 5.4). Furthermore, animal viruses can evade leaving the viral envelope glycoprotein on the plasma membrane by being taken up by endocytosis, which is likely to delay the immune system's detection (Dimmock and Primrose, 1994).

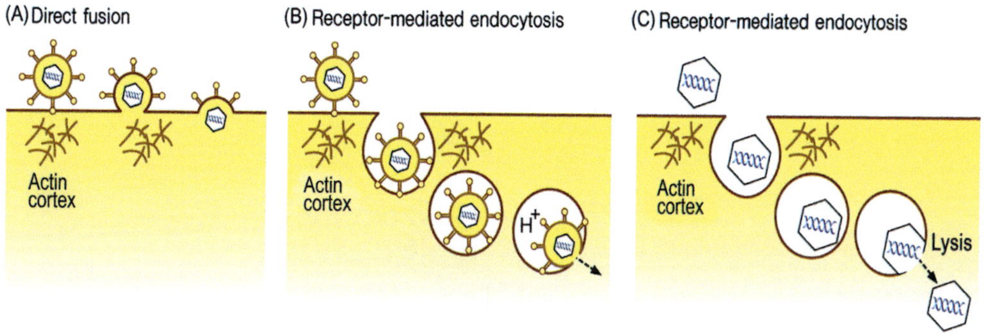

Figure 5.4. *Three modes of viral penetration. (A) Direct fusion. The viral nucleocapsids enter the cell by the fusion between viral envelope and plasma membrane. (B) Receptor-mediated endocytosis for enveloped viruses. The virus particle acts as a ligand for the endocytosis. The virus particle is located inside endosomes. The fusion takes place between the viral envelope and the endosomal membrane. Note that actin cortex (a mechanical support of the plasma membrane), which serves as a physical barrier for the entry, is bypassed through the endocytosis. (C) Receptor-mediated endocytosis for nonenveloped naked viruses. A component of viral nucleocapsid triggers the lysis of endosomal membrane necessary for the release of the viral genome to the cytoplasm.*

Source: Wang-Shick Ryu. Virus Life Cycle. Molecular Virology of Human Pathogenic Viruses. Elsevier. DOI: 10.1016/B978-0-12-800838-6.00003-5.

Generally, clathrin-dependent processes are used in receptor-mediated endocytosis (Figure 5.5). The process by which extracellular molecules are taken up by cells is known as receptor-mediated endocytosis. Numerous viruses, including the influenza virus and adenovirus, use clathrin-mediated endocytosis, which is also the mechanism by which LDL is absorbed. When the virus particle attaches itself to the receptor, clathrins are drawn near the plasma membrane, forming a clathrin-coated pit. The vesicles are fused with early endosomes after an endocytic vesicle forms. The early endosomes contain the virus particles at this point.

Apart from receptor-mediated endocytosis, animal viruses employ several other endocytic mechanisms (Figure 5.6). For example, SV40 and other polyomaviruses enter through caveolin-mediated endocytosis. Otherwise, it is comparable to clathrin-mediated endocytosis; however, caveolin functions as a coat protein in this instance rather than clathrin. Larger particles, like herpes and vaccinia viruses, are admitted through the process of macropinocytosis. Actin-mediated membrane blebbing and ruffling are first caused by the activation of signaling pathways by the virus particle. The internalization of virus particles and the viruses' or their capsids' penetration into the cytosol occurs after the formation of large vacuoles (macropinosomes) at the plasma membrane (Dimmock and Primrose, 1994).

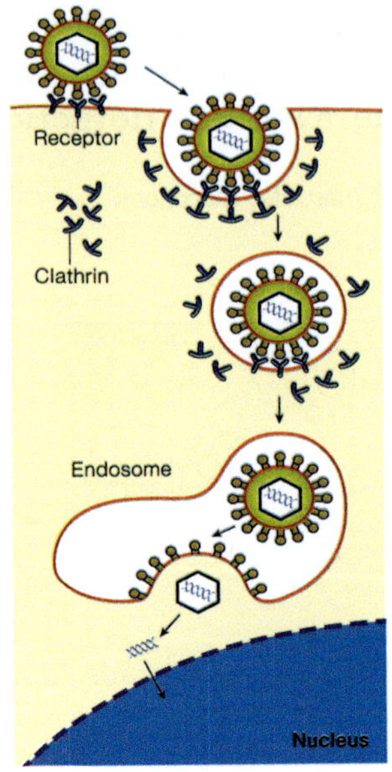

Figure 5.5. *Clathrin-mediated endocytosis. Upon the binding of virus particles to the receptor, clathrins are recruited to form the clathrin-coated pit via its interaction with the AP-2 adapter. Clathrin-coated pits are pinched off by dynamins. After the vesicle coat is shed, the uncoated endocytic vesicle fuses with the early endosome. The capsids are released from the endosome by membrane fusion between the viral envelope and endosome, which is triggered by low pH inside the endosome.*

Source: Wang-Shick Ryu. Virus Life Cycle. Molecular Virology of Human Pathogenic Viruses. Elsevier. DOI: 10.1016/B978-0-12-800838-6.00003-5.

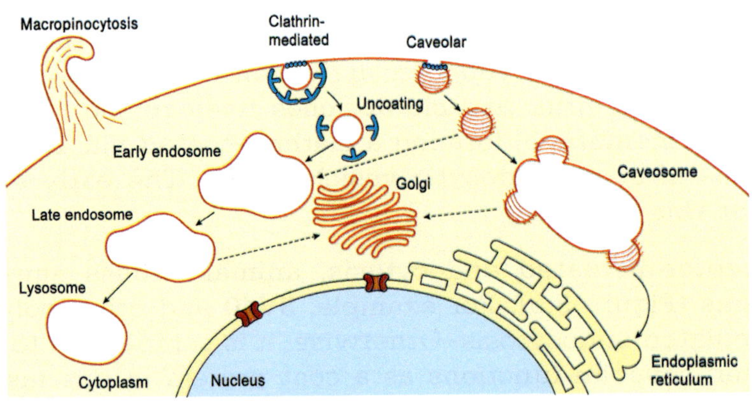

Figure 5.6. *Diverse pathways for receptor-mediated endocytosis. Clathrin-mediated endocytosis. This pathway is the most commonly observed uptake pathway for viruses. The viruses are transported via the early endosome to the late endosome and eventually to the lysosome. Caveolin/lipid-raft-mediated endocytosis. The caveola pathway brings viruses to caveosomes. By the second vesicle transport step, viruses are transported to Golgi, and then to ER. Macropinocytosis is utilized to enter particles with larger size, such as vaccinia and herpes viruses.*

Source: Wang-Shick Ryu. Virus Life Cycle. Molecular Virology of Human Pathogenic Viruses. Elsevier. DOI: 10.1016/B978-0-12-800838-6.00003-5.

5.2.2.2.1. Fusion Peptide

Membrane fusion between the viral envelope and plasma membrane is required for enveloped viruses to enter a cell. The viral capsid's entry into the cytoplasm is aided by membrane fusion. The viral envelope proteins (e.g., fusion proteins) contain a "fusion peptide," which initiates the membrane fusion process. This raises the question of how membrane fusion, which appears to be a challenging biochemical reaction, is accomplished. The fusion protein is present in the virion particles in its prefusion conformation and is intrinsically "metastable." How does the fusion peptide facilitate membrane fusion? The fusion peptides are buried inside the viral fusion proteins in this state. The fusion peptide is anchored in the opposing cellular membrane by a variety of conformational rearrangements that are triggered by various factors, including receptor binding and an acidic pH. In the prehairpin extended intermediates, a complementary amphipathic domain forms concurrently with anchoring (Dimmock and Primrose, 1994). These recently revealed domains are unstable and refold to create structures that are more energetically advantageous. The enthalpy linked to these conformational alterations drives the viral membrane's outer leaflet to mix with the cellular membrane's outer layer, forming the "hemifusion" stalk. Once in contact and starting to mix, the inner leaflet of the lipid bilayers causes a pore (fusion pore) to open between the membranes of the virus and the cell. At the same time, the trimeric structures refold into a highly stable postfusion conformation (Figure 5.7).

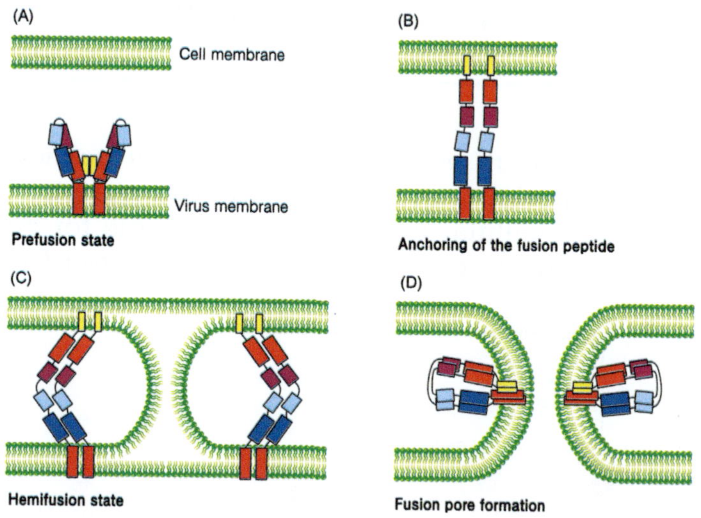

Figure 5.7. *The fusion process between viral and cellular membranes. The fusion peptide (yellow) is buried inside the fusion protein in an energetically unfavorable metastable prefusion state (A). The conformational change induced by triggers (e.g., acidic pH and receptor binding) results in anchoring the fusion peptide in the cell membrane (B), forcing the viral membrane and cell membrane into a hemifusion state (C). Subsequently, the fusion protein is refolded into a highly stable postfusion conformation, which leads to a fusion pore formation (D). For simplicity, only dimers are represented, but the fusion proteins are always trimeric: HA of influenza virus and F (fusion) protein of paramyxovirus.*

Source: Wang-Shick Ryu. Virus Life Cycle. Molecular Virology of Human Pathogenic Viruses. Elsevier. DOI: 10.1016/B978-0-12-800838-6.00003-5.

Several aspects of membrane fusion are noteworthy. First, no cellular factors are involved in membrane fusion; instead, only one viral factor (i.e., the fusion protein) drives membrane fusion. In contrast, budding also involves membrane fusion, so this feature is different. In particular, budding is influenced by a number of cellular components of the MVBs pathway. Second, membrane fusion is driven by thermodynamics and is essentially the mixing of two lipid bilayers. Stated differently, the conformational changes of the fusion protein, which is ready to transition from a metastable prefusion state to a highly stable postfusion state, occur in tandem with membrane fusion.

5.2.3. Intracellular Trafficking

After entering cells successfully, the virus particles must find the right location inside the cell to begin replicating its genome. We call this process intracellular trafficking. The biological significance of cytoplasmic trafficking was actually not understood until live cell imaging technology was developed. The viral nucleocapsids must be directed to the replication site for cytoplasmic viruses. The mechanism for the transport is receptor-mediated endocytosis in conjunction with microtubule-mediated transport (Figure 5.8). Moreover, the viral nucleocapsids must enter the nucleus for viruses that replicate there. Many DNA viruses use microtubule-mediated transport to deliver their viral nucleocapsids to the perinuclear region. The motion of the virus particles is propelled by a dynein motor during this process. The viral nucleocapsids can be thought of as a train in a railroad, to use an analogy (Granzow et al., 1997).

5.2.4. Uncoating

The process known as "uncoating" occurs when the virus particles get closer to the replication site, which is the perinuclear space to the cell periphery. At this point, the viral genome is exposed to the cellular machinery responsible for viral gene expression. Uncoating and cytoplasmic trafficking are frequently associated (*refer to* Figure 5.8).

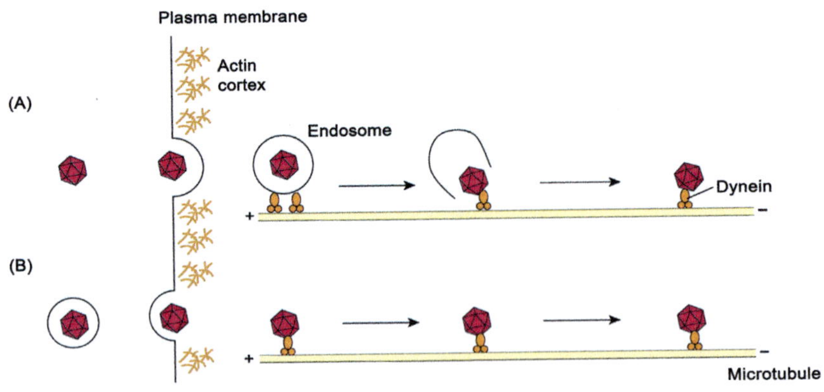

Figure 5.8. *Cytoplasmic trafficking.*

Source: Wang-Shick Ryu. Virus Life Cycle. Molecular Virology of Human Pathogenic Viruses. Elsevier. DOI: 10.1016/B978-0-12-800838-6.00003-5.

Two distinct viruses are used to explain how the entry is linked to cytoplasmic trafficking: (A) adenovirus (naked) and (B) herpes virus (enveloped). Incoming viruses can enter cells by endocytosis (A) or direct fusion (B). Following penetration into the cytoplasm, either endocytic vesicles or viral capsids exploit dynein motors to traffic toward the minus ends of microtubules. Either the endocytic vesicles (A) or the capsids (B) interact directly with the microtubules. The virus can also lyse the endocytic membrane, releasing the capsid into the cytosol (A).

A nuclear pore is required for the viral genome to enter the nucleus in order for the virus to replicate in the nucleus. Several approaches are used, mostly based on the size of their genomes (Figure 5.9). In viruses with smaller genomes, like polyomavirus, the viral capsid enters the nucleus. When nucleocapsids of larger viruses dock with a nuclear pore complex, the capsid (adenovirus, for example) is partially disrupted or undergoes a slight alteration (herpes virus, for example), allowing the DNA genome to pass through and enter the nucleus.

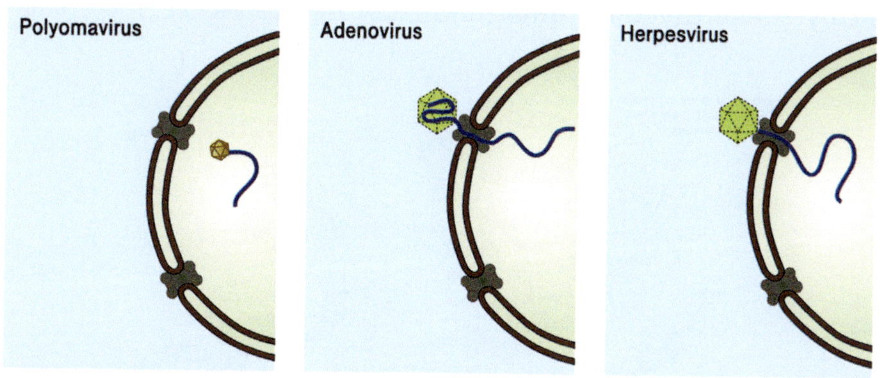

Figure 5.9. *Strategies for the nuclear entry of DNA viruses. (A) Polyomavirus capsids are small enough to enter the nucleus via the nuclear pore complex without disassembly. Uncoating of the polyomavirus genome takes place in the nucleus. (B) The adenovirus capsids are partially disrupted upon binding to the nuclear pore complex, allowing the transit of the DNA genome into the nucleus. (C) For herpesvirus, the nucleocapsids are minimally disassembled to allow transit of the DNA genome into the nucleus.*

Source: Wang-Shick Ryu. Virus Life Cycle. Molecular Virology of Human Pathogenic Viruses. Elsevier. DOI: 10.1016/B978-0-12-800838-6.00003-5.

5.3. ANIMAL VIRUS ENTRY CELLS—THE ROLE OF THE CELLULAR RECEPTOR

Animal viruses need to correctly enter cells via a complex plasma membrane made of a lipid bilayer with membrane-associated proteins floating on either the upper or lower surface (Figure 5.10). In order to move ions and other small molecules through the membrane, certain integral membrane proteins create pores or channels. The addition of sugar residues (glycosylation) modifies other proteins that protrude from the cell surface. Cell adhesion, signaling, cellular recognition, immunity, and other processes are all mediated by these glycoproteins (Granzow et al., 1997).

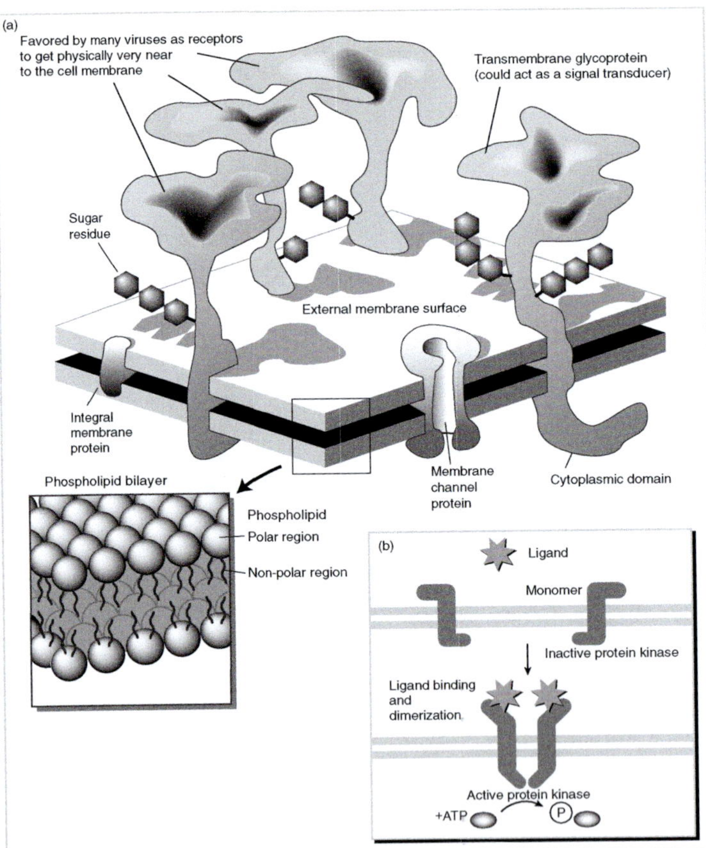

Figure 5.10. *(a) The surface of a "typical" animal cell.*

Source: https://www.blackwellpublishing.com/content/BPL_Images/Content_store/Middle_Sized/Wagner9781405103466/060to076.pdf.

The lipid bilayer plasma membrane is penetrated by cell surface proteins of various functions. The proteins that extend from the surface (mainly glycoproteins) can be utilized by different viruses as "tether points" or "anchors" for bringing the virus close enough to the cell surface to initiate the entry process. This interaction between a cell surface protein serving as a virus receptor and the virus itself is highly specific between proteins. Integral membrane proteins, such as those mediating the transport of small molecules and ions across the plasma membrane, tend not to project as far into the extracellular matrix and are not as commonly utilized by viruses as receptors. Some viral receptors are listed in Table 5.2. (b) The interaction between a cellular surface protein (receptor) and a ligand or co-receptor can lead to chemical and structural changes that transmit signals between the exterior and interior of the cell. This is the process of signal transduction. Here, for example, ligand binding with two monomeric receptor proteins leads to dimerization, activating a protein kinase in the cytoplasm. This in turn leads to phosphorylation of a target protein, leading to further changes in the cell.

Interaction between proteins on the virion's surface and particular proteins on the cell's surface is necessary for virus infection (Granzow et al., 1997). Receptors are the name for cell surface proteins. Remember that the physiological roles of a cell surface protein acting as a virus receptor serve purposes other than allowing the virus to attach and enter; Table 5.2 lists some known viral receptors and their functions. The word "receptor" simply refers to the protein's effect, in this case, the virus's entry into a cell that is being studied.

Table 5.2. *Some Cellular Receptors for Selected Animal Viruses*

Name	Cellular function	Virus receptor for
ICAM-1	Intracellular adhesion	Poliovirus
CD4	T-lymphocyte functional marker	HIV
MHC-I	Antigen presentation	Togavirus, SV40
MHC-II	Antigen presentation/stimulation of B-cell differentiation	Visnavirus (lentivirus)
Fibronectin	Integrin	Echovirus (picornavirus)
Cationic amino acid transporter	Amino acid transport	Murine leukemia virus (oncornavirus)
LDL receptor	Intracellular signaling receptor	Subgroup A avian leukosis virus (oncornavirus)
Acetylcholine receptor	Neuronal impulse transducer	Rabies virus
EGF	Growth factor	Vaccinia virus
CR2/CD21	Complement receptor	Epstein-Barr virus
HVEM	Tumor necrosis factor receptor family	Herpes simplex virus
Sialic acid	Ubiquitous component of extracellular glycosylated proteins	Influenza virus, reovirus, coronavirus

Source: https://www.blackwellpublishing.com/content/BPL_Images/Content_store/Middle_Sized/Wagner9781405103466/060to076.pdf.

The type and distribution of receptors a virus uses determines its ability to replicate in particular cells, a phenomenon known as tissue tropism or host range. For instance, HIV recognizes specific surface proteins on CD4⁺T lymphocytes involved in the immune response and permits an infection of these lymphocytes. The virus has adapted to identify the particular protein and "subvert" its intended function. The major intercellular adhesion molecule (ICAM) is a tool that the poliovirus uses to infect cells. The rabies virus enters neurons through acetylcholine receptors, which allows it to slowly move up the neural net and into the central nervous system. Because most of these receptors are found at the synapses that connect neurons, the virus can "jump" from one neuron to the next without harming any individual neurons. By minimizing tissue damage and inflammation, this pattern of progression allows the virus to "leak" into the host's circulatory system and trigger an immune response. Lastly, sialic acid residues are a specific target for the influenza virus and several other respiratory viruses. Enzymatic addition of sialic acid residues and widespread modifications to the glycoproteins of secretory cells, particularly those of the respiratory system and nasopharynx, occur (Granzow et al., 1997).

Whether or not a receptor is available for the virus to recognize at a particular cell surface influences the tissue tropism of a particular virus. HIV interacts, for instance, with two proteins present in a specific subset of body cells: specific T lymphocyte subclasses and specific neuronal cell types. The poliovirus, on the other hand, only infects primates and is limited in its ability to bind to and penetrate certain motor neurons and small intestinal lining cells. Although poliovirus-specific ICAMs are present on many primate cells, they appear to be "masked" by other membrane proteins on the surface of those cells, making them resistant to infection. Nonprimate cells lack the poliovirus receptor.

Entry-mediated tissue tropism in viral infections is influenced by an additional crucial factor. Several viruses use co-receptors, or proteins, on the surface of cells, in addition to the primary receptor. An essential co-receptor in the HIV instance is one of a class of surface proteins that act as receptors for chemokines, which are proteins that typically serve as receptors for chemical messengers transmitted between cells. For HIV infection to be effective, the co-receptor and the CD4+ receptor need to interact molecularly. The co-receptor for HIV also controls tissue tropism. Macrophages express a protein called CCR5, in addition to CD4, which enables HIV variants that recognize it to exhibit a notable tropism for these cells. As an alternative, T cells express both CD4 and CXCR4, a second HIV co-receptor; certain HIV strains exhibit a notable affinity for these cells.

Lastly, co-receptor proteins are expressed by peripheral mononucleocytes that circulate in the blood and are susceptible to infection by both HIV strains. Therefore, a particular virus may use a major receptor protein but also need one or more additional proteins to be present. Cell and tissue tropism are changed if a particular cell has the major receptor but not the co-receptor. Infection either cannot happen or occurs with reduced efficiency.

It's also critical to realize that certain viruses have different ways of infecting a cell next to the one that was first infected through the receptor-mediated pathway. For instance, a cell's membrane may alter due to infection, enabling fusion with one or more nearby cells. The virus can then easily enter the uninfected cell's cytoplasm without crossing the plasma membrane; this is a well-known characteristic of infections with certain HSV strains that result in the formation of sizable clusters of fused cells, or syncytia.

Complete fusion is not necessary for the contact between cells that allows a virus to spread. The intimate relationship between macrophages and other lymphatic system cells during the induction of the immune response may make it easier for viruses that the macrophages took up but did not destroy to spread. This is undoubtedly a crucial aspect of the HIV pathogenesis (Granzow et al., 1997).

The virus itself might have a surface protein that, in some circumstances, is not necessary for recognition and receptor-mediated entry. The case of HSV1 mutants lacking glycoprotein C (gC) on their envelope is a prime example. This glycoprotein allows the ultimate receptor to get in close proximity by interacting with heparan sulfate on the cell surface.

While mutant viruses lacking gC exhibit notable changes in the specifics of their infection and pathogenesis in lab animals, they replicate exceptionally well in various laboratory-cultured cells. Here, the repeated passage of the cells and their culture cause changes in the cell surface, making it easier for gC-negative viruses to "find" their receptors.

In cells lacking the effective receptor protein, viruses can ineffectively use other surface proteins to infect the cell. These proteins can replace the effective receptor if the right conditions are met. One explanation for some viruses' ability to infect specific cells in culture despite lacking the perfect receptor is this substitution. One instance is the SV40 virus's ineffective ability to infect specific hamsters and murine cells in culture.

Such a typical infections are easily observed in the laboratory, and there is strong evidence to suggest that they can occur occasionally in natural environments. When a virus infects a host that it had not previously affected, it is common for new infectious viruses to emerge in the environment.

Some of these events can be assumed to be the consequence of the novel virus adapting to use a previously unknown receptor after an initial "inappropriate" infection. A rare "inappropriate" animal virus infection in humans that led to modifications in the virus's genetic makeup has been proposed as an explanation for the comparatively abrupt emergence of HIV in human society. The sudden appearance in Hong Kong in 1997 of a novel chicken influenza virus that appeared to have adapted for human transmission could be explained by another instance of such an occurrence.

5.4. MECHANISMS OF ENTRY OF NONENVELOPED VIRUSES

Translocation across the lipid bilayer is the process by which non-enveloped virus particles must be integrated into the cell. During this process, the capsid, or a capsid that has been altered by the cell, physically crosses the cell plasma membrane. Figure 5.11 illustrates it for the poliovirus and entails receptor-mediated endocytosis. The endocytotic vesicle's acidic environment modifies the poliovirus capsid in a particular way, allowing the internal genome (positive-sense RNA) to be released into the cytoplasm, where it can be translated and start the expression of genes.

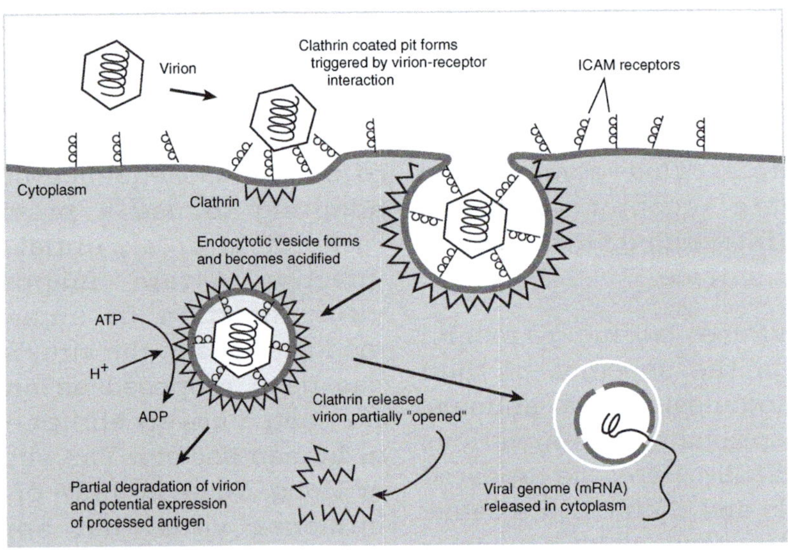

Figure 5.11. *Schematic of receptor-mediated endocytosis utilized by poliovirus for entry into the host cell. The endocytotic vesicle forms due to close association between the poliovirus-receptor complex and the plasma membrane.*

Source: https://www.blackwellpublishing.com/content/BPL_Images/Content_store/Middle_Sized/Wagner9781405103466/060to076.pdf.

A nuclear replicating nonenveloped virus, like the papovavirus SV40, similarly enters the cell, but it is able to move its modified virion to the nuclear membrane through the interaction of viral capsid proteins with the vesicle and other intracellular trafficking proteins. Once there, the virus releases its genome and starts to express its genes through interactions between its DNA and transcription factors found in cells. It is established that the process is virus-controlled because certain genetic changes (mutations) in the SV40 capsid protein will obstruct this transport.

5.5. ENTRY OF ENVELOPED VIRUSES

Viral glycoproteins linked with membranes that extend past the viral envelope allow enveloped viruses to communicate with cell receptors. During viral maturation, the cell Golgi apparatus glycosylates the viral glycoproteins with sugars. The procedure is comparable to what the cell does with its own glycoproteins.

Viral entry can occur through receptor-mediated endocytosis or viral membrane fusion at the cell's surface. Figure 5.12a provides a schematic representation of these two processes. Similar to viruses that lack an envelope, the endocytotic vesicle's acidic pH can cause the viral envelope to change, allowing for the possibility of fusion between the viral envelope and the vesicle's membrane. The electron micrographs in Figure 5.12b depict the fusion of the pseudorabies virus (a close relative of HSV) with the plasma membrane of the infected cultured cells (Mercer et al., 2010).

A straightforward fusion interaction between a single viral glycoprotein and a single cellular receptor can occur during the viral-plasma or vesicular-membrane fusion, or it can involve an intricate network of interconnected protein interactions. For instance, a herpesvirus like HSV must interact with a particular cellular surface receptor to enter the cell, after which five or six viral glycoproteins bring the virus close. The first interaction seems to be a connection between the glycoproteins of the virus and sulfated sugar molecules, or polyglycans, such as heparan sulfate, which is a substance that is attached to a variety of cell surface proteins. The virion can only then be pushed near enough to the plasma membrane to permit interaction with the receptor itself.

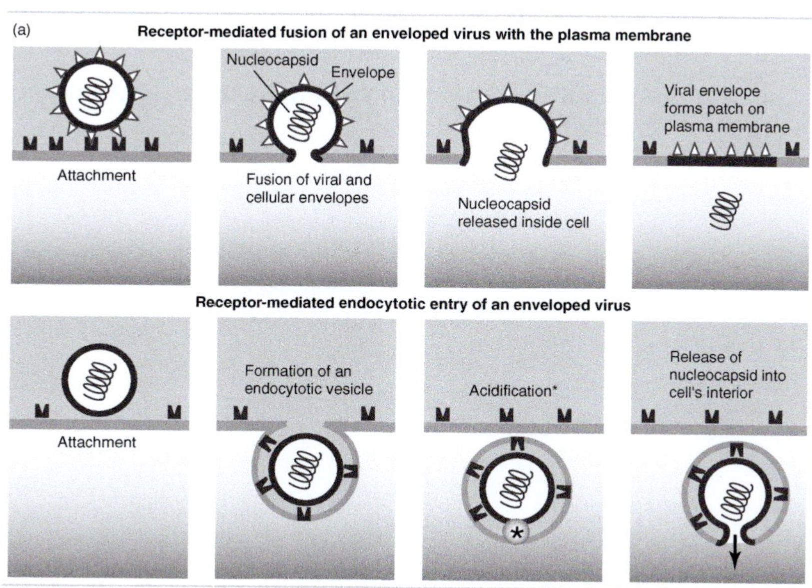

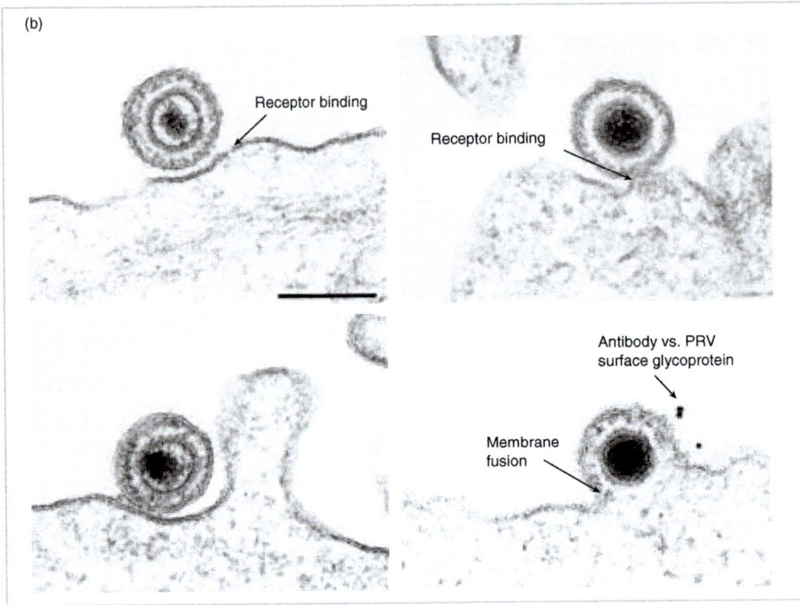

Figure 5.12. *(a) The two basic modes of entry of an enveloped animal virus into the host cell. Membrane-associated viral glycoproteins can either interact with cellular receptors to initiate a fusion between the viral membrane and the cell plasma membrane or can induce endocytosis. The fate of the input virus membrane differs in the two processes. (b) The fusion of pseudorabies virus with the plasma membrane of an infected cultured cell is shown in this series of electron micrographs (the bars represent 150nm). Although each electron micrograph represents a single event "frozen in time," a logical progression from the initial association between viral envelope glycoproteins and the cellular receptor on the plasma membrane through the fusion event is shown. The final micrograph contains colloidal gold particles bound to antibodies against the viral envelope glycoproteins (dense dots). With them, the envelope can be seen clearly to remain at the infected cell surface. (Micrographs reprinted with the kind permission of the American Society for Microbiology from Granzow, H. et al., 1997)*

Source: https://www.blackwellpublishing.com/content/BPL_Images/Content_store/Middle_Sized/Wagner9781405103466/060to076.pdf.

5.6. ENTRY OF PLANT VIRUS INTO CELLS

The unique architecture of plant cells, specifically the presence of a fairly thick and rigid cell wall, creates a special barrier for virus entry. The first penetration into the plant cell must take advantage of a breach in the cell wall's integrity. The virus can enter a plant cell without interacting with particular receptors when it gets into a break like this and positions itself close to the plasma membrane.

Most often, organisms that consume plants or mechanical forces cause breaks or lesions in the plant wall. Aphids, leafhoppers, whiteflies, and thrips are among the invertebrates known to carry various plant viruses above ground. Nematodes are another source of viral infection, as they feed on the plant's roots. Sometimes, the virus spreads to plants from the invertebrate without developing in the vector, as is the case with whitefly-borne geminivirus transmission. Conversely, some viruses can replicate in both their plant and invertebrate hosts. The tomato spotted wilt virus, a plant bunyavirus, demonstrates this with its thrip vector. In each case, the viruses enter the cytoplasm of the plant cell after the insect begins feeding on plant material (Mercer et al., 2010).

Plant viruses may enter a plant through mechanical damage to its cell wall. In experimental settings, this method is most frequently applied when the leaf surface is abraded or scratched before being inoculated with a virus suspension. This can also occur naturally as a result of agricultural practices like harvesting. During cutting operations, the plant may also become infected with the Brome mosaic virus, which is spread by beetles.

5.7. THE INJECTION OF BACTERIOPHAGE DNA INTO ESCHERICHIA COLI

To properly start replication, bacteriophages need to engage with a receptor on the surface of the bacterial cell. Prokaryotic cells exhibit a range of external features on their outer surface, such as glycoproteins and lipopolysaccharides for structural purposes, transport machinery such as sugar or amino acid transport complexes, and the F or sex pilus for cell-to-cell interaction. Bacteria utilize sex pili to facilitate conjugation and genetic material exchange with other bacteria belonging to the opposite sex. Depending on the specific virus, the phage may attach itself to host cells using any one of these structures. Table 5.3 displays some characteristics of bacteriophages replicating in *Escherichia coli* (Mercer et al., 2010).

Table 5.3. *Some E. coli Bacteriophage Receptors*

Virus	Structure	Normal function
T2	OmpF	Porin protein
	Lipopolysaccharide	Outer membrane structure
T4	OmpC	Porin protein
	Lipopolysaccharide	Outer membrane structure
T6	Tsx	Nucleoside transport protein
T1 and T5	TonA	Ferrichrome transport
λ	LamB	Maltose transport protein
MS2	F pilus	Conjugation

Source: https://www.blackwellpublishing.com/content/BPL_Images/Content_store/Middle_Sized/Wagner9781405103466/060to076.pdf.

There are situations when the physical rearrangement of the virus particle occurs during phage attachment to the host cell. For instance, bacteriophage T4 adhering to the surface of vulnerable *E. coli* cells are produced in two steps, as Figure 5.13 illustrates. First, lipopolysaccharide residues on the surface of the cell's outer membrane interact with the tips of the phage tail fibers comparatively weakly. This sets off a second, more intense, and irreversible exchange. This requires a change in the conformation of the tail fibers due to interaction between structures in the outer membrane and tail pins on the virion's base plate. In the end, this causes the contractile sheath around the phage tail to compress and phage DNA to be injected into the host cell. During this process, the phage tail tube enters the cell wall, but the phage DNA still needs to pass through the inner cell membrane. A viral gene product known as a pilot protein aids in the completion of this final step (Mercer et al., 2010).

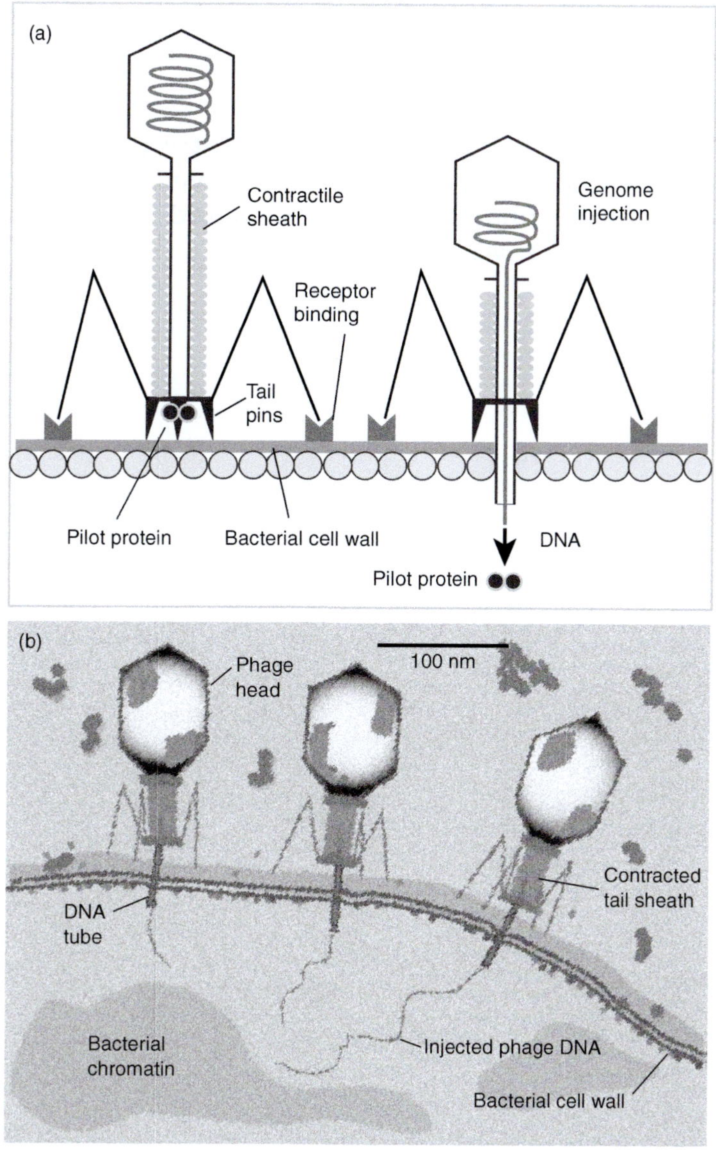

Figure 5.13. *Entry of T4 bacteriophage DNA into an Escherichia coli cell. Initial attachment is between the fibers to the ompC lipopolysaccharide receptor on the bacterial cell wall (a). The binding of protein pins on the base plate to the cell wall leads to the contraction of the tail fibers and sheath proteins, leading to the insertion of the tail tube through the cell wall. As shown in the electron micrograph (b), phage pilot protein (arrow) allows the highly charged viral DNA genome to penetrate the bacterial plasma membrane and enter the cell. Phage DNA can be seen as shadowy lines emanating from the tail tube. (From Dimmock, N.J., and Primrose, S.B., 1994.)*

Source: https://www.blackwellpublishing.com/content/BPL_Images/Content_store/Middle_Sized/Wagner9781405103466/060to076.pdf.

The attachment of bacteriophage l to LamB is one example of a phage where the interaction between the virion and the cell does not immediately change the phage structure. To reiterate, the virus's structure remains unchanged as a result of the MS2 bacteriophage's attachment to the F pilus. It is sometimes said that MS2 and related phages are male-specific because cells bearing a pilus structure (the result of an F-plasmid) are referred to as male.

A bacteriophage's actual number of intracellular entry points varies greatly. Merely the phage DNA and specific auxiliary proteins are allowed to enter the host cell in the case of tailed phage. An entire phage particle, uncoated in the cytoplasm, enters the cell when a contained phage like MS2 enters it.

5.8. NONSPECIFIC METHODS OF INTRODUCING VIRAL GENOMES INTO CELLS

Evidently, the fundamental step in a virus's ability to infect a cell is its ability to successfully insert its genome into the right place within the cell to allow for the expression of its genes. There may be other, albeit rare, ways for viral genomes to be introduced into plant cells given that viruses can internalize themselves into cells without the help of receptors. Cells can be made permeable in the lab, for instance, using chemical or physical techniques, enabling them to absorb fairly large particles. Viral uptake can be caused by high virus particle concentrations added to cells and by appropriately treating those cells. This will be an ineffective process that might destroy the majority of the virus particles. Nevertheless, suppose enough virus particles are ingested and an intact viral genome or two can reach the right part of the cell to start infection. In that case, it is frequently possible to start a productive infection in a small number of cells.

Transfection is a similarly nonspecific and inefficient process that is frequently used to introduce viral genomes, particularly DNA genomes, into cells. By precipitating isolated genomes into aggregates with calcium phosphate ($Ca_3(PO_4)_2$), one can aggregate them into the right-sized particles, which can then be easily incorporated into cells through treatment. As an alternative, viral genomes can be concentrated inside lipid vesicles known as liposomes in solution, which are easily assimilated by cells that have undergone a particular mild detergent treatment to allow the liposome to fuse with the cell's plasma membrane. Once more, productive infection can be started even though the process is inefficient if a few viral genomes are delivered to the right intercellular location (Mercer et al., 2010).

Figure 5.14 provides an illustration of how transfection can be used to study a viral protein's characteristics. Here, a DNA fragment encoding the gene for the varicellazoster virus (herpes zoster virus) glycoprotein, gL, was transfected into cells. The transcriptional machinery of the uninfected cell can express the promoter that controls this gene. Figure 5.14b displays three micrographs that were taken shortly after, twelve hours after, and twenty-four hours after transfection. An anti-gL fluorescent antibody was cultivated on the cells. Later on, it becomes clear that this protein is expressed in the cytoplasm.

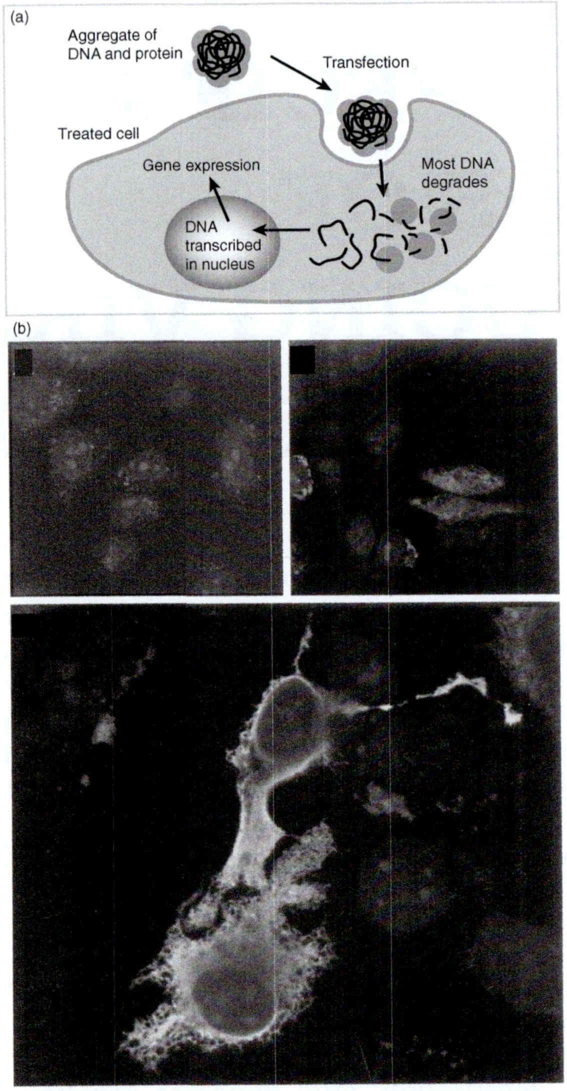

Figure 5.14. *Expression of a varicella-zoster virus protein following transfection of a cell with the viral gene under the control of an active promoter in the uninfected cell. (a) The basic process. The cell membrane is treated with agents that allow it to readily take up large aggregates of protein and nucleic acids by phagocytosis. The transfecting DNA is caused to form aggregates with the use of calcium phosphate (Ca3(PO4)2), and then mixed with cells that have been appropriately treated. While most of the DNA taken up by the cell is degraded, some gets to the nucleus by nonspecific cellular transport of macromolecules, and this DNA can be transcribed, and any genes within it expressed as proteins. (b) An actual experiment. Cells were made permeable and transfected with DNA containing the varicella-zoster virus glycoprotein L gene. The protein encoded in this gene was expressed following its transcription into mRNA. Cells were treated with a fluorescent antibody reactive with the glycoprotein at (clockwise from the top left) 0, 12, and 24 hours after infection. The expression of the glycoprotein in the cytoplasm is clearly evident from the green fluorescence.*

Source: https://www.blackwellpublishing.com/content/BPL_Images/Content_store/Middle_Sized/Wagner9781405103466/060to076.pdf.

5.9. EXIT

Exit can be divided into three steps: capsid assembly, release, and maturation.

5.9.1. Capsid Assembly

When the viral proteins and genome proliferate, the capsid assembly happens next. The two processes that makeup capsid assembly are genome packaging and capsid assembly. These two processes can happen simultaneously or coupled, depending on the viruses involved. Figure 5.15 illustrates that adenovirus is an example of the latter type, and picornavirus is an example of the former. The RNA genome is not assembled until the capsids (i.e., immature capsid or procapsid) are assembled in the case of picornavirus (Sattentau, 2008). The procapsid structure then forms a pore through which the RNA genome is inserted or packaged. In contrast, the assembly of the capsid and the packaging of the DNA genome occur together in the case of adenovirus. The question then becomes, "How does the virus package the viral genome selectively?" The viral capsid proteins specifically recognize a packaging signal, a cis-acting element present in the viral genome, and package either RNA or DNA selectively in response.

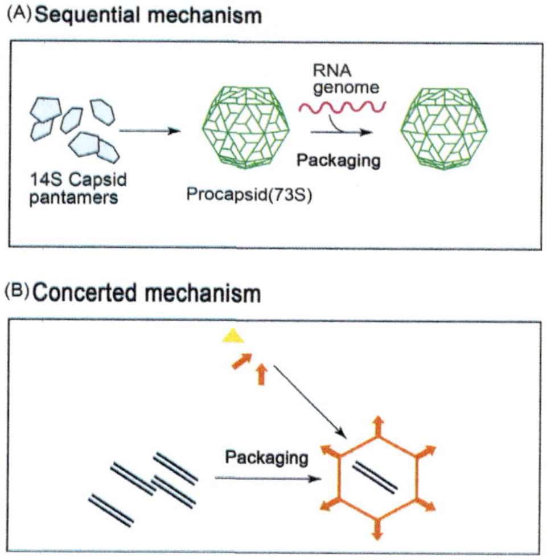

Figure 5.15. *Relationship between capsid assembly and genome packaging. (A) Sequential mechanism. For picornavirus, the procapsid, a precursor of the capsids, is preassembled without an RNA genome. Subsequently, the RNA genome penetrates the procapsid via a pore. (B) Coupled mechanism. For adenovirus, the DNA genome is packaged into the capsid during capsid assembly.*

Source: Wang-Shick Ryu. Virus Life Cycle. Molecular Virology of Human Pathogenic Viruses. Elsevier. DOI: 10.1016/B978-0-12-800838-6.00003-5.

5.9.2. Release

Cell lysis of the infected cells releases the virus particles for naked viruses. Because the cell membrane that contains the assembled virus particles is broken, no special exit mechanism is required. Adenovirus and polyomavirus (i.e., SV40) are two examples of naked viruses. On the other hand, envelopment—the process by which the capsids are encased in a lipid bilayer—occurs before the release of enveloped viruses. Two mechanisms explain how related the capsid assembly is to the envelopment. Once capsid assembly is complete, the envelopment can start (Figure 5.16A). Through interaction between the viral capsids and the viral envelope glycoprotein, the fully assembled capsids are recruited to the membrane in this sequential mechanism. Hepatitis B and herpesvirus are two instances of this. As an alternative, envelopment and capsid assembly may happen simultaneously (Figure 5.16B). An example of this coupled mechanism is the retrovirus (Sattentau, 2008).

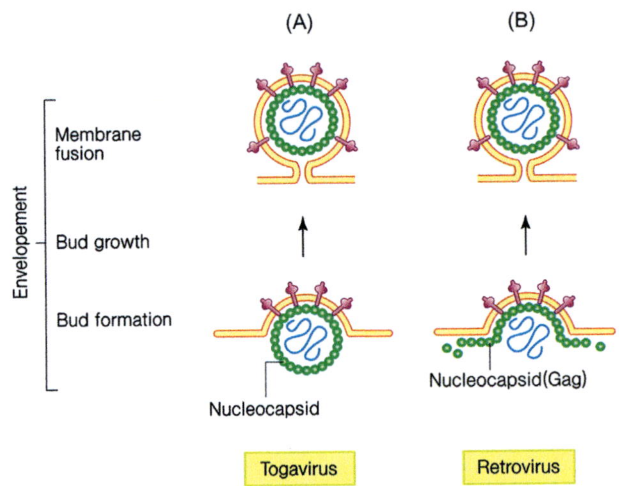

Figure 5.16. *Relationship between capsid assembly and envelopment. (A) Sequential mechanism. The capsid assembly occurs prior to the envelopment. The assembled capsid is then targeted to the membrane for envelopment. Togavirus constitutes a family of positive-strand RNA viruses. (B) Coupled mechanism. Capsid proteins and the viral genome are recruited to the budding site on the membrane. Capsid assembly and the envelopment of the capsid proceed simultaneously. The envelopment process can be divided into three steps: bud formation, bud growth, and finally, membrane fusion.*

Source: Wang-Shick Ryu. Virus Life Cycle. Molecular Virology of Human Pathogenic Viruses. Elsevier. DOI: 10.1016/B978-0-12-800838-6.00003-5.

However, two cellular membranes are utilized in relation to the envelope membrane. Certain viruses, like influenza and retrovirus, envelope in the plasma membrane, while other viruses, like the herpesvirus and hepatitis B virus, envelope in endosomes, such as the Golgi apparatus and the endoplasmic reticulum (ER).

Then, how are the viruses released from the infected cells? The majority of enveloped viruses are released extracellularly through a process known as exocytosis, which is sometimes referred to as "budding" with reference to the plant buds. The envelopment advances in tandem with extracellular release through budding. Then, the question of how budding is triggered mechanistically emerges, and a clue to this was provided by the discovery of a peptide motif known as the late (L) domain, which plays a crucial role in initiating the budding process. For example, the "PTAP" motif is encoded as a late domain by the retroviral Gag protein. In short, the Gag protein subverts the MVB pathway for budding and recruits cellular components involved in the multivesicular bodies (MVBs) pathway through its late domain. Ultimately, it is fascinating to discover how viruses utilize cellular processes to generate their own offspring extracellularly (Sattentau, 2008).

5.9.2.1. Late Domain

The late domain is involved in the budding process of the enveloped viruses. It was initially found in the Gag polyproteins of retroviruses and the M (matrix) proteins of rhabdoviruses. It is made up of four amino acid residues (i.e., PTAP, PPPY, and PPEY) that are either encoded in the HIV Gag polyprotein's p6 subdomain or the rhabdovirus matrix protein (panel A). The nomenclature reflects the intriguing phenotype of virions attached to the plasma membrane without viral release, as if the viral release is inhibited at a late stage of budding caused by a substitution mutation of the late domain motif. It was discovered that the MVBs pathway's cellular factors are recruited by the late domains (panel B). For example, the interaction between the Gag polyprotein and Tsg101, a part of ESCRT-I, depends on the PTAP motif. Crucially, the budding of retrovirus particles is topologically analogous to MVB biogenesis: budding in both instances is oriented in the opposite direction of the cytoplasm. Stated differently, viral egress from the plasma membrane is mediated by the retroviral late domains, which coopt the cellular machinery involved in MVB biogenesis (Figure 5.17).

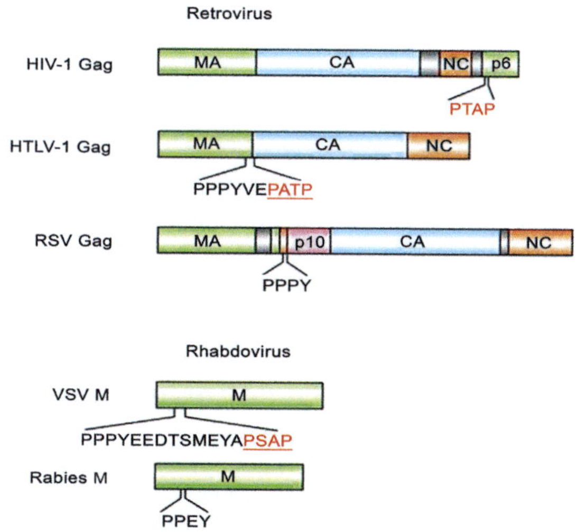

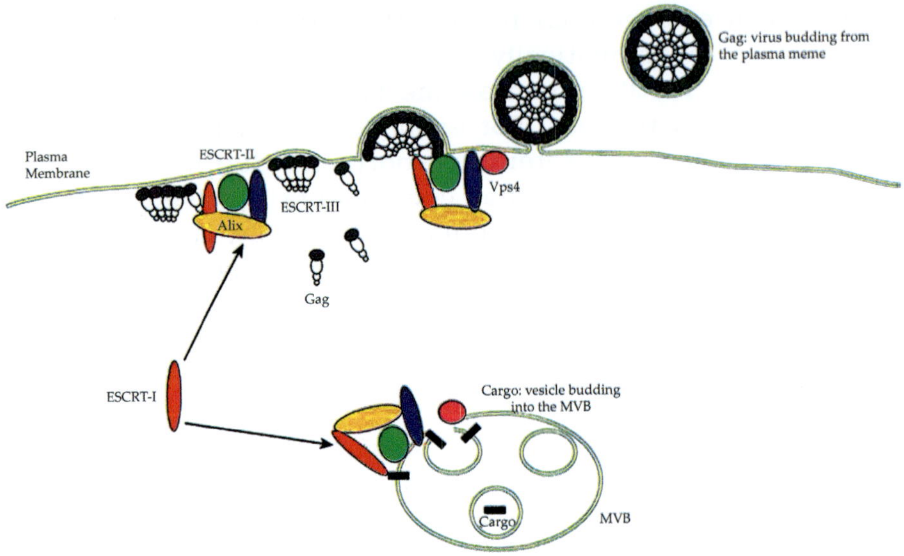

Figure 5.17. *Late domain and its function in MVBs pathway. (A) Three kinds of L domain found in enveloped viruses such as retroviruses (i.e., HIV, HTLV-1, and RSV) and rhabdoviruses (i.e., vesicular stomatitis virus and rabies virus): PTAP motif, PPPY motif, and PPEY motif. A (alanine), E (glutamic acid), P (proline), T (threonine), and Y (tyrosine). (B) L domain and ESCRT complexes involved in MVBs pathway. The budding of retroviral Gag is facilitated by ESCRT complexes, which are normally involved in the MVB pathway. In HIV, Gag interacts with Tsg101 and Alix, an adapter protein, leading to recruitment of the additional components of the MVB pathway, that is, ESCRT-II (green), ESCRT-III (purple), to assemble into a functional complex. Vps4 (red) is involved in recycling the MVB machinery.*

Source: Wang-Shick Ryu. Virus Life Cycle. Molecular Virology of Human Pathogenic Viruses. Elsevier. DOI: 10.1016/B978-0-12-800838-6.00003-5.

5.9.3. Late Events in Viral Infection: Capsid Assembly and Virion Release

5.9.3.1. Assembly of Helical Capsids

Helical viruses require their capsids to coalesce around the genome. The tobacco mosaic virus (TMV) in plants is a relatively well-studied example of this process. All helical viruses seem to follow a similar basic process. This similarity arises from the fact that, when linked to the appropriate protein, single- or double-stranded RNA (or DNA, for that matter) can easily form a helical structure.

Figure 5.18 illustrates how the TMV RNA genome and helical capsid are combined. Capsomers self-assemble to form disks, and the disks that are formed interact with a particular genomic sequence known as pac (packaging signal). The disk assumes a "lockwasher" conformation upon interaction with the RNA, and subsequent capsomer assemblies thread onto the expanding helical array to form the full capsid. It should be noted that in the case of TMV, the RNA resembles a "screw" that breaks through the capsomer disk assembly. Translocation to a helical configuration that expands through ongoing association with the genomic RNA is made possible by this penetration (Sattentau, 2008).

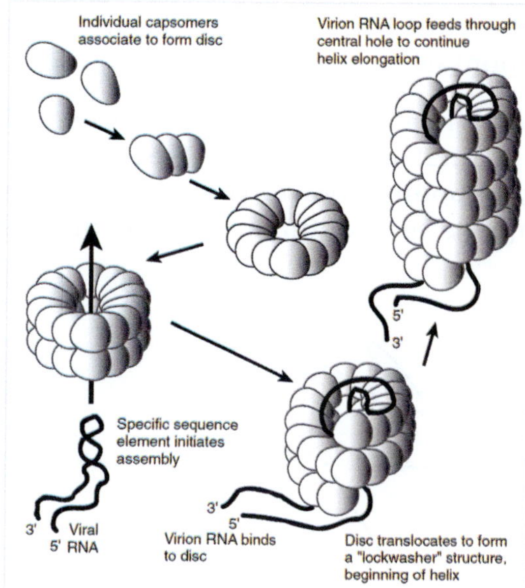

Figure 5.18. *Assembly of the helical tobacco mosaic virus. Steps in the preassembly of the capsomer disk, insertion of viral RNA, and the translational "screwlike" helix assembly process with sequential addition of more capsomers are shown. (Adapted from Dimmock, N.J., and Primrose S.B., 1994.)*

Source: https://www.blackwellpublishing.com/content/BPL_Images/Content_store/Middle_Sized/Wagner9781405103466/060to076.pdf.

5.9.3.2. Assembly of Icosahedral Capsids

The majority of cases that have been thoroughly examined show that particular proteolytic cleavage of one or more capsid proteins that were assembled into the immature virus particle is necessary for the icosahedral capsid to mature from an immature procapsid to the final state. This cleavage, which frequently occurs in conjunction with the inclusion of the viral genome, causes minor structural changes or increases capsid stability. There are very few cleavage steps. Proteins called maturational proteases encoded by viruses hydrolyze one or a small number of distinct peptide bonds. As a result, it is generally accepted that the assembly of icosahedral capsids involves both the specific covalent modifications of the virion proteins through proteolytic processing and the preassembly of procapsids. Protease inhibitors of HIV have been shown to have significant therapeutic value. Maturational proteases are an appealing target for antiviral therapy due to their high specificity and the viral genome encodes them.

Preliminary research on the small RNA-containing virus poliovirus served as the basis for several general models for assembling an icosahedral capsid. The development of empty capsids is one sign of poliovirus infection in the lab. It is evident that the viral capsomers have the ability to self-assemble. This finding was taken to mean that before the genome enters the virion, empty capsids assemble. Ironically, it has been suggested in some recent studies on the assembly of poliovirus and related viruses that empty capsids are a byproduct of the assembly process and that the procapsid assembles directly around the viral RNA. Nevertheless, empty capsids have the ability to form a stable structure spontaneously.

The assembly of capsids in larger icosahedral viruses is a complicated process involving the formation of a final capsid "mold" or pattern by scaffolding proteins. Either way, capsid assembly happens prior to the viral genome entering the capsid, and the production of empty capsids is a defining feature of icosahedral virus maturation (Sattentau, 2008).

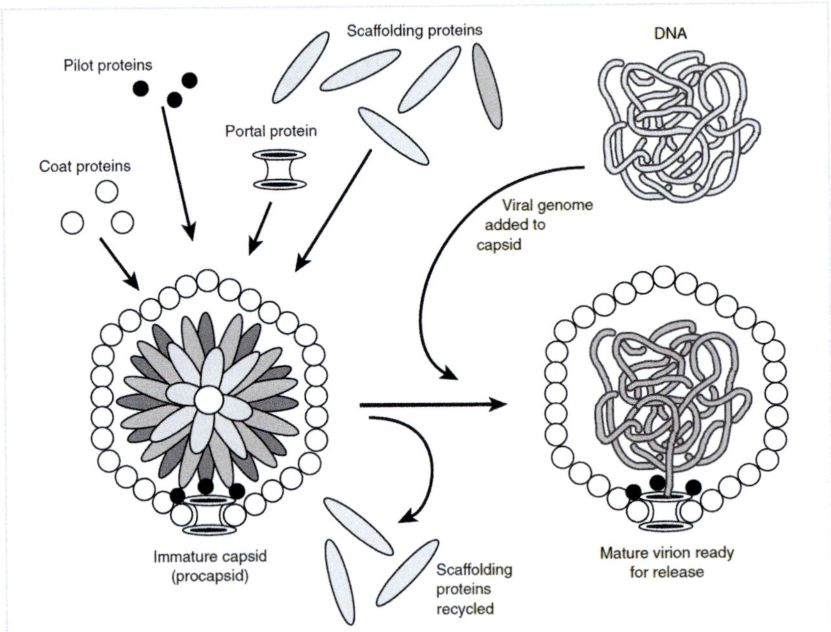

Figure 5.19. *Assembly of the phage P22 capsid and maturation by insertion of viral genomic DNA. Individual capsomer subunits preassemble into a procapsid around scaffolding protein. This latter protein is recycled with phage P22 but can be proteolytically removed with a maturational protease with other icosahedral viruses. The empty head then associates with viral genomes. Genome insertion requires both energy and a conformational change in the procapsid.*

Source: https://www.blackwellpublishing.com/content/BPL_Images/Content_store/Middle_Sized/Wagner9781405103466/060to076.pdf.

As an example of this process, Figure 5.19 depicts the assembly of the bacteriophage P22 head. The process is similar to how herpesvirus capsids are assembled. It should be noted that pilot proteins, which are crucial for the injection of the genome, may also assist in the assembly of capsid proteins. The scaffolding proteins are recyclable and participate in the assembly of multiple capsids. Additionally, note that the term "pilot protein" in this context differs significantly from its meaning in the Teven bacteriophage infection discussed earlier.

Virion-associated enzymes are "activated" by retrovirus proteases in the last stages of virion maturation after release from the infected cell. The structural protein of the virion includes these retrovirus proteases. Other viral proteases are targets for drug development because they are specific to the virus encoding them, and antiviral medications that target the HIV protease have demonstrated notable therapeutic benefits.

5.9.4. Maturation

"Maturation" is the final stage of the virus particle assembly process, which occurs extracellularly after release. Maturation is a necessary stage for picornaviruses and retroviruses to acquire infectivity. In the case of a retrovirus, a significant morphological change, such as the condensation of the capsid structure, occurs concurrently with the cleavage of the Gag polyprotein by the viral PR protein (aspartate protease) in the released virion. Significantly, the particle's infectivity is conferred by this maturation process (Vigant et al., 2015).

5.10 GENERATION OF THE VIRION ENVELOPE AND EGRESS OF THE ENVELOPED VIRION

The infected cell provides the lipid bilayer that comprises the membrane envelope of the viruses that carry them. As of yet, very few (if any) viral genes involved in membrane assembly or lipid biosynthesis have been found. The envelope becomes virus-specific through the insertion of one or more virus-encoded membrane proteins, which are produced during the replication cycle, even though the lipid bilayer is wholly cellular.

Figure 5.20 illustrates some envelope patterns at the plasma membrane for viruses that assemble in the cytoplasm. Viral glycoproteins have their amino termini outside the cell and their carboxy termini inside the cytoplasm when they reach the budding site. They are first synthesized at the rough endoplasmic reticulum and subsequently processed through the Golgi apparatus.

A similar arrangement holds true for viruses that budder at other subcellular locations (e.g., herpesviruses, which bud from the nuclear membrane and then into exocytotic vesicles, or bunyaviruses, which bud into the Golgi itself). Each time, trafficking signals found in the viral glycoproteins guide the protein to its intended location by utilizing the resources of the host cell. Many cells in organized tissue have asymmetric plasma membranes, which certain viruses have evolved to take advantage of. Therefore, some viruses (e.g., influenza viruses) emerge from these cells' apical surface, but other viruses (e.g., the virus that causes vesicular stomatitis) bud from the basolateral surface. The trafficking signals in these cases were found to be located in the amino-terminal region of the viral glycoprotein, which was achieved through the use of sophisticated recombinant DNA techniques to create hybrid versions of the relevant proteins (Yamauchi and Helenius, 2013).

The specifics of envelope formation and virion release are intricate for nuclear replicating enveloped viruses, such as the herpesviruses. As previously mentioned, capsid formation occurs in the nucleus, and complete capsids are thought to attach to tegument (matrix) proteins in the vicinity of the nuclear membrane, which has been altered by the presence of viral glycoproteins that have been glycosylated in the cellular Golgi apparatus. Figure 5.21a shows the outline of the HCMV envelope. Empty capsids can also be enveloped, and some enveloped particles—often referred to as dense bodies—only contain matrix proteins instead of capsids at all. None of these faulty virus particles can start a typical infection because none has a genome.

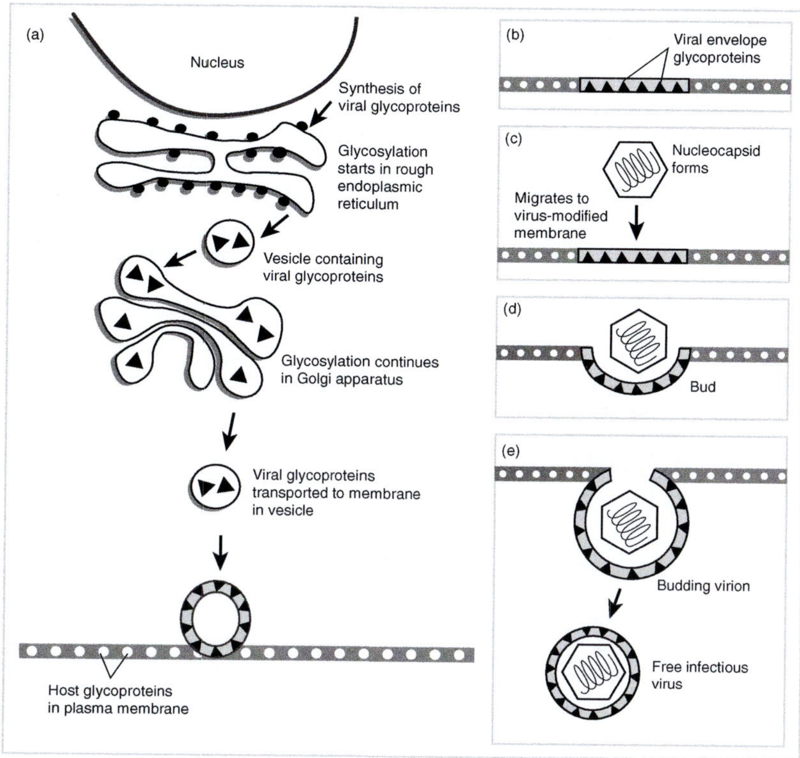

Figure 5.20. *Insertion of glycoproteins into the cell's membrane structures and formation of the viral envelope. The formation of viral glycoproteins on the rough endoplasmic reticulum parallels that of cellular glycoproteins, except that viral mRNA is translated. (a). Full glycosylation takes place in the Golgi bodies, and viral glycoproteins are incorporated into transport vesicles for movement to the cell membrane where they are inserted (b). At the same time (c), viral capsids assemble and then associate with virus-modified membranes. This can involve the interaction with virus-encoded matrix proteins that serve as "adapters." Budding takes place (d, e) as a function of the interaction between viral capsid and matrix proteins and the modified cellular envelope containing viral glycoproteins.*

Source: https://www.blackwellpublishing.com/content/BPL_Images/Content_store/Middle_Sized/Wagner9781405103466/060to076.pdf.

Regarding the fate of HSV after being enveloped in the nuclear membrane, two theories have been put out. Virion nuclear membrane loss occurs either by fusion with the endoplasmic reticulum before exocytosis or the virion retains its nuclear membrane upon egress. The latter theory suggests that the virus reacquires an envelope on the plasma membrane from the trans-Golgi network. Even though experimental data has been interpreted differently to support the two models, Mettenleiter and colleagues have recently offered strong evidence in favor of the second model, or double envelopment. By fusing with the outer nuclear membrane, viral capsids emerge into the lumen that exists between the inner and outer nuclear membranes and enter the cytoplasm (Yamauchi and Helenius, 2013). After budding into exocytotic vesicles, the capsids obtain their mature envelope, and the enveloped virus is then delivered to the cell surface for release. The pseudorabies virus exocytosis electron micrographs included in Figure 5.21b elegantly illustrate the process.

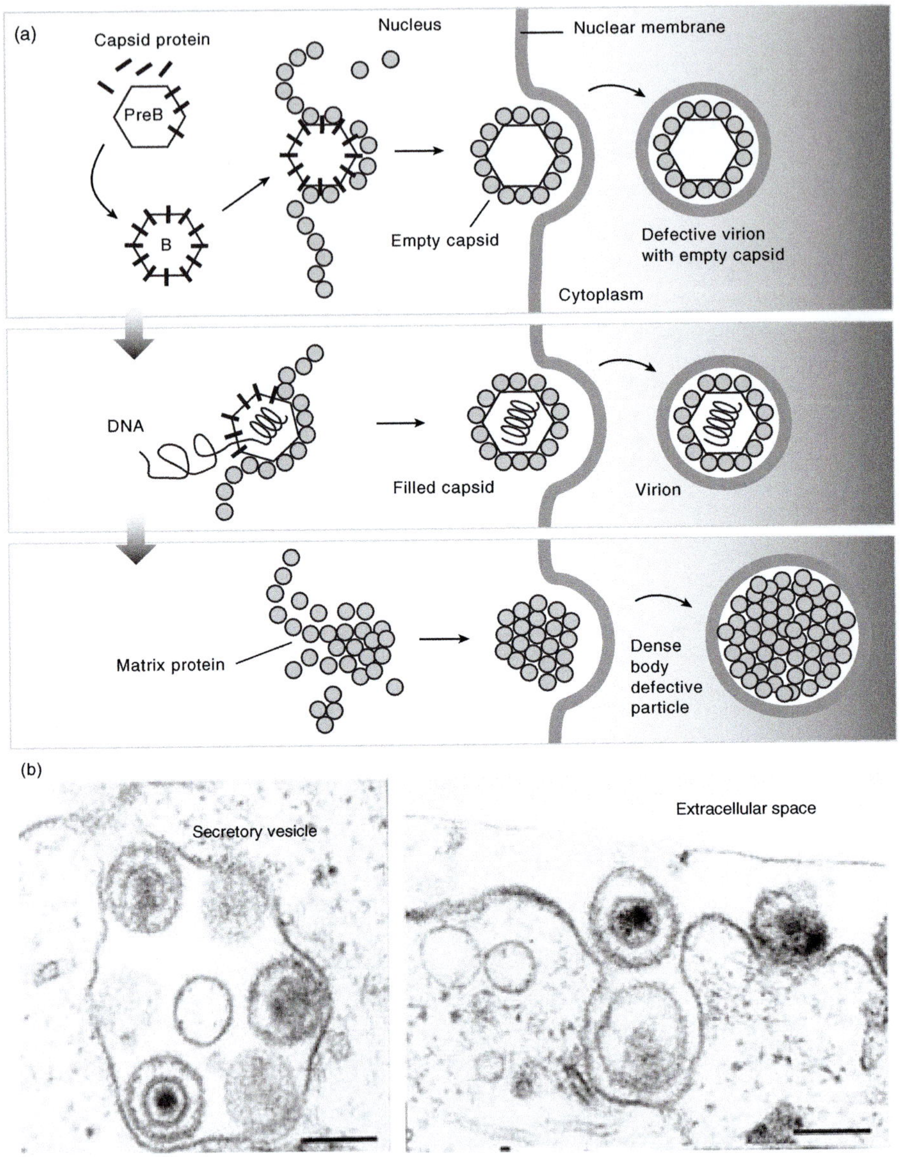

Figure 5.21. *The envelopment and egress of herpes virus. (a) A schematic representation of the nuclear envelopment step for human cytomegalovirus (HCMV). Matrix (tegument) protein serves as the adapter or "glue" to allow the association between capsid and virus-modified envelope, and mature virions bud through the nuclear membrane into the infected cell's cytoplasm. The process is inefficient, and empty capsids can be enveloped. Further, matrix protein can aggregate and itself become enveloped, forming "dense bodies." (Based on a drawing by D.W. Gibson.) (b) Electron micrographs of exocytosis of pseudorabies virus in the cytoplasm of the infected cell; release of enveloped virions is clearly shown. The bars represent 150nm. (Micrographs reprinted with the kind permission of the American Society for Microbiology from Granzow et al., 1997).*

Source: https://www.blackwellpublishing.com/content/BPL_Images/Content_store/Middle_Sized/Wagner9781405103466/060to076.pdf.

CASE STUDY

The campus daycare was recently closed during the peak of the winter flu season because many of the young children were sick with a lower respiratory tract infection. An email announcement was sent to all college students, faculty, and staff that stated the closure was due to a metapneumovirus outbreak. The announcement briefed the campus community with information about human metapneumonoviruses (hMPVs).

The announcement stated that hMPV was a newly identified respiratory tract pathogen discovered in the Netherlands in 2001. New tests confirm that it is one of the most significant and common viral infections in humans. It is clinically indistinguishable from a viral relative known as respiratory syncytial virus (RSV). Both RSV and hMPV infections occur during the winter. hMPV may account for 2% to 12% or more of previously unexplained pediatric lower respiratory infections for which samples are sent to diagnostic laboratories and a lesser percentage in adults.

Both hMPV and RSV cause upper and lower respiratory tract infections associated with serious illness in the young, immunosuppressed, elderly, and chronically ill. Common symptoms include cough, fever, wheezing or exacerbation of asthma, and rhinorrhea (runny nose). It can produce severe enough symptoms to cause intensive care admission and ventilator support. Healthy adults can get a mild form of the disease characterized by a cough, hoarseness, congestion, runny nose, and sore throat.

Members of the campus day care staff were doing their best to limit the epidemic spread of the hMPV outbreak at the center. Primary care physicians need up-to-date knowledge and heightened awareness to recognize this new viral disease in patients.

Questions

1. hMPV and RSV belong to the Paramyxoviridae family. What type of nucleic acid genome do they have? Do they encode their own viral polymerase?
2. Draw a flowchart of the viral life cycle of hMPV.
3. How many genes does hMPV have in its genome?
4. The first report of a fatal encephalitis case associated with hMPV was published in the CDC's Emerging and Infectious Diseases publication in 2005. The authors recommended screening for patients, especially children with encephalitis symptoms of unknown origin. What is encephalitis?
5. Have neurological symptoms been associated with other viruses in the Paramyxoviridae family? If so, which virus(es)?

CLASS ACTIVITY

The viral genome needs to get to the nucleus for the virus to replicate in the nucleus. In other words, the viral nucleocapsid has to overcome two barriers (i.e., plasma membrane and nuclear membrane). Compare and contrast the mechanisms by which the viruses penetrate the two membranes.

SUMMARY

- Viruses are obligate intracellular parasites. Viruses must gain entry into target cells and usurp the host cellular machinery to produce a progeny virus. The multiple steps in the virus propagation inside cells are collectively termed the "virus life cycle." After entering the cell and localizing to an intracellular milieu, the virus sheds its capsid, transcribes its RNA, translates its RNA to the viral proteins, replicates its genome, assembles the viral components, and finally exits from the cell.
- All viruses share the same basic replication cycle. However, the time involved depends on a number of factors, including the size and genetic complexity of the virus itself as well as the nature of the host cell.
- The virus life cycle can be divided into three stages—entry, genome replication, and exit. The virus life cycle can be described in analogy with a businessman's life; the entry to his way to work, the genome replication to his task at work, and the exit to his way home.
- Entry, the first step of virus infection, involves the recognition of viral receptor by a virus particle. The viral entry can be divided into four steps: attachment, penetration, cytoplasmic trafficking, and uncoating. These steps are often linked to each other, so the division into four steps is obscure, but it serves the purpose of explanation.
- The attachment refers to the first encounter of virus particles with host cells, which involves two kinds of host proteins on the plasma membrane: (1) attachment factors and (2) viral receptors. The attachment factor on the cell surface recruits and holds the virus particles, thereby facilitating the interaction of the viral particle with the entry receptor.
- Nonenveloped virus particles must be incorporated into the cell via a process called translocation across the lipid bilayer. In this process, the capsid or a cell-modified capsid physically crosses the cell plasma membrane
- Enveloped viruses interact with cell receptors via the action of membrane-associated viral glycoproteins that project beyond the viral envelope. The viral

glycoproteins are glycosylated with sugars in the cell Golgi apparatus during viral maturation. The process is similar to that carried out by the cell on its own glycoproteins.

- The capsid assembly follows as the viral genome, as well as the viral proteins, abundantly accumulates. The capsid assembly can be divided into two processes: capsid assembly and genome packaging. Depending on viruses, these two processes can occur sequentially or simultaneously in a coupled manner.
- For naked viruses, the virus particles are released via cell lysis of the infected cells. Thus, no specific exit mechanism is necessary because the cell membrane that traps the assembled virus particles is dismantled.

REVIEW QUESTIONS

1. Name the three stages of virus life cycle.
2. Briefly describe the two modes that enveloped viruses use for entry into their host cells.
3. What are the four steps viral entry?
4. How do plant viruses enter their host cells? What feature of the plant cell's architecture dictates these modes of entry?
5. What are the three types of exit?
6. List two types of productive infection.
7. List three types of nonproductive infection.
8. Describe how the T-even bacteriophage attaches and enters the host cells. Which part of the virus particle enters the cell?
9. Simple virus capsids are found in two types of structural arrangements: helical and icosahedral. What are the key features in the assembly of these two kinds of particles?
10. How do enveloped viruses acquire their membranes during their maturation in animal cells?

REFERENCES

1. Connolly, S. A., Jackson, J. O., Jardetzky, T. S., & Longnecker, R. (2011). Fusing structure and function: A structural view of the herpesvirus entry machinery. *Nature Reviews Microbiology*, *9*(5), 369–381.
2. Dimmock, N. J., & Primrose, S. B. (1994). *Introduction to modern virology* (4th ed.). Blackwell Science.
3. Granzow, H., Weiland, F., Jöns, A., Klupp, B., Karger, A., & Mettenleiter, T. (1997). Ultrastructural analysis of the replication cycle of pseudorabies virus in cell culture: A reassessment. *Journal of Virology*, *71*, 2072–2082.

4. Mercer, J., Schelhaas, M., & Helenius, A. (2010). Virus entry by endocytosis. *Annual Review of Biochemistry*, *79*, 803–833.

5. Sattentau, Q. (2008). Avoiding the void: Cell-to-cell spread of human viruses. *Nature Reviews Microbiology*, *6*(11), 815–826.

6. Vigant, F., Santos, N. C., & Lee, B. (2015). Broad-spectrum antivirals against viral fusion. *Nature Reviews Microbiology*, *13*(7), 426–437.

7. Yamauchi, Y., & Helenius, A. (2013). Virus entry at a glance. *Journal of Cell Science*, *126*(Pt 6), 1289–1295.

CHAPTER 6

Host Immune Response to Viral Infection – The Nature of the Vertebrate Immune Response

LEARNING OBJECTIVES

After studying this chapter, you will be able to:

- Describe the innate immune response – early defense against pathogens;
- Discuss the adaptive immune response and the lymphatic system;
- Explain the control and dysfunction of immunity;
- Identify the measurement of the immune reaction.

KEY TERMS FROM THIS CHAPTER

Adaptive immune response	Amino acid clusters	DNA viruses
Bacteria	Defensins	Fungi
Glycoproteins	Host Immunity	Immune Memory
Humoral antibody response	Immune reaction	Protein
Innate immune response	Specific Viral Responses	Surface antibody
Toll-like" receptor (TLR)	Transporter proteins (TAPs)	

INTRODUCTION

The innate immune response and the adaptive immune response are the two distinct components of the immune system that defends the body against invasive pathogens. The innate immune response refers to a broad type of immune response that is triggered by specific molecules or proteins that are produced by or present on bacteria, viruses, or fungi. The initial defense against a pathogen, this reaction causes a generalized inflammation. Furthermore, the innate response aids in alerting the body to the presence of an invasive pathogen and supports the activation of the more focused and powerful adaptive immune response, which involves B- and T-cell responses (Arankalle et al., 2010).

6.1. THE INNATE IMMUNE RESPONSE – EARLY DEFENSE AGAINST PATHOGENS

It is reasonable to assume that at first, the odds favor the virus when it infects a host that lacks immune system function. Ultimately, most viruses have brief replication cycles, enabling them to quickly release hundreds of new virions from a single cell. It takes time for the body to produce enough virus-specific B and T cells, even locally, to kill infected cells and stop the infection from spreading to other host cells and tissues, even though the body starts producing specific antibodies and cellular immune responses immediately. One of the immune system's oldest and most rudimentary defense mechanisms, the innate immune response, is vital in preventing viral replication in the initial stages of infection. This innate reaction gives the host the vital time it needs to create an adaptive immune response that is more focused on controlling the infection (Figure 6.1).

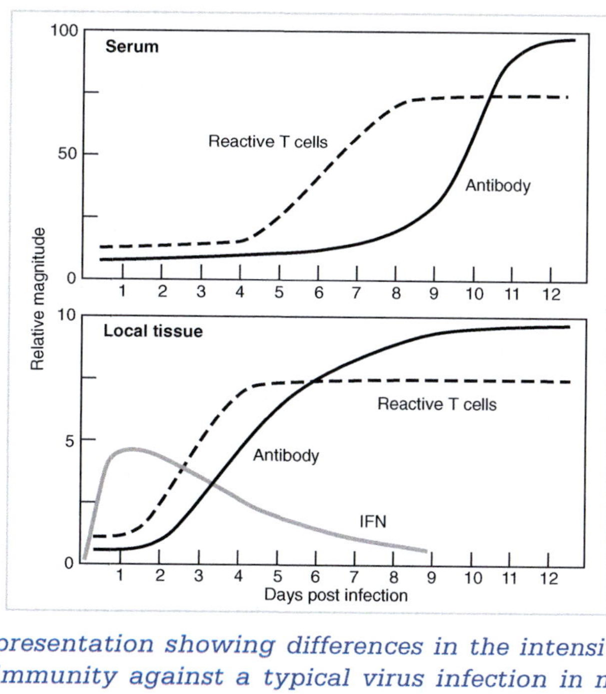

Figure 6.1. *Schematic representation showing differences in the intensity and time of appearance of local versus systemic immunity against a typical virus infection in mice.*

Source: https://www.blackwellpublishing.com/content/BPL_Images/Content_store/Middle_Sized/Wagner9781405103466/077to095.pdf.

6.1.1. Toll-like Receptors

The initial identification of components of the innate immune response was made in Drosophila mutants (Toll), which were found to be particularly vulnerable to fungal infections. These mutants were deficient in a crucial protein that has subsequently been demonstrated to play a role in the cell-signaling response that triggers a broad antifungal response. Years later, a mouse strain with a high susceptibility to gram-negative bacterial infections was discovered to lack a Drosophila protein-related receptor protein. When this receptor protein comes into contact with LPS in normal mice, it sets off a cell-signaling reaction that results in a generalized inflammatory process. The local cellular environment is changed by this inflammatory process, which inhibits bacterial growth until an antigen-specific immune response is triggered. The original Drosophila mutant's name, "toll-like" receptor (TLR), was applied to the LPS-specific mouse receptor. Vertebrates have been found to have multiple TLRs (humans have at least ten) that bind to various molecules linked to bacterial, fungal, viral, and protozoan infections. The innate antiviral defense is primarily mediated by TLR3 and TLR9. For instance, TLR3 can identify and bind to dsRNA, which is produced during the replication cycles of numerous viruses, including DNA and RNA viruses. Activation of TLRs results in the activation of adaptor proteins, which then release pro-inflammatory cytokines. Every TLR triggers distinct pathways by activating distinct adaptor proteins. For instance, type 1 interferons are induced by TLR3 (activated by dsRNA); this process is covered in Chapter 8. It's interesting to note that the cytokines produced by the innate immune response, interferon-mediated antiviral effects and the inflammatory responses triggered by particular TLRs, are crucial in directly stimulating and enhancing the development of particular cellular and humoral immune responses. For instance, TRL4 generates cytokines in response to LPS activation that specifically activate helper T cells that have been stimulated by particular antigens. Furthermore, it is now thought that a key factor in determining the kind of helper T response triggered in reaction to a particular pathogen is the kind of cytokines generated by TLR innate responses. As will be seen below, the T-helper response's selectivity is a key factor in determining whether an antiviral response is primarily humoral or cellular, a process that has undoubtedly aided in the evolution of viruses.

6.1.2. Defensins

Defensins are a group of cellular proteins that together mediate innate immunity, which is another recently identified aspect of the immune system. Numerous cells in the respiratory and gastrointestinal systems secrete these tiny (30–50 amino acid) proteins, which bind to various pathogens such as bacteria, fungi, and certain viruses. This type of binding improves the removal of fungi and bacteria. Research has demonstrated that defensins obstruct the entrance of influenza and HIV into human cells through the process of cross-linking proteins found in cellular membranes. By preventing the virus-induced removal of cellular membrane proteins from the area next to the viral envelope, this cross-linking prevents membrane fusion. Defensins, like TLRs, improve humoral and cellular immunity stimulation. Defensins also seem to be involved in aiding the host's immune system in producing reactions against tumor cells. This serves as another example of the intricate relationships that exist between the various immune system arms and the ongoing development of these arms' constituent parts toward mutually beneficial relationships (Deshmukh et al., 2007).

6.2. THE ADAPTIVE IMMUNE RESPONSE AND THE LYMPHATIC SYSTEM

Part of the general circulatory system, the human lymphatic system is depicted in Figure 6.2. It is essential for the development of the immune response to foreign proteins in the body. Both B-cell immunity (also known as humoral immunity) and T-cell immunity (also known as cell-mediated immunity [CMI]) are activated when an antigen-presenting cell (APC) exposes the immune system to any protein that is not present in the extensive repertoire of proteins that comprise the vertebrate host. Such an alien protein is commonly referred to as an antigen. It may originate from an invasive pathogen (bacterium, virus, or parasite) or a newly expressed cellular protein resulting from aberrant cell growth characteristics—a tumor antigen. Generally speaking, an antigen that does not belong to the host's normal protein composition can be identified as foreign by the immune system of the host and can trigger an immune response. The spleen, thymus, and bone marrow are among the specialized tissues where lymphocytes are generated, differentiate, and mature. They travel throughout the body via the lymphatic and circulatory systems and, in response to an infection, can move from cellular junctions into tissue. They are mostly found in lymph nodes, which are frequently the site of the initiation of a systemic immune response. Antibodies are proteins secreted by B cells that have the specific ability to attach to antigenic determinants on proteins. Due to the antigen-binding sites on their surfaces, activated T cells interact with target cells that are infected by viruses or other foreign antigens, causing the infected cells to lyse. B-cell and cytolytic $CD8^+$ T-cell immunity can both be supported by specific T cells, known as $CD4^+$ T helper cells. These two components of the adaptive immune system work together to enable the host to recognize and eliminate free viruses and virus-infected cells that have viral proteins on their surface or to render them noninfectious (inactivate or neutralize). An overview of how an antigenic pathogen and the adaptive immune system interact.

6.2.1. Two Pathways of Helper T Response – The Fork in the Road

The presentation of virus-specific antigens to cells of the adaptive immune response is a preliminary step in developing a particular immune response. These consist of B cells, cytotoxic T cells, and helper and regulatory T cells. The presentation of antigens to CD4+ helper T cells is a crucial step in this process. Following the presentation of this antigen, some of the cells will differentiate into Th1 or Th2 T helper 1 cells, respectively. The main secreted substance by Th1 cells is γ-interferon, which helps $CD8^+$ T cells differentiate and produce a cytolytic cellular immune response. On the other hand, Th2 cells mainly produce IL-4, IL-5, IL-10, and IL-13, which encourage B cell differentiation and the following humoral antibody response (Couper et al., 2008).

It has been demonstrated that several of the TLRs covered above play a part in stimulating a Th1 or Th2 response. This means that the innate immune response helps to direct the development of the kind and intensity of response that is elicited, in addition to buying time for developing a particular immune response against a viral pathogen (Litman et al., 2005). This is just one of many instances where different immune system components work together to form a functional whole (Figure 6.3).

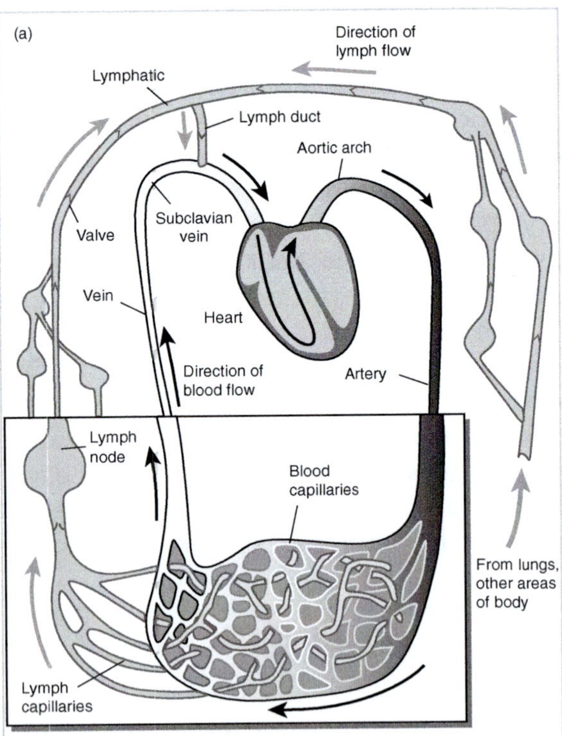

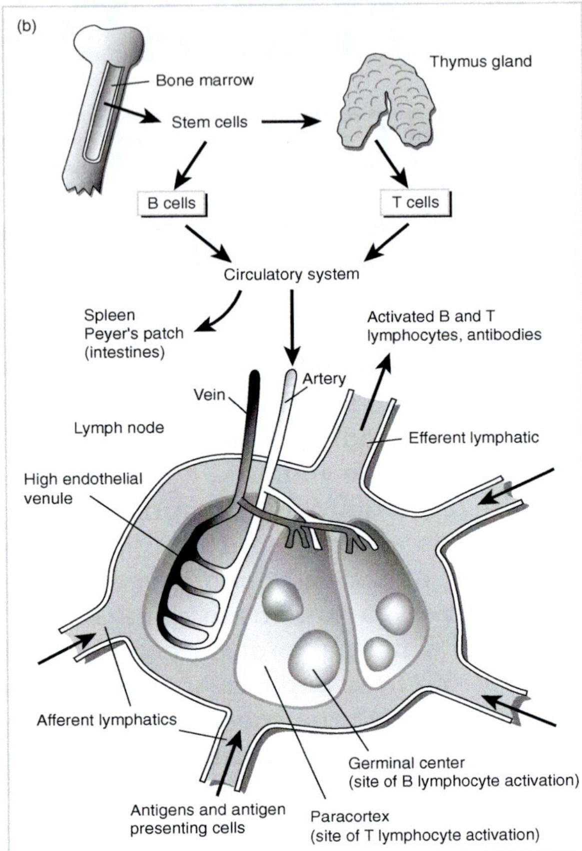

Figure 6.2. *The human lymphatic system. The lymphatic system is the principal organ of the immune system. (a) The relationship between the lymphatic circulation and that of the blood. (b) Some of the important components of the lymphatic system as related to the immune response.*

Source: https://www.blackwellpublishing.com/content/BPL_Images/Content_store/Middle_Sized/Wagner9781405103466/077to095.pdf.

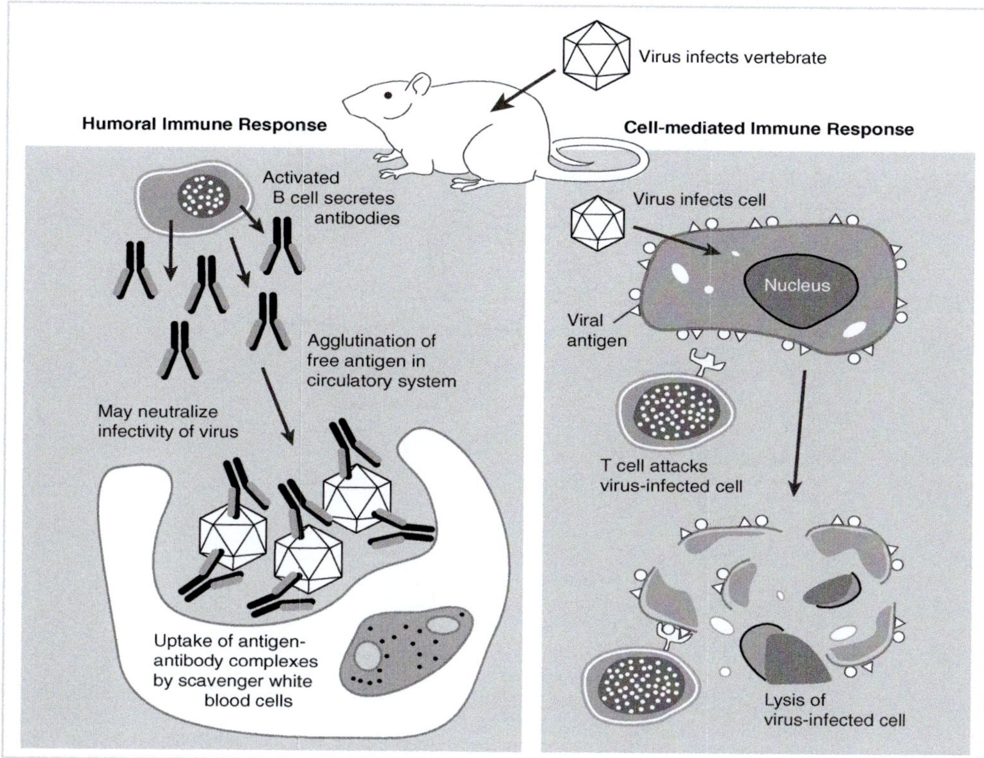

Figure 6.3. *T and B cells in immunity. T lymphocytes play the central coordinating role in evoking the immune response. Upon activation by interaction with a specific antigenic determinant with which they can interact, they proliferate and carry out the functions shown. B cells reactive with specific antigens require reactive T cells for their maturation. Upon maturation, they secrete antibody proteins that bind to antigenic determinants.*

Source: https://www.blackwellpublishing.com/content/BPL_Images/Content_store/Middle_Sized/Wagner9781405103466/077to095.pdf.

6.2.2. The Immunological Structure of a Protein

Any protein has specific amino acid clusters (usually between 10 and 12) that can interact with the right T cells or B cells that produce antibodies to recognize antigens and cause those cells to proliferate. These groups are referred to as epitopes or antigenic determinants. Reactive B-cell epitopes are typically hydrated and hydrophilic. Depending on its protein structure, amino acid sequence, sequence relationship to cellular proteins, degree of glycosylation, and other characteristics, a viral protein may have one, several, or many antigenic determinants. Certain determinants may be shared by two proteins; the more closely related the proteins are to one another, the more determinants they have in common. As a result, viral serotypes that are closely related exhibit strong immunological reactivity. Figure 6.4 shows a schematic representation of the different types of epitopes found in proteins (Restifo and Gattinoni, 2013).

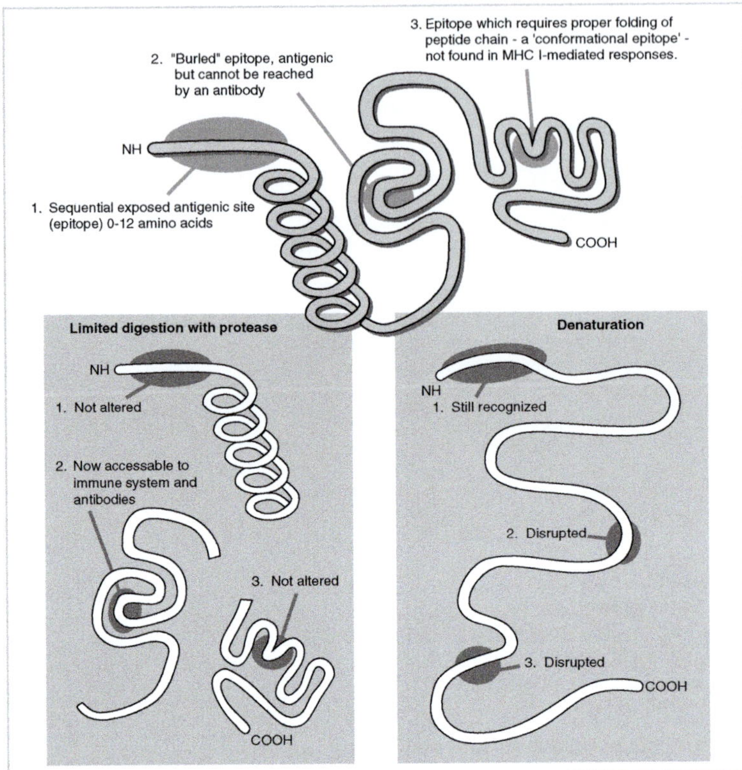

Figure 6.4. *The antigenic structure of a protein. Specific groups of amino acids (usually hydrated) serve as specific antigenic determinants, or epitopes in an antigenic protein. Some of these are insensitive to the protein's physical structure; others require a specific conformation for presentation.*

Source: https://www.blackwellpublishing.com/content/BPL_Images/Content_store/Middle_Sized/Wagner9781405103466/077to095.pdf.

A particular arrangement of amino acids frequently makes up epitopes. Denaturing the antigenic protein won't significantly alter its characteristics or the way it is presented to the immune system when it contains such an epitope. Sequential epitopes are determinants expressed in a protein in either its native or denatured state. Additionally, epitopes may be responsive to the protein region's structure in which they are found. For instance, they might consist of amino acids that have been arranged in close proximity to one another through protein conformation or folding. These are epitopes with a specific conformation that are susceptible to denaturation, or disruption, of the protein structure. The interior of a protein may contain conformational or sequential epitopes that are normally invisible to the humoral immune system. These are factors that are hidden. The protein's denaturation can reveal many of these, which are sequential. A hidden conformational determinant may become visible through appropriate, controlled protein degradation or by denaturing the protein and then refolding it to reveal the epitope (Agerberth and Gudmundsson, 2006).

6.2.3. Role of the Antigen-Presenting Cell in Initiation of the Immune Response

Any protein and a wide variety of other macromolecules can be antigenic. However, in order to trigger an immune response, antigens need to be "processed" and then presented in the right way on the surface of the cell that carries them, known as an antigen-presenting cell. In this instance, the major histocompatibility proteins are in a complex with one of two closely related heterodimeric cell surface glycoproteins. These MHC glycoproteins ensure that only antigen-presenting cells (e.g., the immune system with antigens that are presented by the same organism (macrophages and dendritic cells, for example). Cells present antigens via two fundamental pathways (Figure 6.5). The first is the presentation of endogenously expressed antigens on the surface through the type I major histocompatibility complex, a function of almost all cells (MHC-I). Proteins are partially broken down into epitope-sized peptides in proteolytic vesicles (proteosomes) by a class of cellular proteins known as ubiquitins, which bind to the proteins during synthesis. Transporter proteins (TAPs) subsequently transfer these peptides into the Golgi apparatus, where they attach to freshly made MHC-I glycoproteins and are displayed on the cell surface. When $CD8^+$ T cells are surveyed using these MHC-I complexes as targets, the antigen-bearing cells are eliminated if they become reactive. The immune system searches every cell in this manner to synthesize foreign or aberrant proteins. This endogenous antigen presentation is undoubtedly a key component of local immunity and plays a significant role in the early immune detection of virally infected cells.

A comparatively large population of freely circulating, relatively short-lived effector T cells that are able to recognize the target antigen is necessary for the development of both immune memory and systemic immunity. Long-lived, specialized dendritic cells that originated in the bone marrow and moved to the epithelium are mainly responsible for this. Because of their fundamental function in inducing systemic immunity, dendritic cells and some other immune system cells are frequently referred to as professional antigen-presenting cells. Receptor-mediated endocytosis internalizes and partially digests antigenic proteins or complexes in ways that are not entirely understood. When type II major histocompatibility complex (MHC-II) proteins are present on the cell surface, antigen fragments that contain epitopes are reexpressed. This is schematically depicted in Figure 6.5(b), where the antigenic fragment and major histocompatibility complex (MHC) molecules combine to form a surface structure that can be recognized by $CD4^+$ T and some B cells in lymph nodes to initiate the amplification of cells able to recognize the antigen. In lymph nodes, MHC-II-mediated antigen presentation takes place. It takes time for the process to happen because antigen concentration needs to rise to a certain point in order to trigger the immune response. This lag happens after the initial infection and early viral replication. This latency period is crucial in cases of viral infections, like those caused by HSV, since the virus can infiltrate sensory neurons and create latent infections before a potent immune response is generated. In fact, a particular viral protein expressed soon after infection allows HSV, like some other viruses, to obstruct the MHC-I-mediated early presentation of its antigenic proteins at the infected cell's surface. Certain viruses, such as HIV, have the ability to survive within dendritic cells and be presented to T cells, resulting in the infection of lymphocytes.

HIV is capable of replicating in CD4+ lymphocytes, eventually leading to the destruction of the immune system as the virus continues to replicate within infected lymphatic cells. Additionally, as immature B cells with surface receptors that can bind to antigen proliferate and interact with presented epitopes, they also internalize and process the antigen, providing an alternative method of antigen presentation in the lymph nodes (Agerberth and Gudmundsson, 2006).

Producing effective immunity depends critically on the ability to internalize and process antigens. However, the host can also produce immunological responses to complex conformational epitopes, such as fragments of the virus's dimeric and multimeric proteins, in addition to producing sequential determinants. In fact, the host preferentially develops robust antibody responses against the viral surface proteins. These proteins' abundance and location at the "interface" of the infection and the host's antigenic response contribute to the reactions observed. Several other elements are also at play, such as the proteins' inherent resistance to widespread degradation and the surface antibody (IgG) on immature B cells' capacity to identify native, unprocessed antigens.

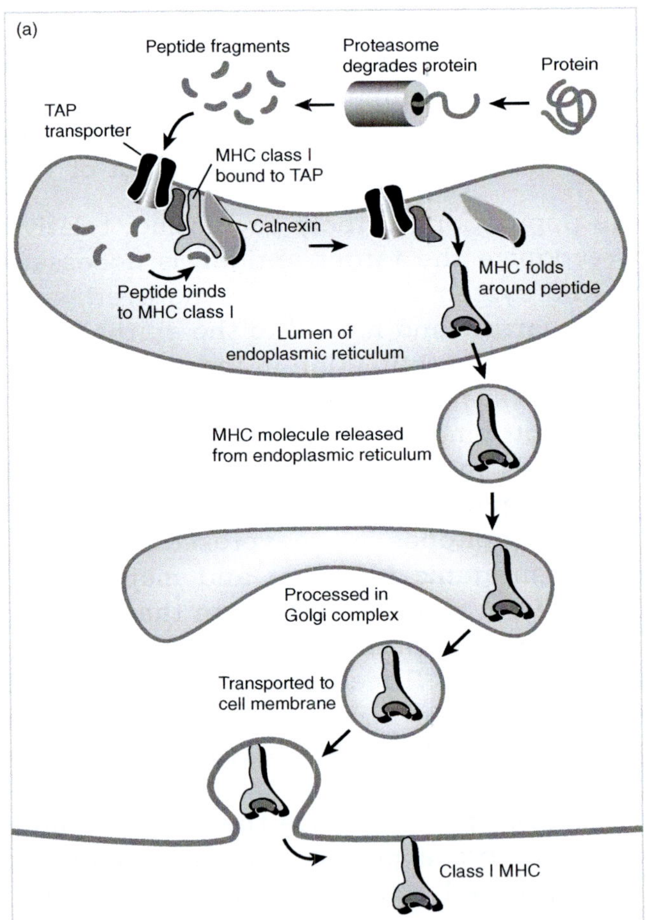

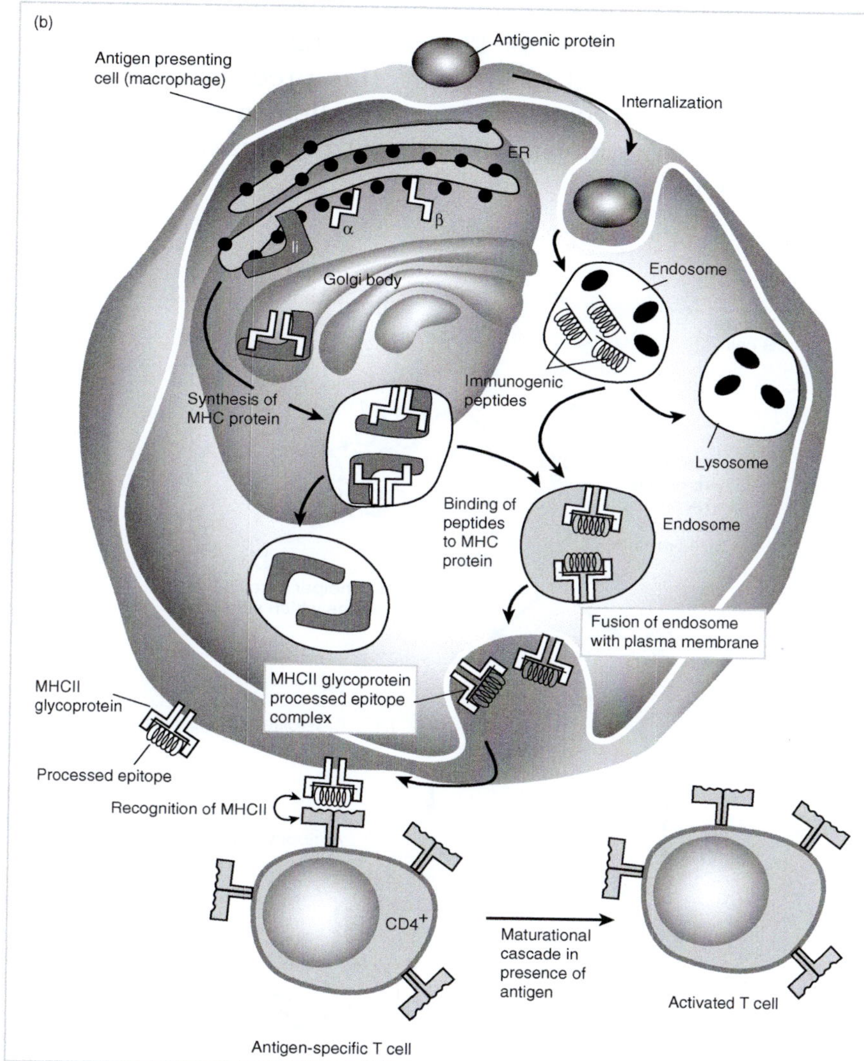

Figure 6.5. *The processing of a foreign antigen and stimulation of the immune response. As described in the text, an antigenic protein can only stimulate the immune response when it is processed by a macrophage and then presented to cells of the immune system in lymph nodes in the presence of histocompatibility antigens. The processing is relatively rapid and involves partial degradation of the antigenic protein and expression of antigenic portions on the surface of the antigen-presenting cell. (ER, endoplasmic reticulum.) (a) MHCI antigen processing and presentation. (b) MHCII processing and presentation.*

Source: https://www.blackwellpublishing.com/content/BPL_Images/Content_store/Middle_Sized/Wagner9781405103466/077to095.pdf.

6.2.3.1. Clonal Selection of Immune Reactive Lymphocytes

T and B cells in the immune system that can identify antigens are stimulated to multiply. Clonal selection occurs because all distinct T cells that recognize epitopes and all distinct B cells that produce specific antibodies are descended from a single reactive cell (i.e., clones of that cell), as Figure 6.6 illustrates. Because so many cell populations in the lymph nodes need to interact, this process mostly occurs there. The ability to produce B cell clones that produce antibodies in the lab has been a crucial tool for researching the interactions and functional organization of different viral and cellular proteins (Kumar et al., 2011).

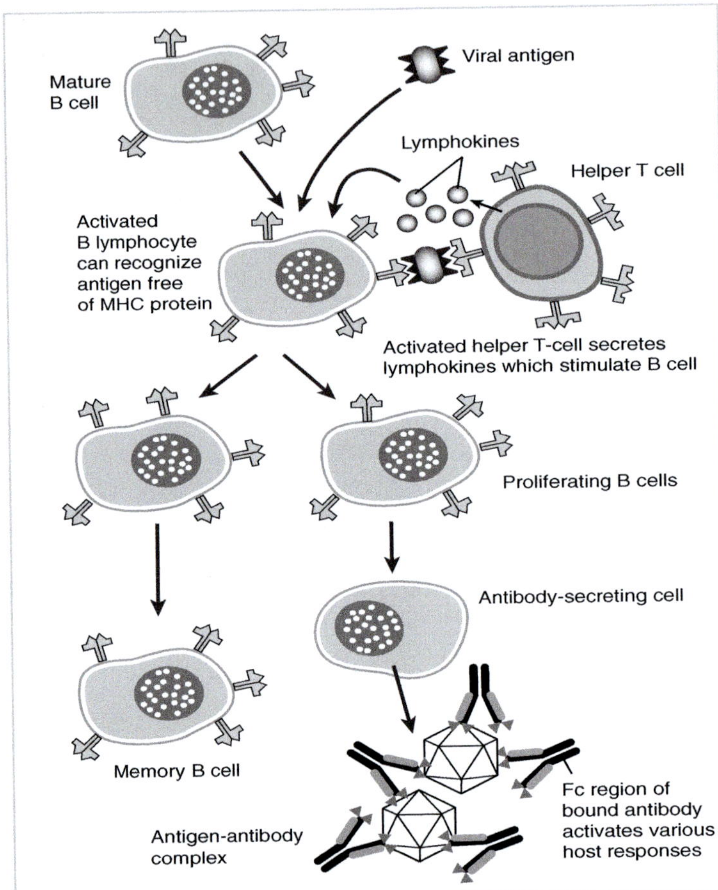

Figure 6.6. *The clonal selection of B lymphocytes. Only the B lymphocytes reactive with a specific epitope can be stimulated to mature by the action of a helper T lymphocyte. Specific mature B cells secrete specific types of antibody molecules, but the same epitope will result in only the stimulation and maturation of B-cell clones reactive with it.*

Source: https://www.blackwellpublishing.com/content/BPL_Images/Content_store/Middle_Sized/Wagner9781405103466/077to095.pdf.

B cells differentiate (mature) when stimulated by the presence of a particular epitope they recognize. Soluble antibodies are secreted by fully differentiated B cells. B cell maturation is mediated by a particular class of effector T cells called $CD4^+$ helper cells. A different subclass of T cells known as $CD8^+$ cytotoxic T cells targets and eliminates any foreign antigen-containing cells, including virus-infected cells. When immunity is at its peak and antigen levels start to drop at the end of the "crisis," a third type of T cells known as regulatory, or Treg, cells suppress the immune response.

6.2.3.2. Immune Memory

The immune system can react quickly when exposed to the antigen again because it "remembers" the antigenic response. Immune memory is mediated by T and B cells with long memory. These memory cells circulate in the blood and lymph and are primarily found in lymph nodes. The antigen-responsive cells keep multiplying as long as the antigen is present. Memory cells are long-lived and remain prepared to react to a second infection with the same or closely related pathogen, unlike most, which have a limited lifespan before going through apoptosis (Beutler et al., 2006). Memory cells also do not function in dealing with the antigen. A second stimulation causes the antigen to interact with these memory cells quickly, leading to a secondary immune response (a remembered immune response) that is both more extensive and quicker than the initial or primary response. Figure 6.7 illustrates how immune memory affects the potency and tempo of the immune response.

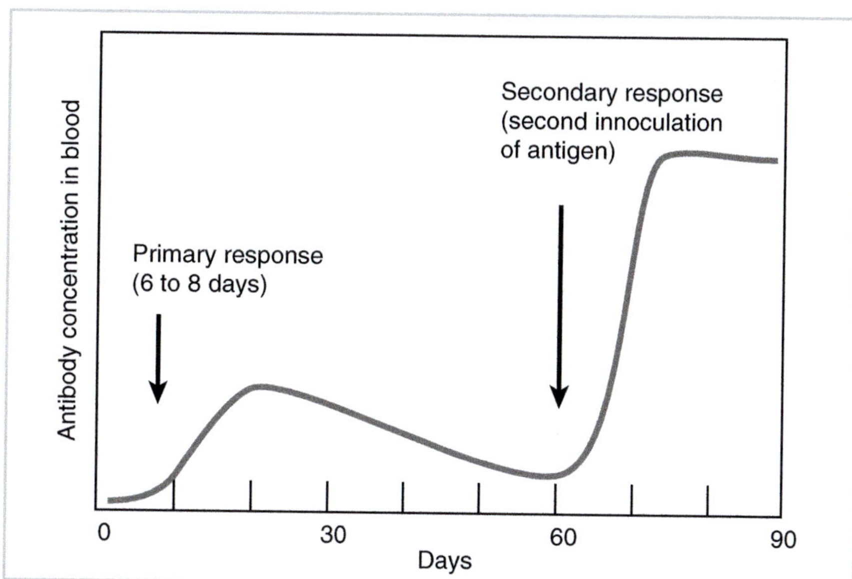

Figure 6.7. *Immune memory. The first exposure to an antigen results in the primary response, which occurs after a week or so. During this time, immune-reactive cells mature. Once the primary response occurs, antibody and reactive T and B cells decline to a low level. Upon restimulation with the same antigen, the memory lymphocytes are rapidly mobilized and a more intense and more rapid immune response follows.*

Source: https://www.blackwellpublishing.com/content/BPL_Images/Content_store/Middle_Sized/Wagner9781405103466/077to095.pdf.

6.2.3.3. Complement-Mediated Cell Lysis

While T cells play a major role in eliminating foreign antigen-bearing cells, B cells can also eliminate antigen-bearing cells by utilizing the complement system, which results in complement-mediated cell lysis (Bermejo-Martin et al., 2010). This system functions by binding antibodies to cells, which set off a series of events with serum complement proteins that ultimately result in the cell's destruction (Figure 6.8).

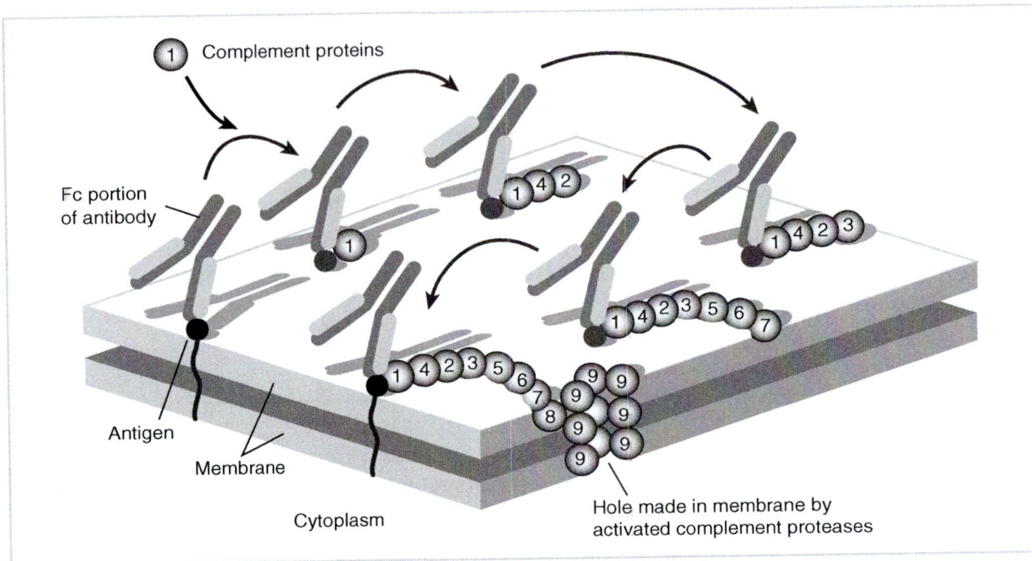

Figure 6.8. *The maturational cascade of serum complement proteins upon binding to an antigen–antibody complex on the surface of a cell. The Fc region, a portion of the antibody molecule that is not involved in binding to the epitope of the antigen, specifically triggers this cascade.*

Source: https://www.blackwellpublishing.com/content/BPL_Images/Content_store/Middle_Sized/Wagner9781405103466/077to095.pdf.

6.3. CONTROL AND DYSFUNCTION OF IMMUNITY

The most effective stimulation occurs in T and B cells with antigenic recognition sites that have the highest affinity for a particular epitope. Lower concentrations of high-affinity interactions can still boost immunity even after general antigen levels decline late in infection and during the healing process. As a result, after infection, the immune system's composition evolves over time. In general, a patient who is recovering will have more specific and higher-affinity antibodies than a person who is just beginning to experience symptoms. Regulatory T cells that are $CD4^+$ and $CD25^+$ mature very late in the immune response and eliminate immunity. These cells play a crucial role in the control of immunity. Hyperimmune reactions, including allergic reactions, could arise if they are not functioning correctly. They may not produce enough immunity if they operate too well. A deficiency in regulatory T cells, which typically make up 1-3% of all $CD4^+$ T cells, is the root cause of several autoimmune diseases.

Other immune pathologies include autoimmune diseases, in which the body's immune system attacks and kills what appears to be healthy tissue. This may be the result of the immune system trying to kill healthy cells that express viral antigens. Subacute sclerosing panencephalitis (SSPE), a pathological reaction to the persistence of measles virus antigen in neural tissue, is an example of an autoimmune pathology caused by viral infection and persistent antigen presentation. Chapter 4 provided a brief description of this. Some autoimmune diseases, like multiple sclerosis, are believed to be brought on by a prior viral infection that seemed to clear up. It has been proposed that immune pathology, in this case, neuronal demylenation, can result from a prior viral infection, possibly years prior. Although the precise mechanism underlying this pathology is unknown, it is thought to be caused by a process known as "molecular mimicry," in which a pathogen's specific epitope resembles one found in the host tissue. Here, normal tissue is also now identified as foreign during a typical immune response against the invasive pathogen. This is the known mechanism for the involvement of group A Streptococcus in rheumatic fever, wherein the bacterial epitope elicits a strong immune response, which in turn causes complications due to its similarity to an epitope present in a protein found in heart tissue. Multiple sclerosis cases require extremely careful statistical analysis of lengthy medical records in order to establish correlations, as this mechanism has not been proven for this condition (Tang et al., 2019).

6.3.1. Specific Viral Responses to Host Immunity

Effective defense mechanisms like the immune system constantly aid in the selection process against viruses that are unable to establish a viable infection. It is evident that many virus infections persist and grow in the context of the host's immune capacity despite the efficacy of the immune response. In fact, the vast majority of DNA viruses that replicate nuclearly form enduring relationships with their hosts. It is evident that they can handle host attempts to eradicate the infection. The ability of viruses to mount numerous potent counterreactions to the immune response is a key component in their ability to survive. While some of these are essentially passive, others involve the blocking of particular immune response components by viruses.

6.3.1.1. Passive Evasion of Immunity – Antigenic Drift

Every animal virus has different serotypes or forms based on their antigens. The type of virus determines how many forms there are. For instance, the measles virus has only one strain, the poliovirus has three main serotypes, the adenovirus has over 40, and the papillomavirus has up to 100 strains. A prior infection with one serotype of a particular virus will result in no protection at all, or only limited protection, from reinfection with another serotype. A serotype is stable and may be limited to a particular geographic area. RNA viruses are more prone to mutations that result in the formation of serotypes than DNA viruses because RNA-directed RNA replication lacks an enzymatic error-correction mechanism, unlike DNA replication. This process is commonly referred to as "antigenic drift," and it is undoubtedly the cause of the drift in influenza virus serotypes as well as the substantial number of rhinovirus serotypes (more than 100).

Antigenic "conservatism" is favored by other factors that counteract this drift mechanism. For instance, a few RNA viruses, such as the poliovirus and measles virus, show many serotypes, and the internal proteins of these viruses remain mostly antigenically stable even in cases of significant drift, as in the case of influenza. The fact that specific structural or enzymatic functions impose significant functional constraints on the amino acid sequence of viral proteins helps to stabilize protein sequences even when they are encoded by highly mutable RNA sequences. These limitations are not as intolerant to changes in the external glycoproteins of viruses that are enveloped.

6.3.1.2. Passive Evasion of Immunity – Internal Sanctuaries for Infectious Virus

Certain viruses have the ability to evade the host's immune response by causing latent or persistent infections in tissue that is not under constant immune surveillance. The capacity of HSV to create latent infection in nondividing sensory neurons is a well-known example. Another illustration is the respiratory syncytial virus's capacity for low-level replication in the nasopharyngeal mucosa, where secretory antibodies offer defense against viral invasion but are unable to eradicate it. Another example of virus infection in a localized area free from intense immune surveillance is the highly localized replication of papillomaviruses, such as those that cause skin warts.

6.3.1.3. Passive Evasion of Immunity – Immune Tolerance

Neonatals and fetuses have immature immune systems. This is a crucial tactic for the fetus' survival as it grows inside its mother, an antigenically unique individual. Neonatals can suffer greatly from fetal and neonatal infections caused by viruses that typically cause relatively mild infections in immunocompetent adults. When rubella infects a developing fetus, it results in severe abnormalities of the nervous system's development. Since the virus does not produce long-lasting immunity in adults, it can still pose a threat to a mother who has already been infected. Neonatal encephalitis with a dire prognosis can result from a primary or reactivating HSV infection in the mother at the time of delivery. Neonatal and uterine cytomegalovirus infections are also strongly associated with neurologically based developmental disorders. The primary method of HIV transmission from mother to child is also active HIV replication during childbirth.

The ability to adapt to the developing immunity of the neonate is used by at least one group of viruses, known as the arenaviruses. These viruses, of which the lymphocytic choriomeningitis virus (LCMV) is the most well-studied laboratory model, are persistent in rodent populations and are passed from mother to offspring. The mouse exhibits a compromised immune response to LCMV and develops comparatively normally despite persistent viremia. A circulating antibody in the tolerant mouse reacts with the virus but is unable to neutralize it. Moreover, T-cell reactivity to the virus is absent. A strong immune response is produced when an immune-competent adult mouse contracts LCMV, but the infection is typically lethal.

The process of building immune tolerance is intricate and includes choosing particular viral genotypes that can infect immune system cells, such as macrophages, in the early stages of an infant's infection. Specific immunity to the virus is suppressed as a result of this infection. It is interesting to note that the virus that spreads between people prefers neural tissue. The viral glycoprotein and viral polymerase of these neurotropic and lymphotropic viruses only differ by one amino acid. The neurotropic variant causes severe disease in an uninfected adult, but it has little effect on immune-tolerant animals. The two variants are produced by random periodic mutations during replication of the resident virus in the animal. Other arenaviruses, some of which are harmful to humans, such as the Lassa fever virus, exhibit similar patterns of infection (Tang et al., 2019).

6.3.1.4. Active Evasion of Immunity – Immunosuppression

Multiple virus infections result in the temporary or permanent suppression of one or more host immunity branches. When EBV is the primary cause of infectious mononucleosis, it results in a self-limiting, widespread infection that is marked by a comparatively high induction of regulatory T cells. In addition to making the virus more successful in spreading its infection, the infected person becomes more vulnerable to contracting new infections. By expressing suppressor proteins, certain retroviruses—HIV in particular—have the ability to specifically prevent T-cell proliferation. Furthermore, AIDS results from a significant loss of immune competence brought on by HIV replication's ongoing destruction of T cells.

The ability of certain wasp polydnaviruses to actively suppress immunity serves as an example of how the virus and host have evolved together. The ovaries and egg cells of parasitic wasps harbor this virus as a persistent genetic passenger. The caterpillars of other insect species serve as food for the developing larvae of these wasps, which lay their eggs in them. In order to prevent the caterpillar from eliminating the embryonic tissue at an early stage of development, systemic, immunosuppressive infection is caused by the polydnavirus that was inserted into it along with the wasp egg. Larval survival greatly decreases if wasps lacking these viruses inject eggs into the caterpillar host.

6.3.1.5. Active Evasion of Immunity – Blockage of MHC Antigen Presentation

MHC-I antigen presentation is specifically inhibited by adenovirus, HIV, and HSV. A particular viral protein that mediates this blockage is expressed in each instance. It takes a moment to realize the significance of HSV's blocking of MHC-I antigen presentation, as the virus replicates quickly and effectively in the cells it infects. The slowly replicating adenovirus will benefit greatly from its ability to impede host immunity. Here, the value is probably found in the early phases of reactivation from latent infection, where the virus needs to be present in small enough quantities to cause infection in a host with a strong immune memory that is biased against HSV replication. Similarly, one of the first genes that HIV expressed codes for the protein Nef, which suppresses MHC-I expression to avoid being recognized by cytotoxic T cells.

6.3.1.6. Consequences of Immune Suppression to Virus Infections

While immune suppression is a useful tool in some medical conditions, certain viruses have the ability to either slightly or significantly suppress immunity during the course of infection and pathogenesis. One example is the requirement for host cell-mediated immunity to be suppressed before organ or tissue transplantation. Some intravenous drug abuse types also cause immune suppression. Reactivation of herpesvirus infections, like varicella zoster (chicken pox), and cytomegalovirus infection are major consequences of immune suppression. Naturally, the same issues can arise from viral infections like HIV that compromise the immune system. Significant populations of people exhibiting immune suppression may have a more serious consequence in that they may act as selective reservoirs for the emergence of drug-resistant and antigenic strains of pathogens. For instance, in critical urban, prison, and Third World populations, incomplete drug therapy, HIV infection, and drug-induced immunosuppression are definitively linked to the current rise in antibiotic-resistant tuberculosis cases.

6.4. MEASUREMENT OF THE IMMUNE REACTION

Measurement of the immune reaction involves various techniques to assess the body's response to pathogens, vaccines, or other foreign substances. This complex process can be examined at different levels, including cellular, molecular, and functional responses. One of the fundamental methods is the Enzyme-Linked Immunosorbent Assay (ELISA), which measures the concentration of specific antibodies in blood samples, providing insight into humoral immunity. Flow cytometry is another crucial technique, enabling the analysis of different cell populations within the immune system by detecting specific cell surface markers. This method can quantify and characterize T cells, B cells, and other immune cells, offering a detailed view of cellular immune responses (Guan et al., 2018).

Utilizing Polymerase Chain Reaction (PCR) and its sophisticated variation, quantitative PCR (qPCR), to identify and measure particular nucleic acids is another crucial method. These methods are very useful for determining the expression of genes linked to immunity as well as the existence and activity of bacteria, viruses, and other pathogens. To assess the immune system's functional capability, assays that measure immune cell proliferation in response to particular antigens or mitogens include the lymphocyte proliferation assay. Cytokine profiling measures the concentrations of different cytokines and chemokines, important immune response mediators, using methods like multiplex bead assays or ELISA. Understanding the inflammatory environment and the harmony between pro- and anti-inflammatory cues is aided by this profiling. Furthermore, assays with high sensitivity for measuring immune responses, such as the enzyme-linked immunospot (ELISPOT) assay, can identify cytokine production at the single-cell level.

The limits of immune measurement are being pushed by emerging technologies like single-cell RNA sequencing and mass cytometry (CyTOF), which offer high-dimensional data on transcriptional and phenotypic profiles of individual cells. A more thorough understanding of the immune landscape and its dynamic changes during infections, vaccinations, and immune disorders is made possible by these cutting-edge techniques. All things considered, measuring the immune response is a complex process that uses various advanced methods to give a thorough and nuanced understanding of how the immune system operates. These metrics are essential for disease diagnosis, assessing the effectiveness of vaccines, and creating novel immunotherapies.

6.4.1. Measurement of Cell-Mediated (T-Cell) Immunity

To measure a particular T-cell response, immune lymphocytes must first be incubated with the target cell. This process is known as cell-mediated immunity. Although it can be challenging and complex, the release of radioactive chromium from target cells is a useful technique for measuring T-cell-mediated cell lysis. Target cells are cultured in a way that allows the radioactive metal to be incorporated. After being rinsed, the cells retain only the radioactivity inside them. As a result, at low gravity (low speeds), radioactivity will settle to the bottom of a centrifuge tube. The target cells are lysed in the presence of reactive killer T cells, and the "hot" chromium enters the solution and cannot be sedimented at low speeds. Measuring the rate of cell replication in response to a particular antigen can also be used to determine the numerical estimate of the quantity of reactive lymphocytes. An antigen and a radioactive nucleoside precursor to DNA are cultured in white blood cells. T cells will integrate this radioactive precursor as they multiply in response to antigens. A lymphocyte stimulation index can be used to quantify the addition of radioactivity relative to a control culture. Incubating antigen-bearing cells with lymphocytes is an additional technique for assessing T-cell immunity. Under a microscope, reactive T lymphocytes can be seen and counted as they form rosettes around the antigen-bearing cell.

6.4.2. Measurement of Antiviral Antibody

Glycoproteins, known as antibody molecules, are secreted and possess the ability to identify and merge with particular segments of viruses or other proteins that are not native to the host. Antibody molecules have a very specific structure where a region of fixed amino acid sequence is found at one location on the antibody molecule, and the antigen-combining sites, which consist of variable amino acid sequences, are at another. The constant region, or Fc region, plays a key role in mediating the secretion of the antibody molecule from the B cell that is responsible for its expression. Serving as a signal to cells and other particular cellular proteins that the molecule bound to the antigen is, in fact, an antibody is another important role of the Fc region (Guan et al., 2018).

6.4.2.1. Enzyme-Linked Immunosorbent Assays (ELISAs)

The Fc region of the antibody molecule is used as a "handle" in several techniques for measuring antibody reactions. "Enzyme-linked immunosorbent assays (ELISAs), a group of incredibly sensitive techniques, employ enzymes that can convert a colorless substrate into a colored product that is attached to the Fc region of an antibody molecule. The enzyme attached to the Fc region will bind with the antibody when it binds to an antigen. The amount of antibody present can be determined by incubating the antigen-antibody complex with suitable substrates for the bound enzyme, which will cause color to develop. Figure 6.9 provides an outline of the method with examples. ELISAs have significant commercial value and are incredibly useful for quick diagnosis. An antigenic peptide bound to an insoluble matrix, such as a flexible plastic strip with dry reagents on it, will, for instance, take on color when the strip is dipped into a plasma preparation containing an antibody against the peptide. As long as the enzyme being used is relatively stable, incubation for a sufficient amount of time will produce some color even in cases where the amount of antibody is very low. it can be modified to work with automated machinery. Commercial kits are currently available that allow one to spot and dry a small sample of bodily fluid that may contain an interesting virus or antibody. After that, the kit is shipped to a lab for quantitative analysis (Arankalle et al., 2010).

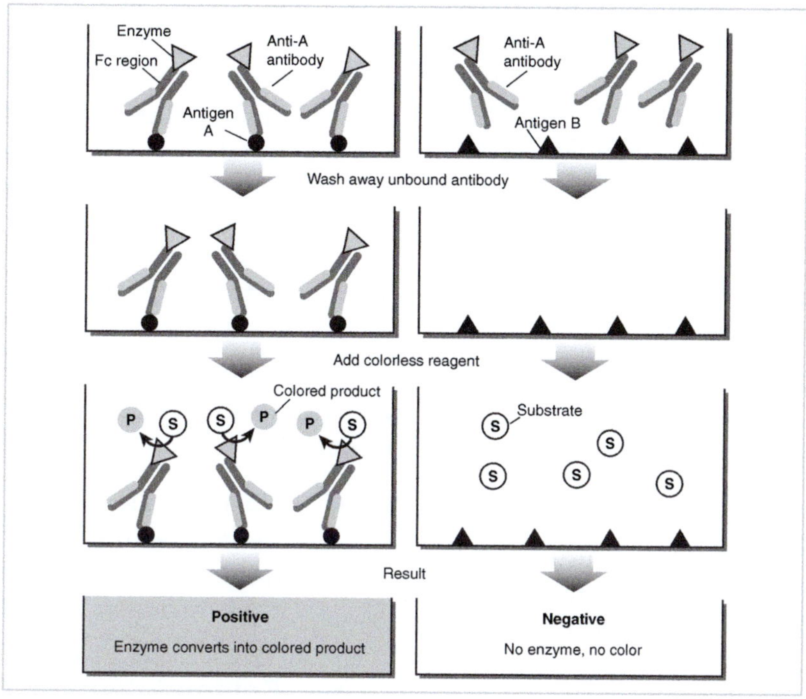

Figure 6.9. *An enzyme-linked immunosorbent assay (ELISA): the method of using a color reaction mediated by an enzyme bound to the Fc region of the antibody molecule.*

Source: https://www.blackwellpublishing.com/content/BPL_Images/Content_store/Middle_Sized/Wagner9781405103466/077to095.pdf.

Our capacity to identify minuscule amounts of viral antigens or antibodies in test material could undergo even more revolutionary changes with the application of lasers and microtechnology developed in the electronics industry. An enormous variety of potential antigens can be bound to a microchip, which can then be incubated with an unidentified antibody and used in an ELISA assay or another technique to produce a fluorescent signal indicating the formation of an antigen-antibody complex. This can be quickly scanned using a solid-state detection device, laser beam, and fluorescent microscope. With the use of these techniques, it may be possible to screen a serum sample in a matter of hours for the presence of antibodies directed against all or almost all known pathogenic agents! Such antibodies indicate exposure to the corresponding pathogen, either recently or in the past.

6.4.2.2. Neutralization Tests

Neutralization tests are a critical component in virology and immunology used to evaluate the efficacy of antibodies in neutralizing viruses. These tests measure the ability of antibodies, whether from an individual's serum or generated through vaccination, to inhibit the infectivity of a virus. The primary goal is to determine the presence and potency of neutralizing antibodies that can block the virus from entering host cells and replicating, a key indicator of protective immunity.

- **Principle and Procedure:** Neutralization tests typically involve mixing a virus with a sample containing antibodies (such as serum) and then assessing the mixture's ability to infect cells in culture. There are two main types of neutralization tests: the plaque reduction neutralization test (PRNT) and the microneutralization assay.

6.4.2.2.1. Plaque Reduction Neutralization Test (PRNT)

- **Virus Preparation:** A known quantity of the virus is prepared.
- **Serum Dilution:** The serum sample is serially diluted to obtain different concentrations of antibodies.
- **Incubation:** The diluted serum is mixed with the virus and incubated, allowing any neutralizing antibodies present to interact with the virus.
- **Infection:** The mixture is then added to a monolayer of susceptible cells cultured in a Petri dish.
- **Plaque Formation:** After incubation, the cells are fixed and stained. Areas where the virus has infected and killed the cells, known as plaques, are counted.
- **Calculation:** The reduction in the number of plaques compared to a control sample without antibodies indicates the neutralizing capacity of the serum.

6.4.2.2.2. Microneutralization Assay

- **Microplate Setup:** This assay uses a microplate format with 96 wells, where each well represents a different dilution of the serum sample.
- **Virus and Cell Addition:** The virus and the serum are mixed in the wells, followed by the addition of susceptible cells.
- **Incubation and Observation:** After incubation, cell viability is assessed, usually through staining methods or other viability assays. The presence of neutralizing antibodies prevents cytopathic effects (CPE) in the cells.

6.4.2.2.3. Applications and Importance

Neutralization tests are vital for several reasons:

- **Vaccine Development:** They help in evaluating the immune response generated by vaccines, determining their efficacy in eliciting protective antibodies.
- **Epidemiological Studies:** These tests are used to assess population immunity, identifying individuals who have been exposed to the virus and have developed neutralizing antibodies.
- **Clinical Diagnostics:** Neutralization tests can aid in diagnosing infections by confirming the presence of virus-specific neutralizing antibodies in patients.

6.4.2.2.4. Advantages and Limitations

Neutralization tests are highly specific and sensitive, providing precise measurements of functional antibody activity. However, they can be labor-intensive, require biosafety facilities for handling live viruses, and may take several days to complete. Despite these challenges, neutralization tests remain a gold standard for assessing immune protection against viral infections.

6.4.2.3. Inhibition of Hemagglutination

The ability of an antibody to block a

particular viral property is the basis for certain techniques used to measure an antibody's antiviral activity. For example, since the early years of this century, it has been known that a variety of enveloped viruses will adhere to red blood cells and cause them to agglutinate. The concentration of viral particles in solution can be roughly determined using this hemagglutination property. The hemagglutination inhibition (HI) test can be used to measure the levels of antibodies against enveloped viruses, as many of them will prevent the virus from causing red blood cells to clump together. The basic mechanism relies on the fact that many antibody molecules bind to the surface of viruses, thereby physically masking them. It was developed long before a thorough understanding of the immune response was available. A virus capable of causing hemagglutination will become coated in antibody and be unable to attach itself to red blood cells if it is preincubated with an antibody against it. This occurs as a result of the virus particle's small surface area, which makes it difficult for certain protein molecules to access certain areas of the surface once they attach themselves to it. The entire surface becomes opaque if enough antibody adheres to it (Steuerman and Gat-Viks, 2016).

Figure 6.10 illustrates an experiment using hemagglutination inhibition, commonly known as an HI test. A standard virus stock and blood serum are all that are needed to assess a patient's immunological response. The basic steps are as follows: Standard red blood cell samples (e.g., guinea pig or chicken red blood cells for influenza virus) are combined with various dilutions of an unidentified antibody that may be present in a patient's serum along with a known quantity of virus stock. Following an appropriate duration, the mixture is softly agitated and exposed to low-velocity centrifugation. Red blood cells that have aggregated form a jelly-like clump and are unable to separate. The characteristic of agglutination is a diffuse red or salmon-pink solution (Steuerman and Gat-Viks, 2016). The red "button" at the tube's bottom is formed by the cell pellet if the red blood cells' inability to agglutinate is due to the antibody's sequestration of the virus. Relative speed, ease of use, and low cost of performance—a crucial factor in small clinical laboratories, particularly in developing nations—make using HI beautiful, rather than accurate.

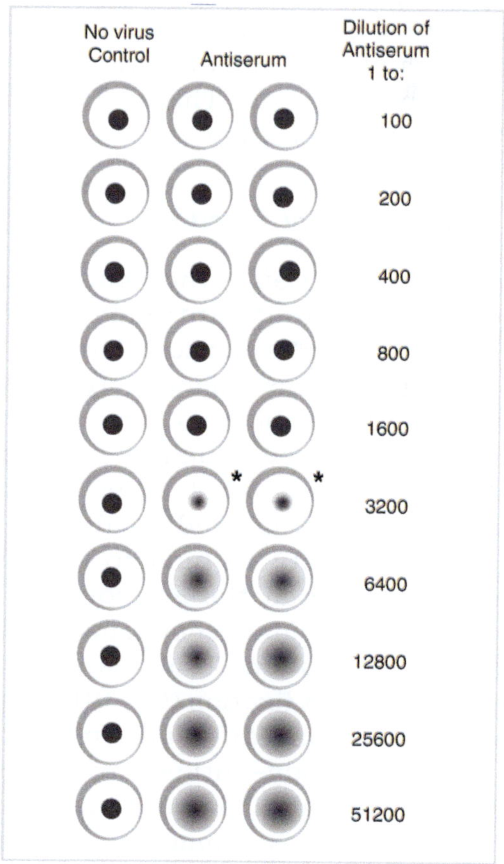

Figure 6.10. *The hemagglutination inhibition assay for measuring antibody against a virus in serum.*

Source: https://www.blackwellpublishing.com/content/BPL_Images/Content_store/Middle_Sized/Wagner9781405103466/077to095.pdf.

The assay is carried out by mixing constant amounts of a known hemagglutinating virus with serial dilutions of serum; then the virus–serum mixture is added to red blood cells. Low dilutions of serum result in sequestering the virus so that it is not available for hemagglutination, and red blood cells in the wells pellet to the bottom under low centrifugal fields. Higher dilutions of the antiserum dilute the antibody concentration to a point where enough virus remains to cause a positive hemagglutinin reaction. If there were more antibody in the serum, a higher dilution would be required to accomplish this. Thus, the hemagglutination inhibition titer of the serum is a measure of how far it can be diluted and still block the hemagglutinin reaction. This is a measure of antibody concentration

6.4.2.4. Complement Fixation

Many soluble proteins that can adhere to cells containing antibody–antigen complexes make up serum complement. The complement proteins change structurally as a result of this binding, and the final protein to bind is eventually activated to become a protease, which lyses the cell. Fixation is the term used to describe a complement's ability to bind to antibody-antigen complexes at the Fc region of the antibody. Once bound, complement is no longer free in solution. This characteristic makes complement fixation (CF) titration—a reasonably easy and affordable method of measuring antibody–antigen reactions—possible. Sheep red blood cells are used in a CF assay, which is commonly performed on a horse, goat, or other large animal, to create an antibody against their surface proteins. After that, the red blood cells are "standardized" so that they will lyse when a particular quantity of antibody is added to them, and the mixture is incubated with guinea pig complement. Red blood cell lysis can be easily measured because, in the event that a lysed red blood cell solution is centrifuged at a low speed, the solution will remain red due to the absence of cells that would otherwise carry the hemoglobin to the tube's bottom and form a pellet.

Red blood cells, complement, and anti-red blood cell serum can all be kept in the refrigerator for a considerable amount of time once they have been standardized. This is what happens when they are used to assay an antibody-antigen reaction. A solution containing an unknown quantity of virus and a fixed amount of known antibody, or a solution containing an unknown quantity of virus and a fixed amount of known antibody, are incubated together in serial dilutions. After that, a predetermined quantity of guinea pig complement is added to them. The complement will be fixed (that is, if an antibody-antigen complex has formed), bound) by it. The complement will remain in solution otherwise. There will be some complement that is fixed and some that is free if there is an intermediate complex level. After the unknown antibody-antigen mixture is incubated with the known quantity of complement, the entire "mess" is incubated with standard concentrations of red blood cells and antibody against red blood cells. There won't be red blood cell lysis if the complement is fully fixed. The red blood cells will partially lyse if some is fixed (Caini et al., 2018). Complete lysis will result if none are fixed. The amount of unknown antibody-antigen reaction can be determined and the CF titer can be obtained by measuring the degree of lysis (by measuring the amount of red color in solution after low-speed centrifugation). Similar to HI, this method is quick, inexpensive, and only needs a few pricey pieces of equipment, but it is not very sensitive or precise. In smaller labs or ones with fewer resources, it's the perfect way to obtain fast results. Today, all hospitals still use it.

CASE STUDY

HOST IMMUNE RESPONSE TO VIRAL INFECTION

The host immune response to viral infection is a multifaceted process, demonstrated here through the case study of a 38-year-old female patient who contracted Influenza (a) The patient, previously healthy with no underlying conditions, experienced typical flu symptoms, including fever, muscle aches, and fatigue, over 10 days, with full recovery in two weeks.

Initial Innate Immune Response

Upon infection, the virus infiltrated respiratory epithelial cells, triggering the innate immune response. This early defense mechanism involved the recognition of viral RNA by pattern recognition receptors (PRRs) such as Toll-like receptors (TLRs). These receptors activated signaling pathways leading to the production of type I interferons (IFNs) and pro-inflammatory cytokines like IL-6 and TNF-⊠. The patient's increased IFN levels were crucial in establishing an antiviral state in neighboring cells and limiting viral replication. Additionally, natural killer (NK) cells were mobilized to destroy infected cells, reducing the viral load during the initial phase of infection.

Activation of the Adaptive Immune Response

Within a few days, the adaptive immune system became actively engaged. The patient's dendritic cells, acting as antigen-presenting cells (APCs), processed viral antigens and presented them to T cells in the lymph nodes. This led to the activation and proliferation of CD4+ helper T cells and CD8+ cytotoxic T cells. The CD4+ T cells produced cytokines that supported the B cell response, while CD8+ T cells directly targeted and killed infected cells displaying viral peptides via MHC class I molecules.

Humoral Immune Response

B cells were activated by viral antigens and received help from CD4+ T cells to differentiate into plasma cells. By day 7, the patient's blood tests indicated the presence of IgM antibodies, signifying an acute response. By day 14, IgG antibodies were abundant, signifying a more mature immune response capable of neutralizing the virus. These IgG antibodies are specifically bound to the viral hemagglutinin and neuraminidase proteins, and prevent the virus from entering host cells and facilitating its clearance.

Memory Formation

As the infection resolved, the formation of memory B and T cells was noted. These cells remained in the body, providing the patient with long-lasting immunity and a rapid response to future exposures to the same virus. Memory B cells retained the ability to quickly produce specific antibodies, while memory T cells could rapidly expand and exert their effector functions upon re-exposure to the influenza virus.

Clinical Implications

This case highlights the importance of both innate and adaptive immune responses in controlling and eliminating viral infections. The insights gained from understanding this immune response have significant implications for designing effective vaccines and antiviral therapies. Vaccines aim to mimic this natural infection process by inducing robust and long-lasting immune memory without causing disease, thereby preparing the immune system for future encounters with the pathogen.

In conclusion, the patient's recovery from Influenza A infection illustrates the coordinated action of the innate and adaptive immune systems, culminating in effective viral clearance and the establishment of immune memory. This comprehensive immune response is crucial for protection against subsequent infections and forms the basis for developing immunization strategies.

CLASS ACTIVITY

EXPLORING THE ADAPTIVE IMMUNE RESPONSE

How do antigen-presenting cells (APCs) activate T cells, and what are the subsequent roles of CD4+ and CD8+ T cells in the adaptive immune response?

Write a short essay or create a presentation explaining the process of T cell activation by APCs and the distinct functions of CD4+ helper T cells and CD8+ cytotoxic T cells in fighting a viral infection.

SUMMARY

- When a virus infects an immunologically naïve host, one might expect that initially the odds are in favor of the virus. After all, most viruses have relatively short replication cycles, resulting in the rapid release of hundreds of new virions from a single cell
- In any protein, certain clusters of amino acids (usually between 10 and 12) are able to interact with the appropriate antigen-recognizing T cells or antibody-producing B cells to lead to proliferation of those cells
- Any protein and many other macromolecules can be antigenic, but antigens must be "processed" and then presented at the surface of the cell bearing them (antigen-presenting cell) in the proper context to be able to evoke an immune response
- Although T cells have a primary role in the destruction of cells bearing foreign antigens, B cells can also destroy antigen-bearing cells by use of the complement system, which leads to complement-mediated cell lysis
- The immune response is an effective one and plays a constant role in selection against viruses that do not mount an efficient infection. Despite the effectiveness of the immune response, it is clear that many virus infections survive and thrive in the setting of the host's immune capacity
- Cell-mediated immunity requires incubation of immune lymphocytes with a target cell and then measurement of a specific T-cell response. This can be difficult and tricky, but for the measurement of T-cell-mediated cell lysis, the release of radioactive chromium from target cells is a convenient method
- Antibody molecules are secreted glycoproteins that have the capacity to recognize and combine with specific portions of viral or other proteins foreign to the host.
- Serum complement is made up of a number of soluble proteins that are able to stick to cells bearing antibody-antigen complexes. As this binding occurs, the complement proteins undergo structural changes, and, finally, the last protein-bound is activated to become a protease, which then lyses the cell

REVIEW QUESTIONS

1. Describe the mechanism through which defensins disrupt viral membranes and inhibit viral replication.
2. Discuss the immunological structure of a protein.
3. What protein structural features are involved in the antigenic nature of epitopes?
4. What steps occur in the immune response following the primary infection of a vertebrate by a virus?
5. Describe the role of major histocompatibility complex (MHC) molecules in presenting protein antigens to T cells.
6. How do APCs activate T cells, and what signals are necessary for this activation?

7. What are some problems arising when considering vaccination strategies for viral diseases?
8. Explain the concept of clonal selection and its importance in the adaptive immune response.
9. How do viruses evade the host immune response, and what strategies do they employ?
10. What are the potential consequences of immune suppression on the outcome of viral infections?

REFERENCES

1. Agerberth, B., & Gudmundsson, G. H. (2006). Host antimicrobial defense peptides in human disease. *Current Topics in Microbiology and Immunology*, *306*, 67–96.
2. Arankalle, V. A., Virkar, R. G., Tandale, B. V., & Ingle, N. B. (2010). Utility of pandemic H1N1-09 influenza virus recombinant hemagglutinin protein-based ELISA for sero-surveillance. *Clinical Vaccine Immunology*, 2010 Jul 14. https://doi.org/10.1128/CVI.00201-10 (accessed on 4th October 2024).
3. Bermejo-Martin, J. F., Martin-Loeches, I., Rello, J., Anton, A., Almansa, R., Xu, L., et al. (2010). Host adaptive immunity deficiency in severe pandemic influenza. *Critical Care*, *14*(5), R167. https://doi.org/10.1186/cc9259 (accessed on 4th October 2024).
4. Beutler, B., Jiang, Z., Georgel, P., Crozat, K., Croker, B., Rutschmann, S., Du, X., & Hoebe, K. (2006). Genetic analysis of host resistance: Toll-like receptor signaling and immunity at large. *Annual Review of Immunology*, *24*, 353–389.
5. Couper, K. N., Blount, D. G., & Riley, E. M. (2008). IL-10: The master regulator of immunity to infection. *Journal of Immunology*, *180*, 5771–5777.
6. Deshmukh, T. M., Lole, K. S., Tripathy, A. S., & Arankalle, V. A. (2007). Immunogenicity of candidate hepatitis E virus DNA vaccine expressing complete and truncated ORF2 in mice. *Vaccine*, *25*, 4350–4360.
7. Guan, W., Wu, N. C., Lee, H. H. Y., Li, Y., Jiang, W., Shen, L., et al. (2018). Clinical correlations of transcriptional profile in patients infected with avian influenza H7N9 virus. *Journal of Infectious Diseases*, *218*, 1238–1248. https://doi.org/10.1093/infdis/jiy301 (accessed on 4th October 2024).
8. Kumar, H., Kawai, T., & Akira, S. (2011). Pathogen recognition by the innate immune system. *International Reviews of Immunology*, *30*(1), 16–34.
9. Litman, G. W., Cannon, J. P., & Dishaw, L. J. (2005). Reconstructing immune phylogeny: New perspectives. *Nature Reviews Immunology*, *5*(11), 866–879.
10. Restifo, N. P., & Gattinoni, L. (2013). Lineage relationship of effector and memory T cells. *Current Opinion in Immunology*, *25*(5), 556–563.
11. Steuerman, Y., & Gat-Viks, I. (2016). Computational methods for immune cell-type subsets. *Nature Immunology*, *17*(1), 4–8.
12. Tang, B. M., Shojaei, M., Teoh, S., Meyers, A., Ho, J., Ball, T. B., et al. (2019). Neutrophils-related host factors associated with severe disease and fatality in patients with influenza infection. *Nature Communications*, *10*(1), 3422. https://doi.org/10.1038/s41467-019-11311-2 (accessed on 4th October 2024).

CHAPTER 7

Replicating and Measuring Biological Activity of Viruses

LEARNING OBJECTIVES

After studying this chapter, you will be able to:

- Understand the outcome of virus infection in cells;
- Measure the biological activity of viruses;
- Discuss the visualization and enumeration of virus particles.

KEY TERMS FROM THIS CHAPTER

Abortive infection	Carcinogenesis	Cell lysis
Chromosome numbers	Embryonic (fetal) tissue	Escherichia coli
Epstein–Barr herpesvirus	Glucose fermentation	Inorganic nitrogen
Mutagenic processes	Nonproductive infection	Virus infection
Plaque assays	Prototrophy	

INTRODUCTION

Cell Culture Techniques

Growing and maintaining cells in the laboratory for molecular biological research is essential. Studying viruses is easier when cells can be cultured where the viruses grow most effectively. This process involves taking cells from their natural environment, identifying their growth and functional properties, and ensuring a constant supply through continuous culture. Depending on the virus being studied, this task can be either routine or challenging (Celis and Celis, 1998).

Maintenance of Bacterial Cells

Because bacterial viruses were easily replicated in bacterial cell cultures, the study of bacterial viruses served as a model for studying all viruses. Certain bacterial cells have extremely slow generation times and are very challenging to grow in culture. However, typical lab cultures of the prokaryotic cells that are most frequently used, like Escherichia coli (*E. coli*), can be cultivated on straightforward, well-defined media that only contain an energy and carbon source (often glucose) as well as inorganic sources of phosphorus, nitrogen, and sulfate (NH_4Cl, $MgSO_4$, and phosphate buffers). Protostrophy, the capacity to grow on media that solely contain inorganic molecules and sugar, enables a complete understanding of all the ultimate sources of biological reactions. A yeast or beef extract broth, perhaps enhanced with required inorganic materials, promotes faster growth.

In liquid culture, bacterial cells can be grown to densities of 10^8 cells/ml, which can be reached during the exponential (i.e., fastest) stage of development. It is possible to cultivate bacterial cells on semisolid or solid surfaces, which enables the creation of colonies or clones in which every cell is descended from a single parent cell. Agar is the material most frequently used for this kind of growth; it is poured into glass or plastic Petri dishes in the form of a thin slab.

These plates are widely used in bacterial virus plaque assays. Plaque assays profit from the fact that a virus's replication causes cells to lyse, leaving a virus infection center cell-free.

Plant Cell Cultures

Studying plant viruses is easier because they can be easily infected in their intact hosts, which are simple to cultivate. This approach enables a basic analysis of various plant virus characteristics. In fact, early structural research on plant viruses was on par with studies conducted on bacterial viruses.

However, until recently, molecular biology research lagged because plant cell culture systems were unreliable. Plant cell architecture makes the manipulation of cells in culture—which is so helpful for the study of animal and bacterial cells—extremely difficult and frequently practically impossible, which is why plant cell culture techniques have not advanced as quickly as those for animal cells. Due to these technical issues, plant cell culture has not significantly influenced the advancement of plant virology. Plant molecular biology has benefited greatly from the ability to cultivate plant cells devoid of their cell walls as protoplasts; however, little virology research has been done with such systems to date.

Culture of Animal and Human Cells

Maintenance of Cells in Culture

In order to keep cells in culture, one needs to use a culture medium that closely resembles blood plasma. This medium includes salts that resemble those in plasma, vitamins, most amino acids (since animal cells cannot synthesize many of these), glucose as an energy source, buffers (usually sodium bicarbonate or carbon dioxide) to keep the medium from becoming too acidic due to lactic acid fermentation from the glucose, and most importantly, blood serum, which is typically taken from calves or horses. Numerous growth factors, such as proteins, are necessary for cell growth in this serum. Antibiotics are also included, as mentioned, to prevent microbial contamination.

The same type of cell (such as a fibroblast or skin cell) can have remarkably different properties depending on its species of origin, age of the donor animal, state of the cell, and the particular culture history, in addition to the uncertainties surrounding the precise requirements for culture. As a result, every cell culture is unique and has its own history (Celis and Celis, 1998).

Figure 7.1 illustrates a general procedure for obtaining mouse mammary epithelial cells. Numerous primary and tumor cell cultures are produced via similar techniques. Typically, cells are cultured in standard-sized dishes with defined sections. Popular sizes vary based on the required number of cells, from 25 cm^2 to 150 cm^2.

Types of Cells

All animal cells are ultimately derived from living tissue. However, some, like HeLa cells, have been cultivated in culture for so long (roughly 50 years) that they no longer resemble the original tissue. Because these continuous cell lines grow quickly and yield large amounts of virus for the study of some fundamental aspects of virus replication, they are very helpful. However, they are not useful for researching the more nuanced effects of virus infection on the development and regulation of cells. Additionally, continuous cell lines are inappropriate for investigating the function of differentiated cells (Celis and Celis, 1998).

Continuous cell lines generally have the following properties:

- They have fragmented and reduplicated chromosomes; that is, they are aneuploid
- They can grow in suspension and in relatively low concentrations of serum, and can overgrow each other; they display no response to neighboring cells
- They are essentially immortal; if periodically diluted and fed with appropriate nutrients, they will continue replicating
- They generally do not display properties of differentiated cells and do not respond to modulators of cell growth or function
- If introduced into an animal (even one of the same species from which they were originally isolated), they

will not grow and will be eliminated by the animal's immune system.

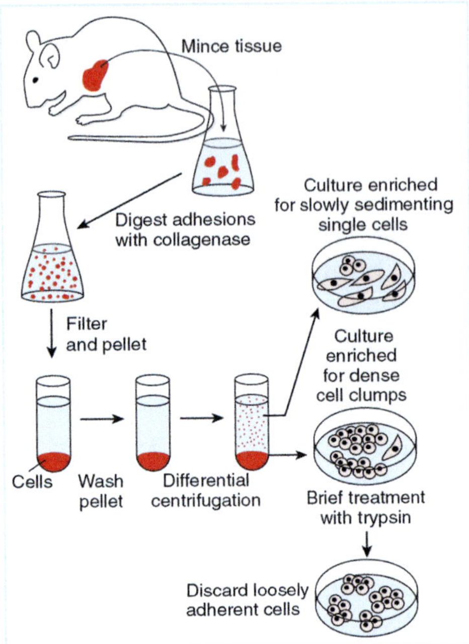

Figure 7.1. *Generating a primary cell culture. Tissue is surgically removed from an anesthetized animal, and then minced and homogenized. The addition of collagenase breaks down extracellular collagen, but the enzyme does not attack intact cells.*

Source: Edward K. Wagner, Martinez J. Hewlett, David C. Bloom, David Camerini, Basic Virology Third Edition, ISBN 1-4051-0346-9.

Primary cells are at the other end of the laboratory cell spectrum. The easiest tissues to isolate primary cells are embryonic (fetal) tissue or the tissue or embryos of newborn animals. Older animal cells are often more difficult to culture and have a shorter time in culture before they die or stop dividing (senescing).

The early stages of culturing primary cells are almost exactly the same as those in the tissue from which they originate, even though the simple act of culturing cells causes rapid changes in the subtle properties of living cells. More quickly growing and replicating cells, like fibroblasts, will outgrow other cells in a mixed tissue source, even though almost any type of replicating cell can be cultured if the tissue containing it is properly isolated. Because of this, primary cell isolation from whole embryos typically results in fibroblast cultures.

Most primary cells have the following properties:

- They have normal chromosome numbers and shape
- They require high serum concentrations containing numerous growth factors
- They cannot divide or even survive for long unless they are maintained in contact with a solid surface
- They are susceptible to contact-induced inhibition of cell migration and growth. They stop growing and moving when they come into contact with other cells in a culture plate, a phenomenon known as contact inhibition. As a result, a certain culture plate area will support the growth of a given number of cells. The cells are able to metabolize energy and remain healthy during contact inhibition. They will proliferate and divide once more when diluted (passaged) and put in a fresh culture dish.
- Their lifespan is limited and expressed in divisions. After being isolated, fibroblasts typically divide 20–30 times before starting to senesce and eventually die. According to recent experimental data, the programmed loss of chromosome end regions (telomeres) during each cell replication may be partially responsible for this finite lifetime. A sufficient loss

of chromosomal DNA triggers the cell's demise.

- They display all properties of differentiated cells and respond to modulators of cell growth or function
- The cells may survive if injected into a member of the same species from which they were originally isolated, but they won't grow tumors. This list of primary cell characteristics is idealized. A specific type of cell isolated from an individual may not have all of the listed properties. For instance, when kept in suspension, lymphocytes separated from the tiny quantity of blood present in a newborn's umbilical cord will live but not multiply. On the other hand, lymphocytes cultured from a large number of adults and adolescents who have contracted infectious mononucleosis will not only live but also proliferate for comparatively long periods of time. These immortalized lymphocytes appear normal on the surface, but they harbor copies of the Epstein-Barr herpesvirus genome, which causes these peculiar characteristics due to the expression of specific viral genes. Although this kind of altered lymphocyte is not a tumor cell, it unmistakably exhibits certain traits that are (Celis and Celis, 1998).

Loss of contact inhibition of growth and immortalization of primary cells

One type of cell found in a lab that has characteristics halfway between continuous cell lines and primary cells is immortalized B lymphocytes. These cell types have undergone changes and now resemble tumor cells in at least some ways. Transformed cells can be produced by long-term primary cell culture. A critical number of growth control genes encoded by the cell are altered by a random accumulation of mutations that occur during the time in culture. Cells experience the cumulative effects of aging (senescence) at a critical point during cultivation (the precise point will depend on the type of cell; for fibroblasts, it is typically 12–15 generations). At this point, nearly all cells enter a crisis period, undergo apoptosis, and ultimately die.

Because chromosomal DNA is linear and cannot replicate fully at both ends, chromosomal replication errors result in senescence. Therefore, the crucial telomeric sequences at the end of chromosomes are lost with each round of DNA replication. The telomeres typically attach to various distinct cellular proteins, hiding the chromosomal ends that structurally resemble breaks in double strands of DNA. Numerous cellular defenses are triggered when the ends are exposed, starting the apoptotic process.

One of the main factors in the development of cancer is when cells fail to undergo normal cell death in response to DNA damage. If these damaged areas are not properly fixed, they can cause mutations and disrupt the function of genes that control cell growth and development. In some cases, a few cells in crisis can protect their chromosomes by activating an enzyme called telomerase, which is typically only active in early development. This process leads to the cells becoming immortalized and sharing characteristics with tumor cells. These immortalized cells quickly take over the culture and can eventually be used to create continuous cell lines.

It is also possible to separate cells with transformed characteristics from an animal's tumors. Various tumor cells in animals exhibit one or more of the same stages of transformation from normal cells that can be seen in primary cell cultures. This provides a crucial hint about the cellular processes that result in cancers. It is crucial to remember that distinct tumor cells may exhibit wildly disparate departures from the typical growth characteristics of the parent cells. Some tumor cells show very little difference from normal cells, maybe just the loss of contact inhibition of growth (especially those isolated early in the course of cancer development). Others have many more changes, particularly long after the cancer was diagnosed. Figure 7.2 schematically depicts the transition from primary cells to continuous line cells and the relationship between these cells and tumors in the original animal. This process of change serves as a compelling experimental illustration of the multiple steps involved in an organism's cellular transformation from normal to cancerous. The alterations intensify when mutations occur in particular growth control and regulatory genes, affecting the cell's capacity to respond to regular signals in the animal and restricting the growth and function of the cell.

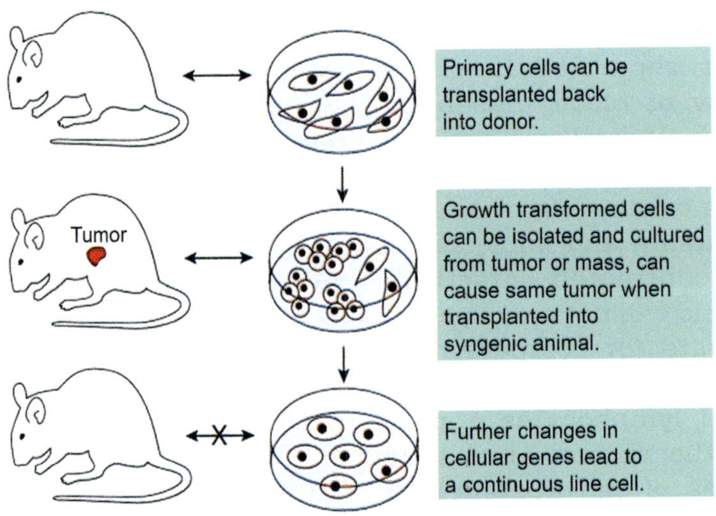

Figure 7.2. *The progression of cells in culture from primary to transformed to continuous lines and their relationship to tissues in the originating animal.*

Source: Edward K. Wagner, Martinez J. Hewlett, David C. Bloom, David Camerini, Basic Virology Third Edition, ISBN 1-4051-0346-9.

While long passage or other mutagenic methods are frequently used to create immortalized cell lines, it is crucial to remember that many tumor cells and certain transformed cells with lost contact inhibition of growth still have a limited lifespan. The genes that regulate response to contact inhibition and life span are distinct and can be mutated independently or in tandem. A highly productive area of research on certain viruses is their ability to convert healthy stem cells into malignant or cancerous cells. Reintroducing these virus-transformed cells into animals can result in tumor development. The study of the process has contributed significantly to our current understanding of carcinogenesis and the characteristics of cancer cells because this transformation necessitates a particular interaction between viral and cellular genes or gene products.

7.1. THE OUTCOME OF VIRUS INFECTION IN CELLS

7.1.1. Fate of the Virus

When a virus infects a cell, the virus's genetic material enters the cell. Whether the virus's outer protein shell, or capsid, stays on the surface of the host cell during infection, like in bacteriophages, or is taken inside the cell, it is altered and then destroyed. If a virus attached to and entered a cell is isolated, it is no longer capable of causing infection and replicating. Therefore, after the initial stages of the virus invading the cell, the only way to isolate infectious virus is to prevent it from entering the cell or wait for new virus particles to be produced (Celis and Celis, 1998).

Certain infections (usually referred to as nonproductive infections) do not result in the production of new infectious viruses. Because this kind of infection stops before the replication process is fully completed, it is also known as an abortive infection. When a virus infects a nonpermissive cell—a cell that lacks the necessary components for virus replication—abortive infections may ensue. Another possibility for a nonproductive infection is to contract a virus that interferes with replication due to a defective gene product. A general rule of thumb in differentiating types of infection is the following:

- Productive infection: more virus out than in
- Abortive infection: no virus out, the virus cannot replicate

The viral genome may be internalized or destroyed during an unsuccessful infection. In the latter case, expression of one or more viral genes may occur. Viral antigens may be expressed by the cell in this scenario, either on the cell surface or in other parts of the cell. It is possible to identify and research these antigens with the appropriate immune reagents (Celis and Celis, 1998).

This explanation makes it evident that an unsuccessful infection can have a significant impact on the host cell and, in the end, the organism. For instance, persistently having non-contagious measles virus in brain tissue can have serious consequences. Furthermore, many DNA viruses only transform cells when they are in an abortive infection state. Comprehending the reason behind a virus's abortiveness can be crucial to comprehending and elucidating the progression of virus replication and the consequences of virus infection on the host. Some questions important to characterizing abortive infections are the following:

- Is the virus genome lost?
- Is part of the genome maintained and expressed?
- If the genome is maintained, is it physically integrated into the host genome, or is it maintained as a separate "mini-chromosome" or episome?

Between the productive and abortive extremes are other forms of infection. For instance, a cell may be a poor host for a particular virus's replication but not completely nonpermissive; this type of cell is frequently referred to as semipermissive. Obviously, the terms "permissive" and "semipermissive" for virus replication are relative and there is no actual, hard cut threshold for them.

There are additional cell-based barriers to viral replication. Mutations in viruses that become lethal only in specific circumstances, like high temperatures, are known as conditional lethal mutations (temperature-sensitive mutations). It is possible for dynamic situations to arise where a virus is gradually released at low levels over time. A chronic, inapparent, or persistent infection can be identified by such a circumstance. Such an infection in an animal or in a cell culture may, in certain circumstances, result in periods of high viral production accompanied by clear disease or cell death. Periods when the host (or cell culture) appears relatively healthy and the virus is hard to detect can alternate with these episodic events.

In certain instances of infection, numerous viruses can create unfinished viral particles that can infect other cells. These incomplete particles are known as defective virus particles and can be generated through various methods, usually occurring during inefficient stages of virus maturation near the end of the infection cycle as the host cell is deteriorating rapidly from virus-induced alterations.

Aside from defective particles forming from empty capsids being packaged, viruses can also generate partial genomes randomly during replication. If these partial genomes have a packaging signal, they can be enclosed, creating a distinct type of defective particles. When a cell is infected with one of these particles, the infection will not be successful as the genome is incomplete.

Interestingly, when a defective virus particle is infected simultaneously with an infectious one, interference can occur. This is because the smaller, fragmentary genome of the defective particle can replicate more quickly than the complete genome, leading to a reduction in the yield of infectious particles. This phenomenon is called a mass effect. Defective interfering particles are classified as such because they can impede the production of high yields of viruses in a research setting. However, these particles can also be utilized to deliver genes to cells in specific circumstances.

It is important to note that herpesviruses and lentiviruses like HIV-1, along with other viruses, have the ability to remain latent in the host's cells, where the viral genome is present but no active virus can be detected.

7.1.2. Fate of the Cell Following Virus Infection

7.1.2.1. Cell-Mediated Maintenance of the Intra- and Intercellular Environment

Cells can change their growth properties through prolonged passage in culture or through mutations in the genetic information they carry. Although such alterations within the animal may result in tumor formation, this is not usually the case. This is because the bodies of vertebrates and the cells that make them up have a variety of "check points" that react to changes in the DNA of specific cells. One of the main purposes of MHC-I mediated antigen presentation is this. A cytotoxic T lymphocyte destroys a cell when an aberrant epitope from a genetically damaged protein that is normally not expressed is displayed on the cell's surface. The apoptotic pathway causes the cell to die as a result of this interaction. Certain cellular genes cause apoptosis, which is a gradual termination of cellular activity and eventual cell death. By causing the death and elimination of highly differentiated cells that are no longer needed (such as immune system effector cells), aging cells, and cells with mutations in genes that normally limit cell division, the process has a protective effect on the body. It is crucial to realize that the apoptotic pathway is a tightly controlled process that only removes cells from the tissue in question that are no longer useful, leading to cell death without releasing cellular contents to the immune system and consequent inflammation and possible pathology. The other main mechanism of cell death, necrosis, should be compared to the apoptotic pathway. In necrosis, the targeted cell swells and bursts, causing inflammation that triggers the immune system. Figure 7.3 contrasts and schematically outlines the two processes (Celis et al., 1998).

A virus that is replicating in a cell needs to be kept in the cell long enough to produce an adequate amount of virus while also preventing the immune system from reacting too strongly to the infection. On the other hand, eradicating contaminated tissue and mounting a controlled immune response as soon as possible is advantageous to the cell and the organism made up of such cells. The conflict between these two processes causes evolutionary change in the virus and the host, and how these processes manifest themselves cause both microscopic and macroscopic changes in virus-infected cells, which is what is known as cytopathology.

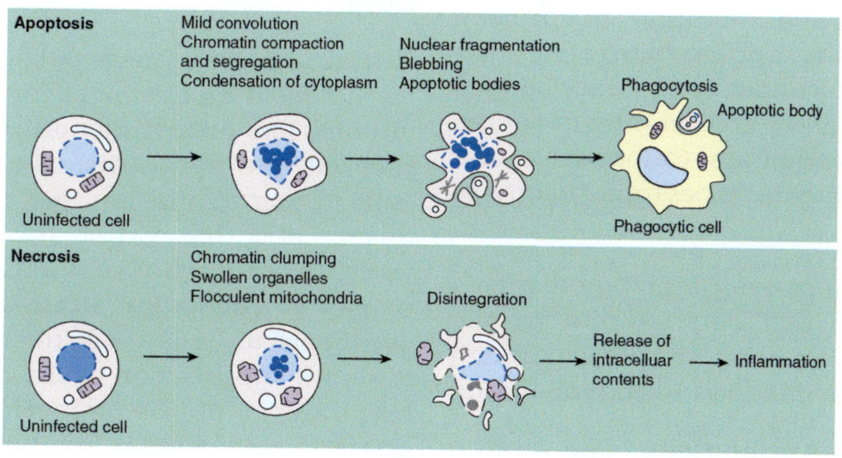

Figure 7.3. *Apoptosis vs necrosis in cell death.*

Source: Edward K. Wagner, Martinez J. Hewlett, David C. Bloom, David Camerini, Basic Virology Third Edition, ISBN 1-4051-0346-9.

7.1.2.2. Virus-Mediated Cytopathology – Changes in the Physical Appearance of Cells

Certain common virus-induced host cell alterations (cytopathology) produce changes easily seen with the naked eye or under a low-power microscope. Every type of cytopathology necessitates a particular interaction between the cell and viral gene products. Even the "exploding" poliovirus or bacteriophage-induced cell lysis is caused by extremely specific alterations to the cell's plasma membrane and lysosomes that are brought on by particular poliovirus gene products. The development of nuclear inclusion bodies, which are indicative of herpesvirus infections, chromosomal abnormalities, and cytoplasmic inclusion bodies, which are diagnostic for poxvirus infections, are less dramatic but still obviously detectable alterations to the cell (Celis et al., 1998)

Cytopathology can encompass changes in cell morphology, growth, and lifespan induced by viruses. Even minor alterations like a virus-induced change in protein expression or the appearance of a new macromolecule fall under cytopathic changes as long as they can be consistently observed with a reliable technique. Cell death is not always a requirement for cytopathology to occur.

A major type of cytopathology involves changes to the cell surface due to the expression thereof viral proteins. Among other things, this can lead to the following:

- **Altered Antigenicity:** The altered cell will stimulate the immune system to generate antibodies to react with viral proteins or previously masked cellular proteins
- **Hemagglutination or Hemadsorption:** Certain viral proteins will stick to red blood cells and cause these cells to stick together
- **Cell Fusion:** Changes to the cell membrane can allow the formation of large masses of fused cells or syncytia. Specific virus gene products are responsible for this. Such fusion induced by the Sindbis virus (a togavirus) can be used to generate somatic cell hybrids.

Another significant form of cytopathology revolves around alterations in cell morphology. For instance, Figure 7.4 illustrates how an HSV-1 infection disrupts the cytoskeleton of the host cell, ultimately altering the cell's shape. In this scenario, viral infection caused the actin fibers to dissociate without degrading the actin itself. This particular biochemical change in the actin subunits leads to significant transformations in the cell's morphology.

7.1.2.3. Virus-Mediated Cytopathology – Changes in the Biochemical Properties of Cells

A virus infection causes particular alterations to the cell's biochemical functions. Certain viruses, like poliovirus and HSV, specifically prevent the synthesis of proteins within cells. Such inhibition has a complicated mechanism that varies depending on the virus. Additionally, a viral infection may cause a particular inhibition of mRNA synthesis in cells. Cell death results from a severe suppression of the macromolecular metabolism of the cell. However, subtle alterations in cellular functions lead to virus production rather than cell death, resulting in complex and multifaceted effects of virus infection on cell function (Celis et al., 1998).

Viral inhibition of apoptosis is a remarkable illustration of how some DNA tumor viruses can evade cell death long enough to enable effective virus replication. As was previously mentioned, modifications to the growth characteristics and life span of virus-infected cells are another significant effect of infection. Tumor suppressor genes, which are named after their ability to prevent the formation of tumors, regulate the growth rate, total number, and lifespan of differentiated cells. To summarize, DNA-tumor viruses impede tumor suppressor genes in order to "activate" the cell for their own replication. This is sufficient information for the purposes of this discussion. The induction of apoptosis would impede the virus's ability to replicate within the cell. Different tumor viruses have different mechanisms of transformation, but in many cases, the inactivation of cellular genes that actively inhibit cell division and the inhibition of apoptosis induced by a particular virus are both significant factors.

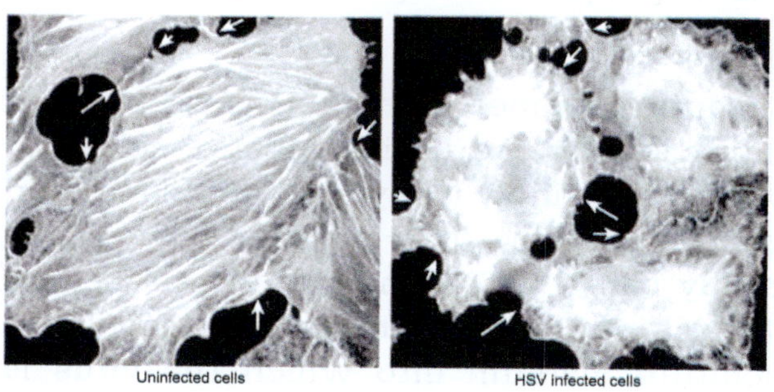

Figure 7.4. *HSV-induced changes in the properties of actin microfilaments of a cultured monkey fibroblast.*

Source: Edward K. Wagner, Martinez J. Hewlett, David C. Bloom, David Camerini, Basic Virology Third Edition, ISBN 1-4051-0346-9.

Interaction between the infected cell and the immune system of the host is another important effect of virus infection. Early on in infection, some viruses, like HSV, have the unique ability to prevent MHC-I (major histocompatibility complex class I)-mediated antigen presentation. This early suppression of antigen processing can give the virus a crucial advantage in infection, even though the cell will eventually express viral antigens as the infection progresses.

When cells are infected with a virus, they can express new genes or increase the expression of existing genes in order to cause a variety of distinct cellular responses. Numerous methods in contemporary molecular biology enable highly accurate identification of the cellular genes activated by viral infection.

One technique is called differential display analysis, and it involves amplifying complementary DNA copies of cellular transcripts using the polymerase chain reaction (PCR), groups of oligonucleotide primers, and retrovirus reverse transcriptase. Changes in the transcription levels of particular cellular genes can be identified by comparing the amplification patterns of products isolated from infected and uninfected cells. Additional techniques utilize microchip technology, where many oligonucleotide probes specific to different cellular genes are attached to a small microarray. These probes are then hybridized with PCR-amplified cDNA samples created from mRNA extracted from both uninfected and infected cells, which have been labeled with distinct fluorescent dyes. Changes in cellular transcription can be identified by comparing the emitted light patterns when the microchip is scanned with a laser beam.

7.2. MEASUREMENT OF THE BIOLOGICAL ACTIVITY OF VIRUSES

7.2.1. Quantitative Measure of Infectious Centers

7.2.1.1. Plaque Assays

Most viruses have cytotoxic effects on their host cells, which result in visible harm or alterations to the cells they replicate in. The alteration of cells in a local area caused by a localized virus infection can easily be seen as a plaque or focus of infection, even if no cells are killed or lysed. A single biologically active virus can cause such a localized infection given the right conditions and dilutions. A plaque-forming unit (PFU) is a virus particle that can start a productive infection.

Plaque formation is a simple process to imagine. Numerous viruses are released by the first infected cell. The viruses will remain close to the initial infected cells and infect nearby cells if they are prevented from spreading widely (they are large compared to even the most complex molecules in the growth medium). This procedure is carried out several times. A single PFU infection will only cause cytopathology in the vicinity of the original infected cell as long as the virus–cell interaction is maintained localized, which is often achieved by making the cell culture medium into which the virus is released highly viscous. It is easy to see and count the resulting plaque a few days after infection. Figure 7.5 illustrates some plaques on cultured cells (Celis et al., 1998).

There are other methods besides cell cultures to obtain infectious centers. Visible centers of infection are created when a plant virus infects a leaf of a susceptible plant and is stimulated by some sort of mechanical abrasion (perhaps rubbing the leaf with carborundum powder). Additional examples can be found in Figure 7.5.

It is possible—indeed, required—to use the chorioallantoic membrane of developing chick or duck embryos for viral assays. Fertilized eggs are carefully opened to reveal the membrane (the embryo is inside the egg) after two weeks of incubation in this assay. After that, the membrane is covered with the virus suspension, the egg is sealed again, and the virus plaques or pocks are allowed to grow.

7.2.1.2. Generation of Transformed Cell Foci

It is possible to quantify additional biological consequences of viral infection using the same titration principles. Certain DNA viruses can change cells' normal ability to control growth under specific circumstances.

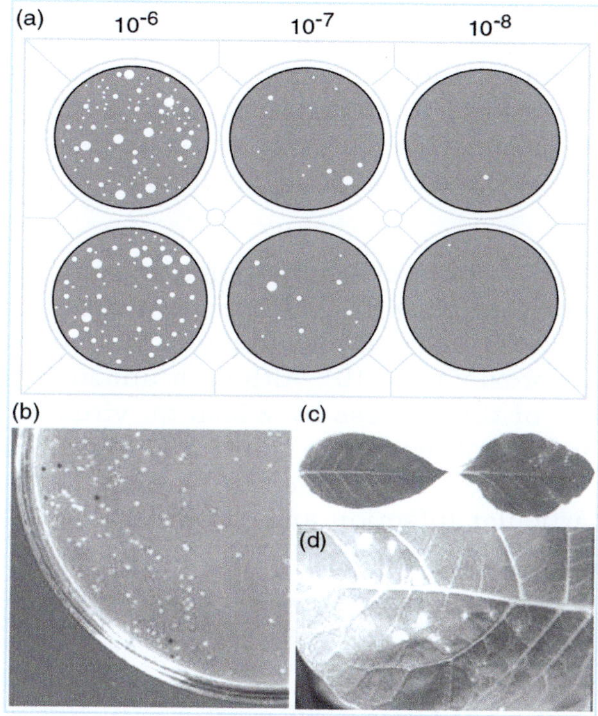

Figure 7.5. *Visualization of virus plaques. (a) A continuous line of monkey cells (Vero cells) was grown in the six-well culture dish. When the cells reached confluence, they were infected in duplicate with a series of 10-fold dilutions of an HSV-1 stock. After 48 hours, the cells were partially dehydrated (fixed) with ethanol and stained. Areas of cell death show as white plaques, each representing a single infectious event with the input virus solution. (b) A portion of the surface of a Petri dish containing agar with bacterial nutrient medium. A "lawn" of E. coli was grown on the plate's surface, and this layer of cells was infected with a solution containing a genetically "engineered" version of bacteriophage λ that can be used to clone inserted genes. Bacteriophages that contain an inserted gene form clear plaques due to the inactivation of an indicator gene (β-galactosidase) and viruses without the insert form dark-colored plaques. (c) Assay of tobacco mosaic virus (TMV). Leaves of a resistant (left) and a susceptible (right) plant that have been infected with small amounts of virus. (d) A higher magnification of plaque development*

Source: Edward K. Wagner, Martinez J. Hewlett, David C. Bloom, David Camerini, Basic Virology Third Edition, ISBN 1-4051-0346-9.

The morphology of transformed cells is distinct, and they frequently outgrow normal cells to create clusters of multiplying cells. Every infectious event will result in the formation of transformed cell clumps, known as a focus of transformation, even if it is abortive and does not produce a new virus. Figure 7.6 depicts an example of a focus on transformed cells. It is easy to see the focal changes in cell morphology, which are a kind of cytopathic effect of transformation. Similar to PFUs, the number of focus-forming units can be counted; however, in this case, the spread of transformed cells, not the spread of the virus, is being counted.

7.2.2. Use of Virus Titers to Quantitatively Control Infection Conditions

There are two key definitions related to infectious virus particles or PFUs. The particle-to-PFU ratio quantifies the ratio of total virus particles to infectious virus particles. To calculate this ratio, the total virus particles must be counted, and an assay must be conducted to determine the number of biologically functional particles (Celis and Olsen, 1998).

PFU ratios for some virus species—bacteriophages and, in extremely rare cases, poliovirus, for instance—are very close to 1. Although viruses like adenovirus and HSV can have ratios of less than 10 in carefully prepared preparations, the optimal ratios for influenza viruses are approximately 10^3. Although unusual, the high particle-to-PFU ratio is a natural byproduct of the process by which flu virus virions are created. Every virus type has a characteristic optimal value that indicates the degree of encapsidation and release of infectious virus from infected cells. Particle-to-PFU ratios can vary based on the specifics of the particular infection and virus.

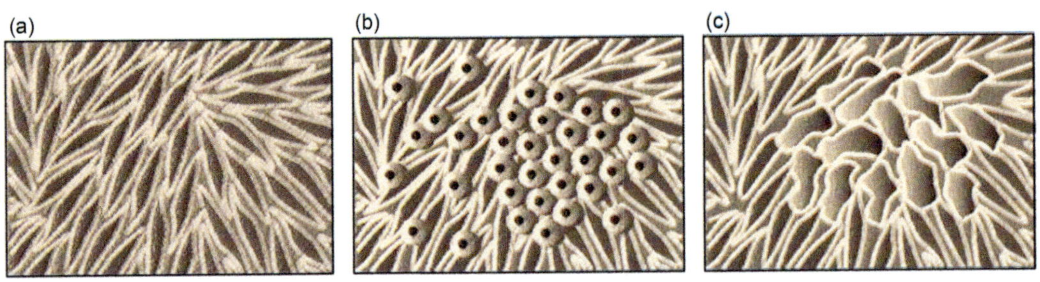

Figure 7.6. *Some representative morphologies of rat fibroblast cells (F-111) infected with different transforming viruses. (a) Shows normal cells with their characteristic parallel orientation. A focus of transformed cells generated by infection with Rous sarcoma virus (an oncornavirus) is shown in (b). Note the rounded morphology and density of these cells. (c) Shows the subtle difference in morphology when normal F-111 cells are infected with SV40 virus, for which they are nonpermissive (Based on portions of a photograph in Benjamin J, Vogt PK. Cell transformation in viruses.*

Source: Edward K. Wagner, Martinez J. Hewlett, David C. Bloom, David Camerini, Basic Virology Third Edition, ISBN 1-4051-0346-9).

The multiplicity of infection (MOI), which is just the average number of PFUs per cell used in the initial infection, is a second quantitative indicator of the circumstances surrounding a virus infection. If 10^6 cells were infected at a MOI of 1, one would need to add 10^6 PFUs of virus. An MOI of 1 indicates that 1 PFU per cell. It is crucial to remember that a MOI can range from zero to a very high value, contingent upon the virus concentration in the initial stock, the nature of the experimental problem under investigation, and other factors. MOI is an average value; statistical analysis shows that when a culture is infected at MOIs greater than 0.1 or so, the number of PFUs interacting with a single cell can vary over a large range (Celis and Olsen, 1998).

7.2.2.1. Examples of Plaque Assays

In the plaque assays depicted in Figure 7.5(a), the cell culture dishes contained 10^6 cells. Therefore, the MOI used to produce the average of 40 plaques observed in the 10^7 dilutions shown was calculated as follows: PFU/cell = $40/10^6 = 4 \times 10^{-5}$. It is clear that when plaques are easily countable, the MOI must be low, and any cell initiating a focus of infection or plaque must have been infected with only 1 PFU. This demonstrates that with normal (wild-type [wt]) animal viruses, a single viral genome delivered to the correct location in the cell is adequate to carry out the entire infection. Moreover, an excessively high MOI may hinder the replication process because the particle-to-PFU ratios can increase quickly with a high MOI. This can occur through the production of defective interfering particles, as discussed earlier.

A virus stock is serially diluted, and aliquots of each dilution are added to a culture dish in order to perform a plaque (or focus) assay. After allowing the plaques to mature, they are tallied. The initial number of PFUs in the solution can be obtained using basic arithmetic.

An illustration of HSV plaque development on Vero cells can be found in Figure 7.5(a). Duplicate serial 10-fold dilutions were added to different plates. Following adsorption, the cells underwent rinsing and were coated with a unique overlay medium designed to prevent the virus from spreading to adjacent cells. After being incubated at 34°C for 48 hours, the cells underwent rinsing, fixing, and staining. Plaques are what are visible. About that, many PFUs were added to each plate at that virus dilution, based on the average number of 40 plaques in the 107 dilutions.

Another example is shown in Figure 7.7. Here, a 100-ml stock of HSV was diluted and infectious units were measured by plaque assay as shown in Table 7.1.

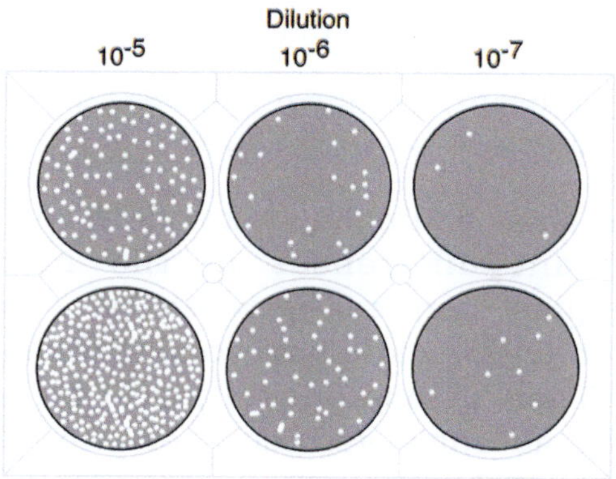

Figure 7.7. *Serial 10-fold dilutions of HSV to determine the titer of the virus in a stock solution. The details of the infection are as described in the legend to Figure 7.5(a), and the calculation of the titer is shown in Table 7.1.*

Source: Edward K. Wagner, Martinez J. Hewlett, David C. Bloom, David Camerini, Basic Virology Third Edition, ISBN 1-4051-0346-9.

One can readily calculate that the original stock was about 6×10^7 PFU/ml or 6×10^9 total units of infectious virus PFU. The following formula is useful to make the calculation:

$$V_f = V_o / D$$

V_f is the final concentration of PFU (units/ml), Vo is the original concentration, and D is the dilution factor.

Table 7.1. *An Example of a Set of Dilutions for a Plaque Assay*

Operation	Dilution of Stock	Plaques Per Dish
0.01 ml of stock diluted into 10 ml of buffer	103	Too many to count
1 ml of the above diluted into 10 ml of buffer	104	Too many to count
1 ml of the above diluted into 10 ml of buffer	105	500–1000 (estimated)
1 ml of the above diluted into 10 ml of buffer	106	(20 +100)/2 = 60
1 ml of the above diluted into 10 ml of buffer	107	(3 +8)/2 = 5.5
1 ml of the above diluted into 10 ml of buffer	108	0
1 ml of the above diluted into 10 ml of buffer	109	0

Source: Edward K. Wagner, Martinez J. Hewlett, David C. Bloom, David Camerini, Basic Virology Third Edition, ISBN 1-4051-0346-9.

In this example, take note of the significant differences in the number of plaques counted between two plates infected with an identical volume of diluted stock. There is inherent statistical variation as well as some variation resulting from experimental error, as it is not possible to predict exactly how many virus particles will be present in a small volume at any given time. This kind of variation is unavoidable when working with samples that contain few particles.

7.2.2.2. Statistical Analysis of Infection

The actual PFU number infecting any given cell will vary greatly at low and moderate MOI values due to chance statistics. At an MOI of 2, for instance, a considerable portion of cells will show no virus, while a greater portion will show 1 PFU. A smaller number of cells will receive three PFUs, while other cells will receive four, five, or more PFUs. A statistical technique originally created for the analysis of gaming results can be used to determine the proportion (or probability) of any given cell being infected with any given number of PFUs (Celis and Olsen, 1998). This is the Poisson analysis, which describes the distribution of positive results in a low number of trials as

$$P_i = (m^i e^{-m})/i\,!$$

P_i is the probability that a cell will be infected with exactly i number of virus and m is the MOI (average number of PFUs added per cell). Using this equation, one can always calculate the probability of a cell being uninfected and, thus, the number of uninfected cells (if you know the number of cells in the sample). Since m^0=1 and 0! ≡1:

$$P_0 = e^{-m}$$

For the MOI of 2 mentioned previously, the proportion (probability) of cells getting i number of PFUs is:

$$P_0 = e^{-2} = 0.135$$

$$P_1 = 2e^{-2} = 0.27$$

$$P_2 = 2^2 e^{-2}/2 = 0.27$$

$$P_3 = 2^3 e^{-2}/6 = 0.18$$

$$P_{4\text{ or more}} = 1.0 - (P_{0\text{–}3}) = 0.14$$

One gets the last number ($P_{4\text{ or more}}$) because the total probability of a cell being infected with no PFUs or any number of PFUs must be 1.0.

7.2.3. Dilution Endpoint Methods

A certain point in the dilution of any virus stock is reached in the hemagglutination inhibition (HI) and hemagglutination (HA) assays, below which a desired property cannot be observed. There is a good chance that an aliquot, or small measured sample, taken after a virus stock has been sufficiently diluted will not contain any infectious virus. The virus has not been eliminated; rather, it has been diluted to the point where it is far below, say, 1 PFU/ml, meaning that there is no virus in any one milliliter.

Since virus stocks can be diluted to the point where any given sample typically contains no PFUs, infection can be measured through dilution rather than titration. This method, known as an endpoint dilution assay, is often referred to as a quantal assay as it involves statistical analysis rather than quantitative measurement. In this assay, a certain number of subjects (such as animals or cell culture wells) are infected with increasingly diluted virus samples and then monitored for signs of illness, death, or cytopathicity (Celis and Olsen, 1998).

Localizing plaques in a quantal assay is not required. One can estimate a virus dilution such that half of the aliquots in that dilution contain the virus and the other half do not by plotting log dilution versus the percentage of infected subjects. This endpoint is known as the ID50 (median infectious dose) or LD_{50} (median lethal dose) in an animal disease assay. It may be referred to as $TCID_{50}$ (median tissue culture infectious dose) for the purpose of measuring gross cytopathology in tissue culture wells.

7.2.3.1. The Relation Between Dilution Endpoint and Infectious Units of Virus

Quantal endpoints relate to the average number of PFUs in the aliquot and are essentially a measure of the infectious virus's dilution. Figure 7.8 depicts a quantal assay in action. Equal aliquots of a diluted HSV stock were added to each well of a 48-well culture plate. The black wells indicate a virus infection (CPE). A table like Table 7.2 can be created for the titration using the tabulated data.

The graph indicates that a dilution at which approximately 50% of the wells would be infected is roughly 4×10^3; consequently, the original sample's $TCID_{50}$ was 4×10^3. A number of basic statistical texts describe statistical methods, such as the method of Reed and Muench, which can be used to obtain more accurate measures of the ID_{50} of a virus stock.

Despite being a dilution measure, ID_{50} directly correlates to PFU; one ID_{50} unit measures the dilution needed to guarantee that 50% of the aliquots in that dilution contain infectious viruses. This will only happen if each aliquot has an average of 0.7 PFU, or in the example above, 7 PFUs in 10 ml.

This conclusion can be drawn from some rough arithmetical calculations: if the initial concentration of PFUs per milliliter (X PFUs/mL) was diluted by a factor D such that there is a 50% chance of PFU infection in each animal or tissue culture well, then the final virus concentration defines a type of multiplicity of infection (called m) where the chance of a positive infection is 50%. This value (m) should have the dimensions of units of infectivity in a standard volume (here, 1 ml). Then:

$$P_0 = 0.5 = e^{-m} = 0.7 \text{ PFU/ml}.$$

Table 7.2. *An Example of a Quantal Assay for Virus Infectivity*

Sample Dilution	Log Dilution	No. of Infected Wells	Total No. of Wells	% Infected
None	0	100	100	100
1/1000	3	39	48	81
1/10,000	4	14	48	29
1/100,000	5	3	48	6
1/1000,000	6	1	100	0

Source: Edward K. Wagner, Martinez J. Hewlett, David C. Bloom, David Camerini, Basic Virology Third Edition, ISBN 1-4051-0346-9.

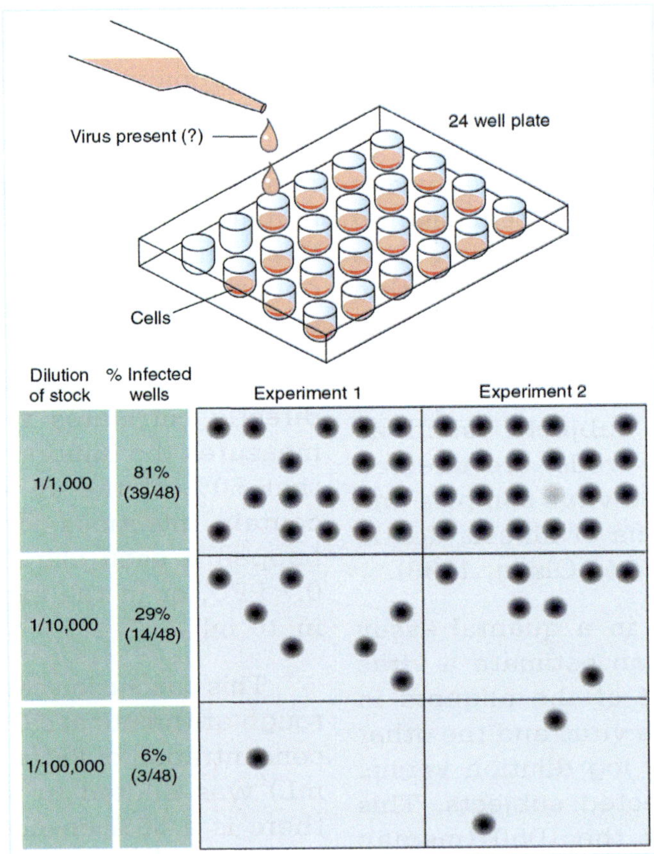

Figure 7.8. *Quantal (endpoint dilution) assay of HSV in tissue culture wells. Replicate cultures of rabbit skin fibroblasts were grown to a density of about 5 ×104 cells per well of a 24-well tissue culture plate. Aliquots of the indicated stock virus dilutions were pipetted into the cultures, and the plate was incubated for 48 hours. The plate then developed a stain that indicated black for virus-infected cells. Any well that received at least 1 PFU of virus stained black (two separate experiments are shown). The percentage of positive (infected) wells is shown at each dilution.*

Source: Edward K. Wagner, Martinez J. Hewlett, David C. Bloom, David Camerini, Basic Virology Third Edition, ISBN 1-4051-0346-9.

7.3. VISUALIZATION AND ENUMERATION OF VIRUS PARTICLES

The majority of viruses are submicroscopic physical particles; although the largest can be identified with an ultraviolet (UV) light microscope, other techniques are needed for detailed visualization. An important driving force behind the application of physical and chemical methods to the study of viral structural properties, including their distinct sizes and shapes, is the development of these techniques for the study of biological processes as a whole.

To conduct a meaningful physical study of virus particles, an investigator needs to know the number of virus particles in a sample and how the sample relates to the biological properties of the virus (measured in other ways).

In the end, counting viruses comes down to being able to see them, which takes unique methods that were not developed until shortly before World War II. Among the most notable of these is the electron microscope (EM), whose creation necessitated an advanced understanding of contemporary electrical and mechanical engineering and particle physics. Without the electron microscope, much of the advancements in molecular biology and medicine that are now taken for granted would not have been possible. It has given scientists unprecedented insight into the workings of cells and biological processes (Cristofalo et al., 1998).

7.3.1. Using the Electron Microscope to Study and Count Viruses

Viruses can only be seen at shorter wavelengths with EM because their dimensions are smaller than those of visible light. Electrons are accelerated to high energy and magnetically focused by the EM. The electrons have a short wavelength due to their high energy, making them substantially "smaller" than the virus particles. Big proteins, RNA, and DNA can all be seen with the EM.

The energy required to achieve short wavelengths presents a challenge, notwithstanding the high resolving power of the EM. The majority of biological materials are transparent to high-speed (short-wavelength) electrons because of their high penetration rate. Therefore, viruses are usually coated or stained with a heavy metal, like osmium or platinum, to be seen. The fundamental protein arrangements and viral structure are preserved during this coating or staining process. The specimen is then exposed to electrons, allowing the particles to be seen on a fluorescent screen. Because the areas are photographed in negative, areas where the heavy metal prevents electrons from passing appear dark on the screen but appear white (light) in prints. Because specimens must be examined in a vacuum due to the physics of electron acceleration and focus, the sample needs to be fixed and completely dry. Because preparation for the EM may change or eliminate minute effects of protein hydration on the arrangement of the polypeptide chains, for example, the EM picture is only a representation of structure. Additionally, sample preparation means that moving objects can only be seen by the EM as frozen in time. Care must be taken when interpreting images showing the virus's entrance, exit, or changes to the infected cell. It is never certain if a virus is biologically active (capable of replicating) or if the observed process is the exact one causing biological effects. (Maunsbach, 1998)

The process of "shadowing" a virus particle with heavy metal is shown in Figure 7.9. Such a shadow-cast can provide exquisite detail of the geometry of the virus. Much of the early development of shadowing and visualization methods of viruses was carried out by Robley C Williams at the University of California, Berkeley.

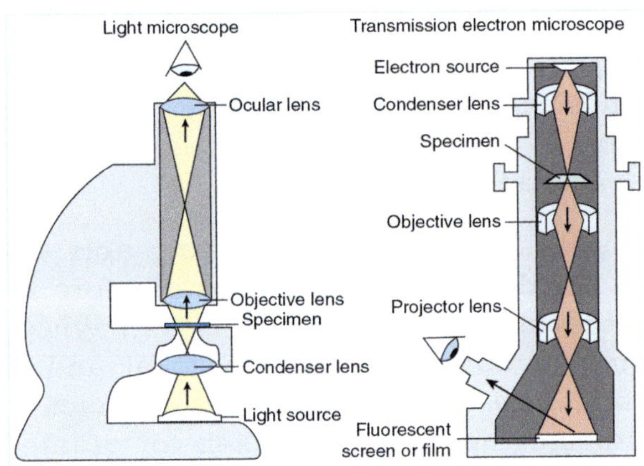

Figure 7.9. *A schematic comparison of light and electron microscopes. The principles behind the focusing of the image are similar except that magnetic fields must be used to focus electrons. The higher energy of the electrons accelerated through high voltage produces very short wavelengths, resulting in high resolving power.*

Source: Edward K. Wagner, Martinez J. Hewlett, David C. Bloom, David Camerini, Basic Virology Third Edition, ISBN 1-4051-0346-9.

Subtle staining techniques heavy metals to protein molecules can yield even more detailed results. Additional shadowing techniques, like carbon shadowing, can also add more detail. By using computer image enhancement, features that are hidden in traditional EM can be resolved, and apparent resolution can be further dramatically increased.

Cryoelectron microscopy is a technique that was developed to circumvent the issues of structural deformation of particles that arise during preparation for conventional electron microscopy, particularly with enveloped viruses. Because this technique does not use any stains or heavy metal shadowing, the particles are better preserved. Rather, virus particles are quickly frozen on the EM grid, ensnaring them in a thin layer of vitreous ice—ice that is incapable of forming large crystals. Unlike the metal ion-stained and dried specimens of conventional EM, the particles are hydrated in what is probably a more normal state within this glass-like matrix. The frozen-hydrated particles are imaged by utilizing the difference in electron density between the surrounding water matrix and the protein or lipid, as no stains are used.

The specimens are examined under a microscope with a cold stage to preserve the ice structure under vacuum. Data are gathered using a very low dose of electrons to minimize damage from the beam's intensity to prevent unintended changes. Computer techniques similar to those used to improve the resolution of X-ray diffraction information can also be applied to the observed images. These methods resulted in the HSV capsid image displayed in Figure 7.11(a). Using this technique, the structure of the recently identified Mimivirus, which has a massive 1.2-Mb genome, has also been determined. The inner virion and genome of the capsid are surrounded by two lipid envelopes, just like those of the distantly related poxviruses. An outer protein shell then encloses these membranes (Figure 7.10).

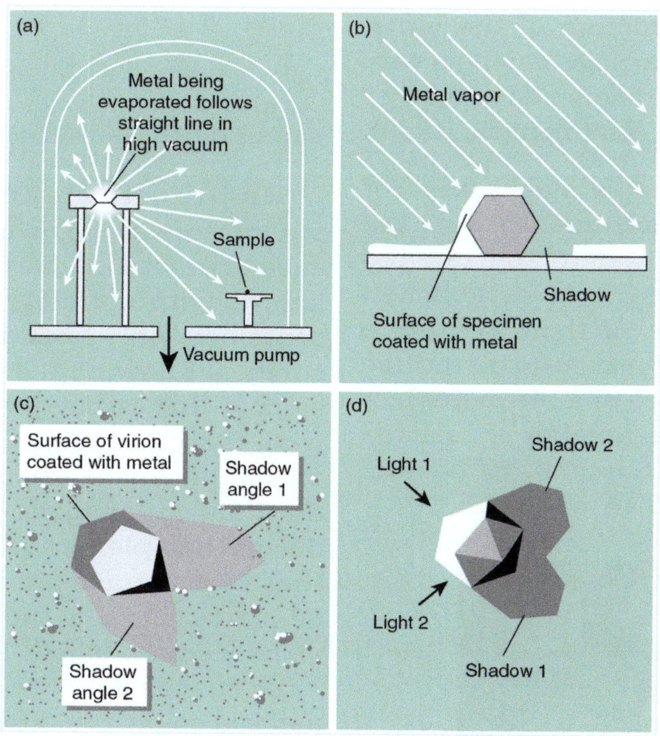

Figure 7.10. *Shadowing specimens for viewing in the electron microscope. (a) A sample of heavy metal is vaporized in a vacuum chamber. This vapor travels in straight lines from the source and forms a layer on all surfaces in its path. (b) Any object in the path will cast a shadow on the grid on which it is supported. (c) A double-shadowed virus in the electron microscope. (d) An icosahedral model is placed in two light beams to show the equivalence of the shadows.*

Source: Edward K. Wagner, Martinez J. Hewlett, David C. Bloom, David Camerini, Basic Virology Third Edition, ISBN 1-4051-0346-9.

7.3.1.1. Counting (Enumeration) of Virions with the Electron Microscope

Virus particles are countable since they can be separated and seen. While a count of particles in a solution free of contaminating cellular material is highly helpful, it does not reveal the proportion of infectious (biologically active) particles. Once the total number of particles in a solution is known, the amount of genome per particle and, consequently, the genome size can be calculated by measuring the total nucleic acid (genomes). Once more, one can determine the precise quantity of protein per capsid by using particle number, which enables one to determine specifics about the structure of the virus. Lastly, counting virus particles can be a very helpful medical tool for diagnostic and other purposes.(Figure 7.11) (Maunsbach, 1998).

While visualization is necessary for all counts, the measurement of these parameters can be linked to the number of virus particles once it is established that a specific number of them contain a particular amount of an enzyme (e.g., reverse transcriptase in the case of a retrovirus), or interact with a specific number of test red blood cells (hemagglutination), or contain a specific amount of DNA or RNA.

(a)

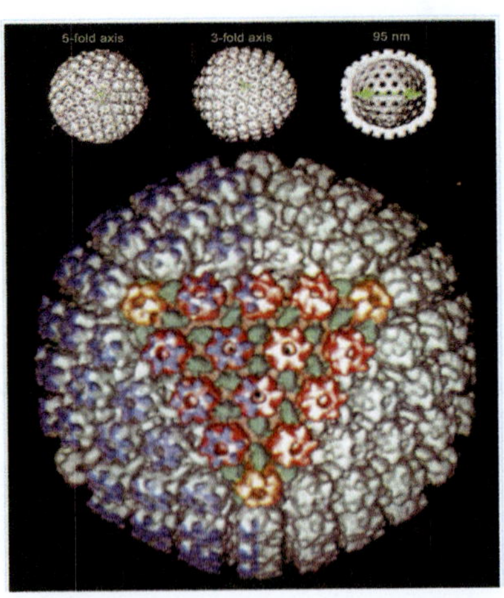

(b)

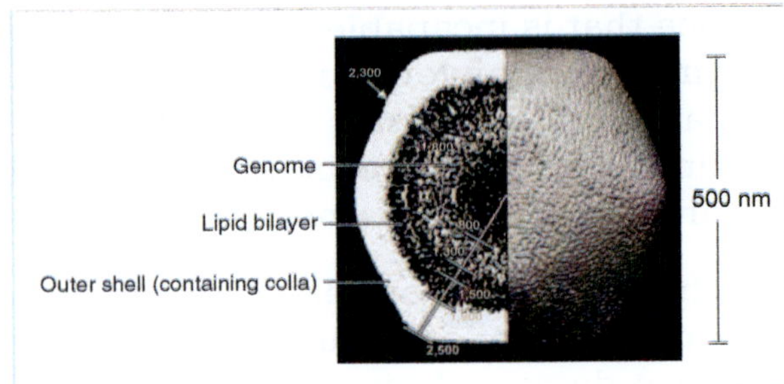

Figure 7.11. *Computer-enhanced three-dimensional reconstruction of viral capsids using cryoelectron microscopy. (a) The HSV-1 capsid reconstructed from 60–80 images, which provides a resolution on the order of 2.5 nm, but many more can be combined, (b) A schematic rendering of the cryoelectron microscopic structure of the Mimivirus genome reconstructed to 7.5 nm resolution*

Source: Edward K. Wagner, Martinez J. Hewlett, David C. Bloom, David Camerini, Basic Virology Third Edition, ISBN 1-4051-0346-9.

In theory, particle counting is straightforward. For instance, it would be easy to determine particle number if it were known with certainty that every EM field contained virus from a particular volume of solution. Knowing the portion of the original sample that is being used for visualization is all that is needed. This fraction depends on the observed sample's volume and any dilution procedures that were applied during sample preparation. In the event that an average microscopic field contained 30 particles and the volume of solution visualized matched 10^{-4} ml of the initial virus suspension, the original suspension's particle count could be approximated at 3×10^5 ($30/10^{-4}$) particles per milliliter. However, this is not a particularly accurate way of measuring particle concentrations. The problem comes from the fact that despite the basic simplicity of the approach, it is difficult to achieve careful dilution and even spread of virus in the field of view of the EM, and many artifacts can arise

(Maunsbach, 1998). By including a known quantity of a standard, like uniformly sized latex beads, in the initial suspension, some uncertainties can be reduced. After that, it is possible to count the quantity of beads and particles in the EM field. The number of particles in the initial suspension can be calculated using the ratio of these and the number of beads used to create the solution. A variety of distinct virus preparations can be processed using the same procedure because it is simple to add a predetermined number of beads from a standard solution.

7.3.2. Atomic Force Microscopy –Visualization of Viruses and Infected Cells, Potentially in Real Time

Although three-dimensional structural information can be obtained using an electron microscope, this method is only an average one; highly detailed structures are dependent on the entire population of particles that are observed. This is an intrinsic drawback of X-ray crystallography, one of the highest-resolution methods for studying molecular structures. Furthermore, these techniques necessitate a great deal of sample fixing and preparation, and only cautious statistical analysis can be used to infer subtle information about the properties of individual virus particles and their structures within a population, as well as dynamic changes. Larger viruses produced from cryoelectron microscopy and X-ray crystallography may, therefore, be somewhat misleading and not totally representative of the population despite their grand symmetry and apparent perfection (Pawley and Centonze, 1998).

When molecules interact at incredibly close (quantum scale) distances, electrons can "tunnel" between atoms, creating a tiny but detectable force between them. This is a rather strange property of the interaction. This quantum force has been used in the atomic force microscopy (AFM) technique, which creates a contour map of the surface as the probe is moved over a sample, such as a cell, subcellular component, or virus. The probe is held at a constant tunneling force over the sample. This technique can, in principle, be used on living cells to provide animated real-time analyzes of changes in cellular surface structure as virus infection progresses and requires little to no sample preparation. It presents a useful addition to the previously mentioned methods. Its most significant application is the ability to study the architecture of a single virus particle or a group of different individuals at a resolution that is extremely close to that of cryoelectron microscopy. This technique has been applied to the imaging of virus interactions with cells as well as the capsid structures of viruses in crystals. Figure 7.12 illustrates an atomic force microscope view of a capsid of the herpes simplex virus.

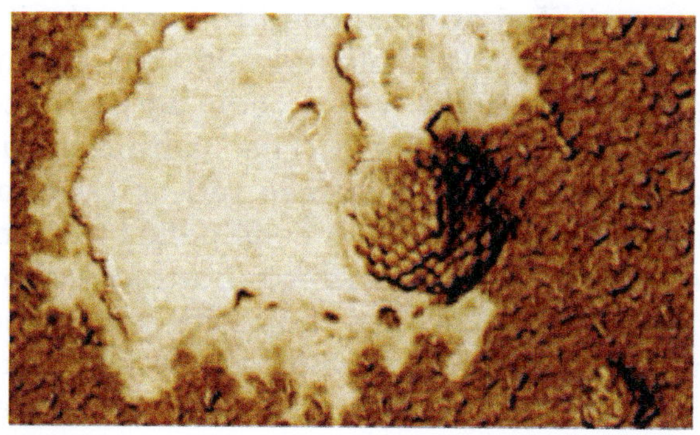

Figure 7.12. *Atomic force microscopy was used to visualize the protein capsid of a herpes simplex virus on a glass substrate.*

Source: Edward K. Wagner, Martinez J. Hewlett, David C. Bloom, David Camerini, Basic Virology Third Edition, ISBN 1-4051-0346-9.

7.3.3. Indirect Methods for "Counting" Virus Particles

It is possible to correlate the quantity of virus particles in a standard solution with other easily quantifiable characteristics of the virion once this information is known. For instance, the number of particles may be connected to the quantity of virus-causing agglutination. Since it takes a specific quantity of virus particles to coat red blood cells and cause agglutination, the ability of many enveloped viruses to agglutinate them can be used as a proxy for virus particle count. The assay's standard operating conditions result in roughly 104 influenza virus particles per hemagglutination unit (HA unit). A HA unit is the minimum amount of virus required to agglutinate the reference sample.

Standard red blood cell samples—chicken or guinea pig red blood cells for influenza viruses—are combined with various dilutions of an unidentified virus stock, which might have come from a patient's serum. Following an appropriate duration, the mixture is softly agitated and exposed to low-velocity centrifugation. Red blood cells that clump together like jelly when they agglutinate are unable to separate. The characteristic of agglutination is a diffuse red or salmon-pink solution. Red blood cells pellet to form a red "button" at the tube's bottom if they do not agglutinate. The simplicity and speed of use of hemagglutination are what make it so beautiful, not accuracy. This is crucial in small clinical laboratories, particularly in developing nations.(Figure 7.13). (Pawley and Centonze, 1998).

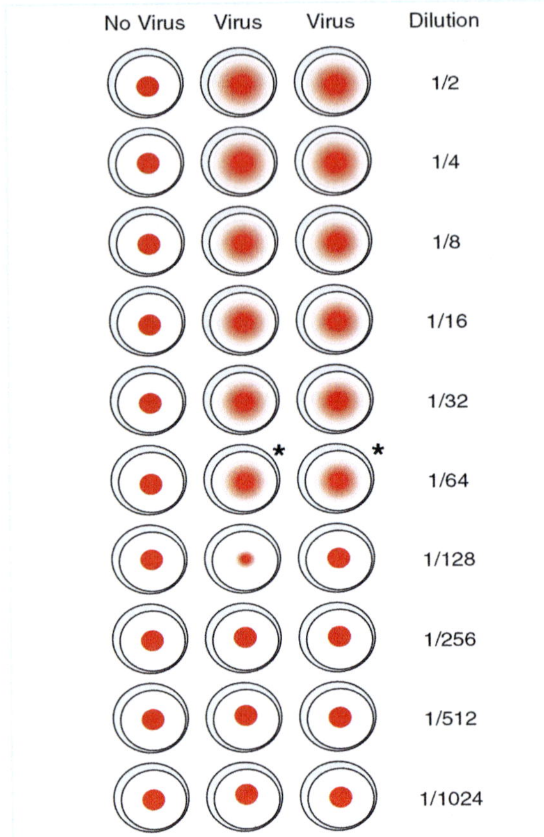

Figure 7.13. *Assay of influenza virus by hemagglutination*

Source: Edward K. Wagner, Martinez J. Hewlett, David C. Bloom, David Camerini, Basic Virology Third Edition, ISBN 1-4051-0346-9.

Particle number can be estimated using similar enzyme tests, but only after the amount of enzyme in each particle is known. For instance, the laboratory can easily assay the enzyme reverse transcriptase, which is present in retrovirus virions. Simple arithmetic can be used to estimate the number of virus particles from the amount of enzyme recovered in an unknown sample. The number of reverse transcriptase units per virus particle is a constant that can only be found once. But keep in mind that the ability to count the particles in the first place is a prerequisite for all of these indirect methods (Osborn, 1998).

CASE STUDY

RNA VIRUSES: A CASE STUDY OF THE BIOLOGY OF EMERGING INFECTIOUS DISEASES

Viruses account for only a small fraction of the 1400 or more species of pathogen that plague humans—the majority are bacteria, fungi, or helminths. However, as both the continuing toll of childhood infections such as measles and recent experience of AIDS and influenza pandemics illustrate, viruses are rightly high on the list of global public health concerns. Moreover, the great majority of newly recognized human pathogens over the past few decades have been viruses, and a large fraction of emerging infectious disease "events" have involved viruses.

There are two kinds of viruses: RNA viruses and DNA viruses. The latter largely consist, with the exception of a handful of pox- and herpesviruses, of viruses that have probably been present in and coevolved with humans for long periods. RNA viruses are very different. The majority of RNA viruses that infect humans are zoonotic, meaning that they can infect vertebrate hosts other than humans. Many of those not regarded as zoonotic are believed to have had recent (in evolutionary terms) zoonotic origins. So, it is the RNA viruses that are of greatest interest in the context of One Health.

In this study, we review current knowledge of how RNA viruses in humans and other vertebrates are related, in terms of both their evolution and their ecology, to try to understand where human RNA viruses came from in the past and where new ones might emerge in the future. Until recently, research on these topics was essentially a series of case studies. Extraordinary work has been done detailing events such as the historical emergence of HIV-1 in Central Africa and the more recent emergence of the Nipah virus in Southeast Asia. But while every emergence event is a fascinating story in its own right, our aim here is to look beyond the specifics and to try to identify any underlying generalities that tell us something useful about the emergence of RNA viruses as a biological process.

Diversity of Human RNA Viruses

The diversity of human RNA viruses was recently surveyed using a formal methodology, and we updated that information here. All RNA viruses known to infect humans were included, with the exception of those only known to do so as the result of deliberate laboratory exposures.

In this study, we use virus species as designated by the Ninth Report of the International Committee for the Taxonomy of Viruses (ICTV) (noting that this differs from earlier ICTV reports used in previous work and that it will doubtless change again in the not-too-distant future). ICTV designations may not always accurately reflect the biological meaning of a "species," i.e., reproductive isolation. The operational criteria used for RNA viruses may include any or all of (i) phylogenetic relatedness based on sequence data, (ii) serological cross-reactivity, (iii) host range, and (iv) transmission route. It is also important to note that any analysis at the level of a virus species implicitly ignores a great deal of biomedically relevant diversity. This point is best illustrated by the influenza A viruses: the epidemiology and public health importance of seasonal influenza A and the H5N1 or H7N9 "bird flu" variants are very different, but all are included within a single species. Less variable virus species than influenza A may still contain multiple serotypes and other functionally distinct subtypes. Despite these limitations, the species remains the most useful unit for studying virus diversity currently available.

RNA Viruses of Humans and Nonhumans

One striking observation is that 160 human-infective RNA virus species (89% of the total) are regarded as zoonotic; i.e., they can also infect other kinds of vertebrate hosts. (The definition of "zoonotic" ignores arthropod vectors; these are regarded as specialized transmission routes rather than alternative host species.) The nonhuman hosts usually (>90% of all zoonotic RNA virus species) include other mammals and less commonly (<40%) birds. Humans rarely, if ever, share their RNA viruses with anything else. Although the bias toward sharing viruses with other mammals is obvious, it is less clear whether we preferentially share viruses with particular kinds of mammals. Many human viruses (both RNA and DNA) are shared with ungulates, carnivores, rodents, primates, or bats (3), but our knowledge of the host range of most viruses is too incomplete for us to be confident about any underlying patterns. The remaining 20 RNA viruses are not known to infect nonhuman hosts naturally. However, most of these have close relatives that can infect other mammals. The only exceptions are hepatitis C, hepatitis delta, and rubella virus.

The overlap between the ability to infect humans and the ability to infect other mammals can be illustrated in other ways, too. Of the 62 recognized RNA virus genera containing species that can infect at least one kind of mammal, 50 (81%) contain species that can infect humans. Of the 19 recognized RNA virus families that contain species reported to infect mammals, all but 2 include species found in humans. The exceptions are the Nodaviridae, essentially insect viruses, and the Arteriviridae, which include species infecting a range of different mammals, notably the simian hemorrhagic fever virus.

Human-adapted RNA Viruses

There is a semantic argument that only those viruses that are capable of persisting in human populations in the absence of a nonhuman reservoir should be described as "human" viruses. In our terminology these are, by definition, the level 4 viruses, comprising 47 species, 20 of which are not known to have any natural hosts other than humans. These 47 viruses—referred to here as "human adapted"—represent 12 families and 29 genera. Their most striking common characteristic is that almost all of them are transmitted by ingestion, inhalation, or direct contact; just 2 are transmitted by vectors.

There are several possible routes for a virus to reach level 4 on the pathogen pyramid. One possibility is that humans are exposed to a virus that is already capable of effective transmission between humans, i.e., the virus is preadapted to humans (noting that this does not preclude further adaptation once the virus has entered the human population). These have been termed "off-the-shelf" viruses. Such viruses may be rare, perhaps extremely rare, variants of the population in the nonhuman reservoir, in which case the main determinants of the rate at which such viruses enter the human population is the amount of genetic variability within the reservoir and the rate at which humans are exposed to the preadapted variants.

Emergence and the Changing Cast of RNA Viruses

New RNA virus species continue to be discovered, identified, or recognized in humans. Recent examples include Nelson Bay orthoreovirus, Irkut virus, primate T-lymphotropic virus 3, human coronavirus HKU1, and human rhinovirus C. Moreover, there is usually a backlog of reports of new human viruses that have yet to be formally recognized as species. Not all of these viruses will have recently invaded human populations; many will turn out to be long-standing human pathogens that have only recently been recognized or accepted as "species."

It is therefore important to understand that the continued accumulation of recognized human RNA virus species may reflect less the possibility that genuinely new viruses are continually emerging, most likely acquired from nonhuman reservoirs, than the fact that we are still getting to grips with the taxonomic diversity of viruses that have been with us for some time. This distinction between viruses that we have only just discovered and viruses that have only just discovered us is, of course, crucial in the context of emerging infectious diseases. If most of the so-called new viruses are not new at all, then this implies that events such as the advent of HIV/AIDS in the early 1980s or the curtailed SARS epidemic in 2003 may be just unusual, one-off occurrences with their own specific causes. If, on the other hand, genuinely new viruses are appearing all the time, then the HIVs and SARS-CoVs are more accurately regarded as just the highly visible tip of a much larger iceberg. Without a much more detailed and thorough understanding of the phylogenies and origins of all human viruses, not just those with high public health profiles, we cannot resolve this question.

Surveillance and Risk Assessment

Our conceptual model has practical implications for disease surveillance and risk assessment, especially in the context of newly emerging infectious diseases. The importance of early detection of potential epidemics or pandemics cannot be overstressed, a point made by several major studies. The early detection through clinical surveillance of SARS and effective intervention based on case isolation and quarantine prevented a potentially catastrophic pandemic. A matter of some debate is whether or not surveillance should be extended into the nonhuman reservoirs of infection from which novel human pathogens are most likely to emerge—a concept sometimes referred to as "getting ahead of the curve."

It helps, of course, if we know what we are looking for and where best to look for it. We currently have only the beginnings of answers to these questions. Viruses, especially respiratory viruses, are often considered the most obvious threat to global public health. New viruses are very likely to have a zoonotic origin, almost certainly acquired from mammals or birds. Emergence events are most likely to occur in regions—so-called hot spots—that combine high human population densities with high densities of domestic animals and/or a high diversity of wildlife

One strategy to increase the likelihood of early detection is to implement sentinel surveillance in people in close, high-risk contact with animal populations, such as bush meat hunters or slaughterhouse workers. In tandem with recent advances in the technologies available for virus detection—especially those based on high-throughput nucleic acid sequencing—such programs should at least improve our knowledge of the diversity of viruses "out there" that humans are exposed to, a process sometimes referred to as "chatter." Pathogen discovery programs, particularly in understudied taxa such as wild rodents and bats, should also add greatly to our knowledge of potential threats to human health.

Conclusion

Emerging diseases caused by RNA viruses are a One Health issue. There is a continuous interchange, over both epidemiological and evolutionary time scales, between viruses in humans and viruses in other animals that we cannot ignore. RNA viruses that pose serious threats to global public health have arisen repeatedly by jumping into humans from other animals. This has been going on for millennia, and it continues today, as fast as ever and perhaps even faster. We have to anticipate that new viral threats will emerge in the coming years or decades, and we need to be prepared to face these new challenges as they appear.

It is worth pointing out that the first virus was discovered in nonhuman animals (foot-and-mouth disease virus at the very end of the 19th century) before they were identified in humans. The same is true for several important viruses, such as retroviruses (and lentiviruses specifically), rotaviruses, papillomaviruses, and coronaviruses. A corollary of this is that veterinary rather than medical expertise may, at least initially, be our best source of knowledge about newly discovered viruses.

CLASS ACTIVITY

The data in the table below show the results of attempting to infect three different cell lines with the La Crosse encephalitis virus (LAC). With electron microscopy, observations were made to detect virus particles on the surface of the cells and virus particles present in endocytotic vesicles (endosomes) inside of the cell. A "+" indicates that the virus was present in the majority (>80%) of the cells observed, whereas "+/–" indicates that the virus was present in only a few (<5%) of the cells observed. In addition, the average yield of virus per cell was measured. Using these data, answer the following questions about these cell lines.

Data for La Crosse Encephalitis Virus			
Cell Line	**Virus on Surface**	**Virus in Endosomes**	**Virus Yield Per Cell**
HeLa	+	+/–	5
CEF	–	–	0
BHK-21	+	+	200

- Which cell lines are susceptible to infection by LAC? Why?
- Which cell lines appear to be permissive for LAC infection? Why? (c) Propose a hypothesis to explain the data for HeLa cells compared to BHK-21 cells. How can you explain the difference in average yield per cell? How would you test your hypothesis?

SUMMARY

- Growing and maintaining cells in the laboratory is necessary for any molecular biological investigation.
- The study of bacterial viruses provided the model for studying all viruses because it was convenient to replicate such viruses in easily grown bacterial cell cultures.
- To maintain cells in culture, a culture medium approximating blood plasma must be used. This medium contains salts similar to those found in plasma; most amino acids (since animal cells cannot synthesize many of these); vitamins; glucose as an energy source; buffers (usually carbon dioxide/sodium bicarbonate) to prevent lactic acid (resulting from glucose fermentation) from making the medium too acidic; and most importantly, blood serum, which is usually obtained from calves or horses.
- When a virus infects a cell, its genome enters that cell. Regardless of whether the virus capsid remains at the surface of the host, as in bacteriophage infection, or is internalized, it is modified and then disrupted.
- Some basic types of virus-induced changes to the host cell (cytopathology) result in changes that are readily observable by eye or with the aid of a low-power microscope.
- Cytopathic effects on the host cell by the great majority of viruses cause observable damage or changes to the cells in which they are replicating. Even if cells are not killed or lysed, the alteration of cells in a local area due to a localized virus infection can readily be observed as a plaque or focus of infection.
- In the hemagglutination inhibition (HI) and hemagglutination (HA) assays, there comes a point in diluting any virus stock below which a desired property cannot be observed.
- Quantal endpoints are simply a measure of dilution of infectious virus, but they relate to the average number of PFUs in the aliquot.
- Most viruses are submicroscopic physical particles, and while the largest can be discerned in an ultraviolet (UV)-light microscope, detailed visualization requires other methods.
- The dimensions of viruses are below the resolving power of visible light, so their visualization requires the shorter wavelengths available with the EM.

REVIEW QUESTIONS

1. You have diluted a 1-ml sample of virus stock by taking 100 µl from the stock solution and adding to it 0.9 ml of buffer. You then take 10 µl of this dilution and dilute it into 1 ml. You then infect two plates that contain 10^5 cells each with 100 µl. One plate had 25 plaques, while the other had 29 plaques. What was the titer in the original stock?

2. Using the Poisson distribution, calculate the proportion (probability) of cells infected with the indicated number of plaque-forming units (PFUs), given the multiplicity of infection (MOI).
3. Virus particles are very carefully isolated from an infected cell stock. You use this material to infect a culture of 10^6 cells with an MOI of 7 PFUs/cell. What is the maximum percentage of cells which could be productively infected?
4. You apply a virus stock solution containing 3×10^6 virus particles to 3×10^5 cells. What is the MOI for this infection?
5. What features of the electron microscope make it an excellent tool for examining virus particles?

REFERENCES

1. Celis, A., & Celis, J. E. (1998). General procedures for tissue culture. In J. Celis (Ed.), *Cell biology: A laboratory handbook* (2nd ed., Vol. 1, pp. 5–15). Academic Press.
2. Celis, A., Dejgaard, K., & Celis, J. E. (1998). Production of mouse monoclonal antibodies. In J. Celis (Ed.), *Cell biology: A laboratory handbook* (2nd ed., Vol. 2, pp. 392–397). Academic Press.
3. Celis, J. E., & Olsen, E. (1998). One-dimensional sodium dodecyl sulfate–polyacrylamide gel electrophoresis. In J. Celis (Ed.), *Cell biology: A laboratory handbook* (2nd ed., Vol. 4, pp. 361–370). Academic Press.
4. Cristofalo, V. J., Charpentier, R., & Philips, P. D. (1998). Serial propagation of human fibroblasts for the study of aging at the cellular level. In J. Celis (Ed.), *Cell biology: A laboratory handbook* (2nd ed., Vol. 1, pp. 321–326). Academic Press.
5. Maunsbach, A. B. (1998). Fixation of cells and tissues for transmission electron microscopy. In J. Celis (Ed.), *Cell biology: A laboratory handbook* (2nd ed., Vol. 2, pp. 249–259). Academic Press.
6. Osborn, M. (1998). Immunofluorescence microscopy of cultured cells. In J. Celis (Ed.), *Cell biology: A laboratory handbook* (2nd ed., Vol. 2, pp. 462–468). Academic Press.
7. Pawley, J. B., & Centonze, V. E. (1998). Practical laser-scanning confocal light microscopy: Obtaining optimal performance from your instrument. In J. Celis (Ed.), *Cell biology: A laboratory handbook* (2nd ed., Vol. 3, pp. 149–169). Academic Press.

CHAPTER 8

The Molecular Genetics of Viruses

LEARNING OBJECTIVES

After studying this chapter, you will be able to:

- Understand viral genomes;
- Discuss genetic manipulation of viral genomes;
- Focus on positive and negative stranded RNA viral genomes;
- Explain segmentation in viral genomes.

KEY TERMS FROM THIS CHAPTER

Ambisense	Archaea	Double-stranded (ds) DNA
Enzyme	Fungi	Genome
Hepatitis delta virus	Map units	mRNA
Proteins	Ribonucleotides	Ribozymes
Viroids	Virus replication	

INTRODUCTION

The expression of information encoded in the viral genome through cellular machinery in the infected cell is necessary for virus replication. Despite this, double-stranded (ds) DNA, the genetic material found in cells, is not present in many viral genomes. Because of these and other factors, viruses have evolved and maintain various distinctive ways to fit into the genetic information flow within cells. The most effective method to integrate all the different aspects is to keep in mind that, in order for the cell to convert the mRNA into viral protein, a virus needs to express a minimum amount of its genetic material as mRNA that is easily translated by unaltered translational machinery. The cell can then be transformed into a factory for the replication of viruses by these proteins, or they can change the cell in ways unique to the virus.

The type of genome a virus uses determines how its messenger RNA is produced. Originally proposed by David Baltimore, the co-discoverer of reverse transcriptase and winner of the Nobel Prize, the way viruses must produce mRNA offers an operationally useful classification scheme. A plan like this also offers a generally trustworthy and scientifically supported way to forecast the relationship between two viruses. It's not hard to see that two viruses that produce mRNA during infection in very different ways or that use very different genetic material might not be all that related!

Viruses can be arranged according to their genetic makeup, which has the additional advantage of making it easier to understand the different ways in which the virus induces a cell to replicate its genome. Since the cell does not use RNA as genetic material or replicate RNA from an RNA template, the patterns for DNA and RNA viruses differ significantly. DNA viruses typically follow a pattern similar to cellular DNA replication (Baker et al., 2014).

8.1. VIRAL GENOMES

It is crucial to represent this genetic information standardized because all the information needed for a virus to replicate itself must eventually be preserved as genetic information in the viral genome. The viral genome is typically depicted as a schematic of the overall genomic structure (i.e., circular or linear, single- or double-stranded nucleic acid, etc.), both cis-acting genetic elements and viral genes represented. Thus, the schematic depicts the virus's genetic and physical maps (a condensed version of the encoded genomic sequence information).

Map units (mu) can also be used to express viral genomes. A map unit is a relative term because its size varies based on the particular genome. The phrase is still widely used to describe genomes, particularly when identifying a particular genetic function within one. Map units can range from 1 to 100 or from 0.0 to 1.0. The size of a map unit for any genome can be calculated if one knows the genome's base or base pair size.

If the genome is linear, the start of the map (0.0 map unit) and the end (1 or 100 map units) are arbitrary definitions of the two ends. If the virus's genome is circular, the start and end points of the map must be conventionally determined at a particular location within the genome. This could refer to a recognized genetic element, or in other situations; it could be the location of a highly specific restriction endonuclease cut in the genome.

Although more difficult to generalize, most proteins found in eukaryotic or prokaryotic cells have more than 100 but fewer than 1000 amino acids. Based on this, one could estimate that a virus with a genome consisting of 7000 bases could encode 10 to 20 "average"-size proteins. These estimations are limited to the information contained in overlapping translational reading frames and do not incorporate protein information. However, these approximations are highly helpful in ascertaining whether or not the virus encodes all of the new or unique proteins observed expressed in a cell after infection. Furthermore, conclusions regarding the possibility of certain viral proteins being expressed unmodified or derived from precursors can be drawn using this information.

8.1.1. Evolution of Viral Genomes

The most prevalent biological organisms on Earth are viruses. According to virology studies, there are at least 10^{33} viruses on Earth, which is approximately ten times more than there are bacteria. They are abundant in plants, soil, oceans, and human bodies, among other places. In a healthy human, there are 10^{13} cells, 10^{14}–10^{18} bacteria, and an undetermined number of viruses. According to these data, bacteria act as our second genome because their genetic complexity is 100 times higher than that of our own. We are 99% bacteria, with over 1.5 kg of various types of bacteria living in our stomachs, making up the entirety of our genetic makeup. More than 200 viruses have been found in human gut samples based on similarities to other known viruses. Viruses make up our third genome. Due to the presence of fungi and archaea in our digestive tracts, humans are both a complex ecosystem and a superorganism. The name "bacteriophage" comes from the fact that these viruses are capable of lysing bacteria.

It all began with the widely acknowledged fact that RNA is the primary building block of life on Earth. The synthesis of RNA may have occurred at hydrothermal vents, which can reach temperatures as high as 400 °C to freezing at the ocean's bottom. Energy from chemical reactions in the rocks, which are made up primarily of granite with a high concentration of metals, and catalytic support from clay may have contributed to the evolution of life. RNA has developed into ribozymes, which are catalytic oligonucleotides. This catalytic RNA can join or cleave the RNA, allowing the molecules to replicate and change. Plant pathogens, like viroids, are a reflection of the early RNA world's characteristics. The ribozymes that are common in plants and capable of causing harm are called viroids. Their non-protein-coding naked RNA and lack of a protein coat give them the appearance of being relics from a world before proteins. They are tiny, made up of a few hundred single-stranded RNA nucleotides folded into a structure resembling a hairpin loop that may protect from external dangers. Given that hepatitis delta virus (HDV) antigens are linked to human proteins and that HDV is the only virus known to be a catalytic ribozyme and human pathogen, it has been hypothesized that a viroid may have entered the human body and evolved into a human virus by obtaining genetic information for a protein from the host.

Plant viruses, like the tobacco mosaic virus (TMV), may represent the next stage in the evolution of viruses. These viruses have rod-shaped structures and are incredibly stable. Many plant viruses utilize their host cellular RNA replicases, which are among the earliest and oldest polymerases, rather than creating their own replicase. Nearly all plant viruses are tiny and contain a vast amount of genetic information. A basic small RNA molecule would have become more unstable if it had been allowed to grow longer. The current viruses with segmented genomes may reflect this accumulation of several similar molecules in a protective compartment. RNA viruses evolved defense mechanisms for the ends of their RNA, and some of them had structures resembling tRNA that could fold back and attach to the target RNA to transfer specific amino acids. This could be the start of synthesizing peptides (Baker et al., 2014).

It is speculative as to how DNA changes from RNA. RNA has many more functions than DNA, but RNA has long-term memory and DNA stabilizes genetic information. On the other hand, RNA is very brittle and inconsistent. Telomerase, a complex of reverse transcriptase, RNA-dependent DNA polymerase, and an endogenous RNA made up of single repeats of a few ribonucleotides, catalyzes the conversion of RNA to DNA in the telomeres of chromosomes. Reverse transcriptase (RTase) is an enzyme that is closely related to telomerase. It is the enzyme that turns RNA into DNA and is a characteristic of retroviruses. Additionally, RTases have been identified in bacteria, which may have been a holdover from earlier retrovirus infections. On the other hand, RTase might also be a precursor to a retrovirus. Phage genomes have been found to contain sequences homologous to RTase, which likely serve to increase phage diversity by facilitating their transmission into new hosts. Lastly, the three domains of life were formed before the invention of DNA by viruses, and the enzyme RTase seems to be a key evolutionary link between RNA and DNA.

8.1.2. Classification of Viral Genomes

Currently, 71 families comprising more than 4000 viruses have been described. Viruses are incredibly diverse compared to plants, animals, and even bacteria, despite having small genomes. Viruses can be classified as DNA or RNA viruses based on their genomes. Both DNA and RNA viruses can have a circular, linear, or segmented structure, and they can be single- or double-stranded. The characteristics of DNA and RNA viruses, such as whether they are monopartite or multipartite, set them apart. The genome of a monopartite contains only one nucleic acid molecule. Few viruses with single-stranded genomes are known to have multiple segments, and all double-stranded DNA genomes contain a single nucleic acid molecule. RNA viruses, on the other hand, typically have multipartite genomes; this is especially true for single-stranded RNA viruses.

Furthermore, the genomes of single-stranded viruses can have two different positive senses: positive sense (+), in which the RNA in the genome has the same polarity as mRNA and encodes genes, or negative sense (–), in which the entire ssRNA genome needs to be copied, and the copied strand is sequenced. Ambisense viruses are a type of single-stranded virus that exhibits both + and – sense. Ambisense viral genomes are found in bacteriophages that belong to the family Leviviridae, including MS2, Qb, and Mimivirus. Compared to RNA viruses, DNA viruses are larger in size. DNA viruses with a maximum size of 305,000 nucleotides are rare. RNA viruses are small, and DNA viruses are referred to as large viruses. Some single-stranded RNA genomes can have up to 31,000 nucleotides in size. The fragility of RNA, which gives large RNA strands a propensity to break, and the fact that RNA viruses are more prone to mutations than DNA viruses may impact the single-stranded virus's ability to remain small. Compared to double-stranded viruses, single-stranded DNA and RNA viruses are more brittle.

The genomes of large double-stranded DNA viruses like herpesviruses and poxviruses are intricate, similar to their host cells. The genomes of Polyomavirus are associated with histone proteins from the host cell, creating a structure like chromatin within the virus. Once inside the host cell, these genomes act like small satellite chromosomes, following the directions of cellular enzymes and the cell cycle. mRNA from the Vaccinia virus is polyadenylated at the 3¢ end. Adenovirus genomes contain split genes with non-coding introns, protein-coding exons, and spliced mRNAs. To enhance the replication process, Adenovirus manipulates various cellular processes, like RNA splicing and polyadenylation, to maximize the coding potential of the limited viral genome.

The genomes of phages are the simplest and least complex. In 1984, the T4 phage's genome yielded the first evidence of introns in prokaryotes. The 160 kbp double-stranded DNA genome of the T4 phage is highly compressed, containing sequences and promoters that regulate translation nested inside the coding regions of overlapping upstream genes. It is made up of three self-splicing group I introns that are found in the genes encoding the anaerobic ribonucleotide reductase (nrdD), aerobic ribonucleotide reductase (nrdB) small subunit, and thymidylate synthase (td). The td and nrdD introns are mobile and can be inserted into new phage genomes, just like any other group I introns. This is because they each have an open reading frame that encodes for a homing endonuclease. Deleting the intron-borne homing endonuclease gene has rendered the nrdB intron immobile. The existence of these introns was hypothesized to provide the phage with a selective advantage by allowing for the control of splicing, which in turn could control the expression of the intron-containing genes. Furthermore, group I intron evolution in T-like phages was thought to have been largely influenced by horizontal intron transfer between phages via homing.

The type of host cell also affects the size of the viral genomes. The minimum genome size of bacteriophages is caused by the compact nature of their genes, which overlap, and the tendency of viruses with prokaryotic host cells to replicate quickly to keep up with their host cells. Viruses that use eukaryotic cells as their hosts exhibit extreme compression during the core's packing into the capsid, allowing only the optimal amount of DNA to be packed. Viruses encapsulate their genomes in capsids made of protective proteins. The type of genome affects how viruses package themselves. A molecular motor with a spool-like structure is used by viruses with double-stranded DNA genomes to pack their genomes into the capsid. On the other hand, viruses possessing single-stranded DNA or RNA genomes utilize a cooperative mechanism wherein the genome is packaged and assembled into the capsid simultaneously, thereby improving the capsid's assembly efficiency. The genomic RNA sequence of bacteriophage MS2 will create a brief term loop to interact with the capsid proteins (Baker et al., 2014).

The symmetric protein dimers required to assemble the capsid will undergo conformational changes brought on by this interaction. Some members of the Myoviridae family of bacteriophages, like T4, have comparatively large genomes—up to 170 kbp. 10% of the 1200 open reading frames in the Mimivirus genome, currently the largest known viral genome at ~1.2 Mbp, exhibit similarities with known function proteins. The genomes of herpesviruses and poxviruses, two eukaryotic viruses, can reach 235 kbp. These genomes contain genes, mainly those of enzymes involved in nucleic acid metabolism, that are involved in their own replication. Since these viruses encode extra biochemical machinery, they can thereby partially evade the host cell's biochemical constraints. However, doing so comes with the drawback of encoding all the information required for a genome to fit into the capsid, which also puts pressure on the capsid's size. All viruses are obligatory intracellular parasites, regardless of the makeup of their genomes. They can only multiply inside the right host cells, and the host cell that is being parasitized needs to be able to recognize the encoded genomes of the viruses. To achieve this, the host cell must either recognize or match the genetic code used by the virus. Likewise, the signals controlling viral gene expression have to be unsuitable for the host.

8.1.3. Size, Structure, and Composition of Double-Stranded DNA Virus Genomes

Every class of cellular life forms has double-stranded DNA in its genome, serving as the standard mechanism for replication and expression. On the other hand, viruses and other detrimental and self-centered entities take advantage of every potential nucleic acid conversion, as their genomes can be either single- or double-stranded DNA or RNA. The genomes of cellular life forms, which were tiny and had unidentified gene functions, are compared with the sequenced and annotated virus genomes. However, the discovery of giant viruses in recent years has led to a rapid expansion in the size of viral genomes, with a range spanning up to three orders of magnitude, from approximately ~2 kb to approximately ~2 Mb. The difference in complexity and genome size between viruses and cells is completely erased by the giant viruses, whose genomes are surprisingly larger than those of many bacteria and archaea. Since their genomes are quite complex and can be divided into small and large genomes based on the type of replication, many viral groups have double-stranded DNA. While large genomes encode a DNA polymerase specific to the virus, small genomes replicate their genomes using a host DNA polymerase. These viruses share genetic similarities with the host cells that they infiltrate. More proteins are encoded by large DNA viruses than by small DNA viruses.

The double-stranded DNA genomes are represented by two main viral groups, namely people who belong to the families Herpesviridae and Adenoviridae. One of the most prevalent pathogens that most patients experience is a mild, self-limiting illness caused by the human adenovirus. One of the viruses that is best understood is the adenovirus, whose fundamental biology has been thoroughly investigated over the previous 60 years. The human adenovirus that causes respiratory infections was first identified in the adenoid tissue in the early 1950s. With as few as five virus particles, this virus can spread remarkably among patients who come into contact with it. In confined populations such as daycare centers, hospitals, retirement homes, and military training venues, adenovirus-induced acute respiratory disease is the most common infection. It accounts for approximately 8% of all childhood respiratory tract infections and can result in pneumonia, bronchitis, or bronchiolitis, with approximately 25% of diagnosed cases necessitating hospitalization.

Other localized illnesses like colitis, hemorrhagic cystitis, nephritis, encephalitis, myocarditis, and disseminated illness with multiorgan failure are also brought on by it. Some of these conditions can be more severe in younger and older people, particularly in those whose immune systems are weakened, like those undergoing transplants or those with AIDS. Adenoviral vectors that are extremely effective at delivering genes into mammalian systems—particularly human cells for transgene expression and therapeutic gene delivery in human gene therapy—have been developed due to the extensive and detailed elucidation of many aspects of the adenoviral life cycle. The nucleoprotein core of adenoviruses is composed of 30–40 kb of linear double-stranded DNA, and it is encased in an icosahedral, non-enveloped capsid with a diameter of 70–100 nm. About thirty to forty genes are present in the genomes of these viruses. Every DNA strand's terminal sequence contains a 100–140 bp inverted repeat. The structures that the denatured single strands form, which resemble,"" are crucial for DNA replication. Each strand has a covalent attachment at the 5¢ end to a 55 kDa protein called the terminal protein. In the process of genome replication, this protein may serve as a primer to start the synthesis of new DNA strands. Adenoviruses have much more complex gene expression, with groups of genes expressed from a small number of common promoters. Each promoter can express a range of polypeptides through the use of alternative splicing patterns and multiply spliced mRNAs (Baker et al., 2014).

Aside from variations in host and tissue tropism, there is Very little variance in the structural parameters and genomes of adenoviral viruses. The human adenoviral serotype 5, with a genome of approximately 36 kb, is one of the well-characterized adenoviruses. Depending on the stage of expression, its coding region is split into early (E1–E4) and late (L1–L5) transcripts. Most adenoviral vectors lack the essential early E1 region, which prevents them from replicating in the majority of cell lines. Adenoviral vectors are more suitable for *in vivo* and *in vitro* studies requiring short-term transgene expression after the deletion of E1, according to a number of studies. In order to enable prolonged transgene expression, deletions in the crucial E2 or E4 regions are made before creating second-generation adenoviral vector constructs. Helper-dependent Adenoviral vectors (hdAd) are produced by eliminating all viral protein-coding genes to enhance cloning efficiency. The overall genome size of these vectors is reduced from 36 kb for the wildtype to 30 kb for E1/E3-deleted Adenoviruses and approximately 500 bp for helper-dependent Adenoviral vectors due to the removal of protein coding sequences.

8.1.3.1. Analysis of Adenovirus 2(AD2) Genome

The genome of adenovirus 2(Ad2) was the first to be fully sequenced, which is of the size ~36 kb, encoding over 40 proteins. Adenoviral coding regions are designated as early or late depending on their expression before or after the DNA replication. The early genes E1A, E1B, E2, E3, and E4 are the first ones to get transcribed. They encode for the proteins involved in activating transcription of other viral regions, altering the cellular environment to promote viral production. E1A proteins also induce mitogenic activity in the host cell, which stimulates the expression of other viral genes. E2 proteins regulate viral DNA replication, while E3 and E4 proteins are involved in altering the host immune responses and cell signaling. Activation of the major late promoter (MLP) followed by the start of viral DNA synthesis, allowing the expression of late genes encoding primarily virion structural proteins. L1–L5 of the late regions are transcribed from an alternatively spliced transcript. The regions encoding the L4-22 K and L4-33 K proteins are initially expressed at low levels from a novel promoter located within the L4 region and these proteins functions in fully activating the MLP. Four small proteins, including the structural protein IX (pIX) and the IVa2 protein produced at intermediate/late stages of infection, help in packing viral DNA into immature virions. The late products VA RNA I and II inhibits the activation of the interferon response, impede cellular micro-RNA processing and also influence the expression of host genes. There are 100 bp inverted terminal repeats (ITRs), located at both ends of the genome, which act as the sites for the origin of replication, with the ~200 bp viral packaging sequence positioned next to the left ITR (Figure 8.1). Even though Adenovirus is being studied in great detail for more than 60 years, our knowledge of the genes encoded by this virus is still expanding. In 2007, a new open reading frame (ORF) was identified to be located between the fiber ORF and E3, termed as U exon. The U exon protein (UXP) is expressed from a unique promoter during later stages of infection and is hypothesized to play a significant role in the transcription (Bitner et al., 1982).

More than two alternatively spliced mRNAs are produced by the majority of the adenovirus's transcription processes. Maintaining the compactness of the adenovirus genome, regulatory events occurring at the RNA processing level are critical for managing the virus's lytic cycle. In the later stages of viral infection, splicing of large introns leads to the production and accumulation of shorter mRNAs. The ORF of the gene is not interrupted by viral introns, with the exception of E1A, L4-33K, and the U exon protein. Furthermore, compared to cellular genes, adenoviral genes have a very small number of introns. One to three introns are cut out of viral mRNAs as they mature. Few and small introns appear to have been preferred by selection pressure because the virus must condense a large amount of its genetic information into a tiny genome. The Adenovirus genome's deep cDNA sequencing has revealed a large number of novel alternatively spliced transcripts, indicating that the virus may produce a large number of new or altered polypeptides in the infected host cell. This indicates that there are still a lot of unanswered questions regarding this genome.

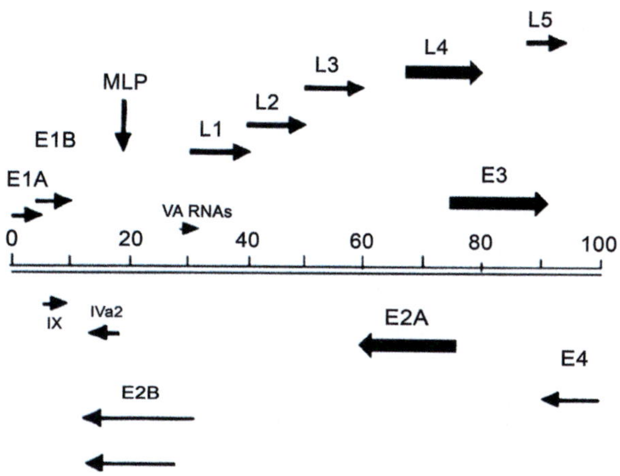

Figure 8.1. *Organization of the Adenovirus genome.*

Source: Chaitanya KV. Structure and Organization of Virus Genomes. Genome and Genomics. 2019 Nov 18:1–30. doi: 10.1007/978-981-15-0702-1_1. PMCID: PMC7119911.

8.1.3.2. Herpesviridae

There is at least one member of the large family Herpesviridae for every animal species that has been studied to date, with approximately 100 members in total. There are eight different human herpesviruses; while they all have a similar overall genome structure; they vary in terms of nucleotide sequence and fine-grained genome organization. A family of enveloped DNA viruses with intricate genomes is called Herpesviridae. Numerous vertebrate and invertebrate hosts, including humans, horses, cattle, mice, pigs, chickens, turtles, lizards, fish, and invertebrates like oysters, serve as their nuclei for replication. Three subfamilies comprise the eight human herpesviruses. Herpesviruses have extremely large, complex genomes comprising approximately 35 virion polypeptides and large, complex virus particles with up to 235 kbp of linear, double-stranded DNA. Although the genome sequences and protein synthesis of Herpesviridae members vary greatly, they exhibit strong structural and genome organization similarities, and nearly all of their genomes encode the enzymes necessary for protein processing, nucleic acid metabolism, and DNA synthesis. Two covalently joined sections known as unique long (UL) and unique short (US) regions, each surrounded by inverted repeats, make up a small portion of herpesviral genomes. Because of these repeats, distinct regions can be structurally rearranged, allowing genomes to exist as a mixture of four functionally equivalent isomers. These genomes have also been reported to have multiple repeated sequences; the size of the genome can vary by up to 10 kbp depending on how many sequences are repeated (Bitner et al., 1982).

The successful completion of the virus lytic cycle requires integration of the viral genome into the host cell's chromosomes. Herpesviruses, on the other hand, do not require integration in order to preserve their genomes as extrachromosomal circular episomes within the nuclei of infected cells. Herpesviruses may also be able to integrate into the host's chromosomes in certain situations, as evidenced by reports of chromosomal integration of herpesviral DNA. The rolling-circle mechanism was proposed as the mechanism for replicating non-integrated Herpesviral DNA, resulting in long DNA concatemers that are cleaved into single genome equivalents during nucleic acid encapsidation. Furthermore, it has been discovered that 1% or so of the global population has Human Herpesvirus 6 (HHV-6) integrated into their germlines. However, it is still unclear how linear CIHHV DNA replicates. Chromosome insertions of α-Herpesvirus, Herpes simplex viruses, and Equine Herpesvirus have been identified after transfection of sheared or subgenomic viral DNA fragments or infection with defective interfering particles. The majority of the integrated viral genome is made up of subgenomic fragments, meaning that infectious viral particles cannot possibly be produced. Several cells exhibiting integrated viral DNA exhibited a changed phenotype, supporting the theories regarding the oncogenic potential of these viruses.

8.1.3.3. Analysis of Herpes Simplex Virus Genome

The Herpes Simplex Virus (HSV), whose genome comprises double-stranded DNA and is approximately ~152 kbp in size, is the prototype member of the Herpesviridae family. The Herpes Simplex Virus's entire nucleotide sequence has now been identified. About 80 genes, closely spaced and with overlapping open reading frames, make up this virus. Under its own promoter, each gene expresses itself. Possibly the most thoroughly researched complex virus genome is HSV-1. Prior to the advent of nucleotide sequencing, traditional genetic analysis and mutant analysis were used to map the HSV genome extensively. A single linear double-stranded DNA strand, approximately ~152,261 base pairs long, makes up the HSV-1 genome. UL and US are the two segments that make up the genome. Between the L and S segments at the ends of the genome are small regions of repetitive sequence. In most HSV-1 wild type populations, the L and S segments invert at a very high rate during DNA replication, resulting in four genome isomers that occur at equal frequencies. This virus synthesizes 75 genes that code for well-known proteins. Out of them, 69 genes have been discovered to have one copy, while three genes have two copies each. 43 of the 75 genes with known functions are thought to be the core genes shared by the α, ⊠, and ⊠ herpesviruses and found in the UL region of the genome. Many essential processes, such as the entry of the core DNA into the host cell and its replication, as well as the assembly into the capsid and its dissemination, are mediated by these genes. Every gene found in the distinct short segment is extremely divergent and non-core. The majority of these are located at the segment's ends. Non-core gene-encoded proteins play a role in species- and lineage-specific processes like immune evasion, transcriptional transactivation, and host cell recognition.

The α-Herpesvirinae subfamily of the Herpesviridae family includes the Type 1 Herpes simplex virus, which can cause encephalitis, cold sores, genital sores, and eye infections. Many HSV-1 strains with different levels of virulence have been identified; these strains may differ in virulence because of base substitutions that alter cis-regulatory elements or amino acids. KOS is a human labial lesion-isolated strain of HSV-1 that is commonly used as a marker to study the pathophysiology and function of the HSV-1 gene. When compared to other HSV-1 strains like McCrae and 17, KOS is less virulent, which has piqued interest in comparative genomics. An unpaired 42-bp Illumina library was created and run at Washington University's Genome Technology Access Center using KOS genomic DNA extracted from African green monkey (Vero)-infected kidney cells (Bitner et al., 1982).

Figure 8.2. *Organization of the herpes simplex virus-1 genome.*

Source: Chaitanya KV. Structure and Organization of Virus Genomes. Genome and Genomics. 2019 Nov 18:1–30. doi: 10.1007/978-981-15-0702-1_1. PMCID: PMC7119911.

Since viral DNA was isolated from Vero cells, potential contaminating host reads that matched the *Rhesus macaque* and/or human genomes were removed using Bowtie. The remaining reads of 16,494,831 bp were assembled into contigs using the Velvet de novo assembler against the reference HSV-1 strain 17 genome (GenBank accession number NC_001806) with SeqMan Pro (DNASTAR, Inc.). Because the HSV-1 genome includes two sets of inverted repeat regions TRL/IRL and IRS/TRS, contigs assembling into one of the repeat units were reverse complemented and also placed into the other repeat unit. The final KOS genome is a linear double stranded DNA of 152,011 bp, consisting of 80 genes and has 13 gaps, exclusively at VNTR regions, totaling up to 1582 bp in length. In the Gen Bank annotation, the sequence and length of each VNTR were copied from strain 17. Using Bowtie to align the filtered reads against the de novo assembly, the average sequence coverage per base pair for the KOS genome was determined to be 4257 (Figure 8.2).

Fast statistical alignment (FSA) was used to align the genomes of strains KOS and 17, and custom Perl and R scripts were applied to identify nucleotide variants between the genomes. KOS is distinct from strain 17 due to 1024 SNPs, of which 320 are non-synonymous alterations in 65 of 77 open reading frames of HSV-1. 172 indels separate the two genomes; the majority of these are single base insertions or deletions in non-coding regions. Nevertheless, 26 indels involve codon deletions or frame additions. The comparative genomics of KOS and other HSV-1 strain genomes are currently undergoing analysis. These investigations will broaden the field's ability to more accurately identify the genetic characteristics of KOS and how they contribute to its pathophysiology.

8.1.4. Locating Sites of Restriction Endonuclease Cleavage on the Viral Genome—Restriction Mapping

By dividing a DNA segment into smaller segments and pinpointing the locations of the breakpoints, restriction mapping is a technique for mapping unknown DNA segments. The technique is based on the utilization of proteins known as restriction enzymes, which have the ability to cut or break down DNA molecules at brief, targeted sequences known as restriction sites. Gel electrophoresis is a laboratory technique that divides DNA fragments based on size after a DNA segment has been digested using a restriction enzyme. The fragments obtained can be analyzed using this technique.

When creating a restriction map, there are three common ways to digest the unknown DNA sample. Here, the DNA sample is divided into two parts, each of which is digested separately using a different restriction enzyme. A third part of the sample is simultaneously double-digested using both restriction enzymes. After that, gel electrophoresis is used to separate each digestion sample and the sizes of the DNA fragments are noted. Each digestion's fragments will have the same overall length. Nonetheless, each restriction site can be mapped in accordance with the lengths of the fragments since the length of each unique DNA fragment is determined by the locations of its restriction sites. For an accurate site map, the double-digestion information is especially helpful. A restriction map is the last illustration of the DNA segment that displays the locations of the restriction sites.

8.1.4.1. Method

Before beginning any digest, the experimental protocol requires a sample of purified plasmid DNA. Next, digestion is carried out using the selected enzyme or enzymes. After that, the samples are run on an electrophoresis gel, usually an agarose gel.

After the electrophoresis is finished, the sizes of the fragments in each lane are added together. Each digest's fragments should add up to the same size as the others, and the total of the individual fragments should match the size of the original fragment. If the sum of the fragment sizes is incorrect, there are two possible issues. It's possible that a few of the smaller pieces rolled off the gel's end in one instance. This is often the result of running the gel too long. Another potential reason for inaccuracy could be that the gel was insufficiently dense, which prevented it from resolving small fragments. This results in a lack of separation between similar-sized fragments. By placing the restriction endonuclease (REN) sites in locations on the original DNA fragment that would satisfy the fragment sizes produced by all three digests, one can infer the location of the REN sites if all three digests produce fragments that add up.

8.1.4.1.1. Rapid Denaturation and Renaturation of a Crude DNA Preparation by Alkaline Lysis of the Cells and Subsequent Neutralization

This technique involves lysing cells in alkaline conditions to denature DNA by disrupting hydrogen bonds between strands. As the pH is lowered during neutralization, genomic DNA may tangle and remain denatured, causing strands to come back together in a disordered manner. However, circular supercoiled plasmids' strands will align closely and renature correctly. Genomic DNA will form an insoluble aggregate while supercoiled plasmids remain in solution. Following phenol extraction to remove proteins and other molecules, DNA can be concentrated through ethanol precipitation.

8.2. GENETIC MANIPULATION OF VIRAL GENOMES

Errors can happen during nucleic acid replication. While this is an extremely uncommon occurrence in living things, it happens quite frequently in viruses because of their fast replication rate and frequently erroneous replication enzymes. Certain viruses, such as HIV, can cause the polymerase to produce one error for every 10,000 bases that are transcribed, resulting in many changes (Essere et al., 2013).

Erroneous base insertions during genome replication are typically the cause of genetic alterations. For instance, during DNA replication, an A:T pair may be miscopied into a G:T pair or an A:C pair may form instead of an A:T pair. Rarely, polymerase slippage may result in the loss of a nucleic acid molecule. These types of genome alterations are known as mutations, and they may have an impact on how a gene functions. Nevertheless, mutations frequently result in minuscule alterations to proteins. These alterations build up over a very long time (often over many generations), resulting in the emergence of related but different organisms. Such alterations in viruses result in the production of minute antigenic variations amongst strains or serotypes.

If a protein is essential for the replication of a virus or possesses a highly important structure, a mutation in that protein can have a significant impact on its function. The virus may occasionally benefit from the alteration. An antiviral medication utilizing a nucleoside analog cannot control a virus if, for instance, a viral DNA polymerase mutates to the point where the enzyme is no longer sensitive to the analogue. Below are some mutations that result in the hypothetical, extremely brief, unspliced open reading frame being altered. Keep in mind that DNA is double-stranded even though only the mRNA sense strand is visible.

Gene: ATG-GTT-GAT-AGT-CGT-TAT-TTA-CCT-CAA-TGG-CAG-TAA

Protein: Met-Val-Asp-Ser-Arg-Tyr-Leu-Pro-Gln-Trp-Gln

A change in the seventh codon from TTA to TTC would result in phenylalanine replacing leucine. This alteration could impact the protein's folding and alter its functionality.

Met-Val-Asp-Ser-Arg-Tyr-*Phe*-Pro-Gln-Trp-Gln

The protein's structure might not be impacted at all by a mutation in the same codon to GTA, which would result in a protein with one aliphatic amino acid substituted for another.

Met-Val-Asp-Ser-Arg-Tyr-*Val*-Pro-Gln-Trp-Gln

A shorter protein would result from a mutation in this codon to TGA, a stop codon; this would most likely eliminate the function of the protein and, if it occurred very near the N-terminal amino acid, would cause the gene to lose all of its protein. For purely historical reasons, geneticists have sometimes referred to these translation termination mutations as amber, ocher, or opal mutations.

Met-Val-Asp-Ser-Arg-Tyr!

A change in the charge of the protein resulting from a mutation in the second codon (GTT) to GAT would replace valine with asparagine. This would most likely drastically change how the protein functions:

Met-*Asp*-Asp-Ser-Arg-Tyr-Leu-Pro-Gln-Trp-Gln

Any such mutation could be reversed by a different base change, or it could mutate again. These changes are referred to as reversible. But if a base is lost, the modification is practically irreversible. In fact, if a mutation essentially has no ability to reverse itself, this type of mutation (addition or loss of genomic material) can be inferred. If the first two bases of the second codon were lost, we would see an example of a non-reversible change. Because the second codon is now a translation stop codon as a result of this frame shift mutation, the protein would be lost.

ATG-*TGA*-TAG-TCG-TTA-TTT-ACC-TCA-ATG-GCA-GTA-A

Other frame shifts can completely change a protein, and thus can completely change or lose the protein's function.

8.2.1. Analysis of Mutations

- **Complementation:** Significant variations in virus plaque morphology, growth rates, host range, cytopathic effects, interactions with the host immune system, and other aspects are caused by numerous mutations. As a result, a mutant virus can be distinguished from its normal or wild type (wt) parent. Mutant viruses are beneficial because they can provide information about the function of mutated genes by changing the function of a particular viral gene, which can change the course of virus replication, among other things.
- **Recombination:** Occasionally, during the infection process, two viruses with distinct mutations may physically entangle one another. This intimate relationship may result in the creation of a recombinant viral genome. While the exact mechanism underlying the formation and resolution of recombinant genomes is still being studied, it is commonly observed that genome replication is the most frequent event in these scenarios. Furthermore, the greater the distance between the mutations on the parental genomes, the higher the likelihood of a recombinant organism developing. Even so, albeit infrequently, closely spaced mutations—even two distinct ones within a single gene separated by merely a few base pairs—can

recombine (Essere et al., 2013).

Two novel viral genomes are produced by a recombination event (one has both mutations, one has neither and is, therefore, wt). Four different virus types are subsequently produced by such an infection: two single mutant input viruses, wt, and double mutant viruses through recombination. A HSV example is displayed in Figure 8.3. A virus that has a mutation in the thymidine kinase gene (U_L23TK-) co-infects with a virus that has a temperature-sensitive (ts) mutation in DNA polymerase (U_L30 Polts). As a result, one parent virus cannot develop at a high temperature (a nonpermissive temperature), while the other parent virus is immune to a nucleoside analogue-induced inhibition of DNA synthesis.

The mixed infection gives rise via recombination to four distinct viral genotypes: wt, double mutant (Polts, TK-), and the two single mutant parents. Plaques unique to their genotype can be identified through appropriate screening or selection methods. All genotypes can be distinguished through virus plaque replication and incubation under various conditions (Essere et al., 2013).

Progeny viruses are plated in the experiment, and plaques are allowed to develop. On different plates, four exact duplicates are then created. These are cultivated in diverse environments that facilitate the differentiation of the virus genotypes:

- At high (nonpermissive) temperature, only wt and TK-will make plaques
- At normal (permissive) temperature, wt, Polts, TK-, and PoltsTK- all make plaques
- At permissive temperature with an antiviral nucleoside analogue drug, only virus lacking the TK gene (TK- and PoltsTK-) will make plaques.
- At nonpermissive temperature with antiviral drug, only the mutant (TK-) virus will form plaques.

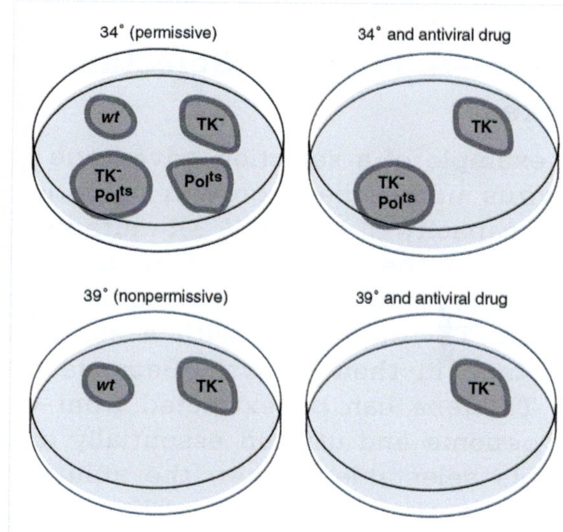

Figure 8.3. *Replica plating of virus plaques to distinguish genotypes produced by recombination following a mixed infection. Mixed infection of two genotypes will (rarely) produce novel genotypes by recombination. The example shows replicas of virus plaques that were developed under different selective and screening conditions as described in the text.*

Source: https://www.blackwellpublishing.com/content/BPL_Images/Content_store/Middle_Sized/Wagner9781405103466/207to228.pdf.

8.2.2. Isolation of Mutants

Choice. A mutant virus can be selected by growing in environments where the original virus will not replicate or will replicate poorly if the mutation results in a notable change in the virus's growth characteristics, resistance to a medication, or resistance to a neutralizing antibody. Although this is a perfect scenario for virus isolation, it is sadly difficult to create suitable selection conditions for many desirable mutations. Similar to bacteria, many mutations alter certain aspects of the virus in animal cells, but they do not significantly alter its growth characteristics.

8.2.3. HSV Thymidine Kinase—A Portable Selectable Marker

One example of a selection advantage for the virus under these growth conditions is the capacity of the HSV TK-mutants to replicate in the presence of an antiviral medication. Additionally, a selection scheme provides TK$^+$ organisms with a selective advantage in their growth. Because the HSV TK gene can be extracted from the viral genome and utilized essentially as a portable selectable marker, the ability to quickly identify cells that are TK$^+$ or TK-is a valuable tool.

Over the past 20 years, the HSV TK gene has been used to harness the power of selection to create recombinant viruses and cells that carry foreign genes. This is because the HSV TK gene is easily isolated and cloned from the HSV genome due to its convenient size, which is 2.3 kilobases of DNA. This gene encodes the structural gene for the enzyme as well as the promoter that regulates the mRNA encoding the structural gene's expression. This promoter's expression in uninfected cells and in viruses other than HSV is equally significant.

In order to employ the HSV TK gene for selection, a constitutively TK$^-$ cell line is generated. The process involves cultivating cells in the presence of an analog of a toxic nucleoside. The cell lines are easily established because only TK$^-$cells are capable of replicating in these circumstances. One can either add a TK plasmid to the plasmid of interest or incorporate the HSV TK gene into a plasmid containing the desired gene for transfection in order to create a recombinant virus or cell line. Both plasmids will be incorporated because transfection strongly favors cells absorbing relatively large aggregates of DNA (Holmes et al., 2005).

Cells or viruses that have undergone recombination with the added genome and are, therefore, TK$^+$ can be chosen once they have recovered from the transfection process. The entire virus recovered from the initial transfection can be used to selectively infect TK$^-$cells in order to select for recombinant viruses; only viruses expressing TK will be able to replicate because only those cells will make it through the selection process.

Recombinant HSV can also be generated via the TK selection technique. This is implied by the observation that in the majority of cells, the TK gene is not necessary for HSV replication. Consequently, wt virus DNA can be cotransfected with a cloned plasmid of recombinant DNA that contains HSV sequences from either side of the TK gene recombined with the gene of interest, which replaces the viral TK gene itself. All cells infected with the TK$^+$ parental virus will perish when the progeny virus of this transfection is used to infect TK$^-$ cells being grown in the presence of the toxic nucleoside analog. Therefore, plaques containing the inserted gene of interest will be significantly enriched for TK-virus.

8.2.3.1. Screening

The virus mutant can be selected or screened if the mutation results in a discernible alteration in the morphology of the plaque or in another aspect of infection. Here, one uses the change to identify the viruses by looking for them. For instance, a single gene mutation may change the virus's ability to package, reducing the virus's burst size when it leaves an infected cell. With a background of numerous other viruses, it would be easy to distinguish between the large and small plaque-producing viruses caused by this mutant virus, which would produce smaller plaques than the wt parent. Of course, other examples include the HSV ts mutants previously mentioned, which are not able to replicate at 39°C. They cannot be chosen for selection because they cannot replicate in environments with restricted growth; however, they can be chosen or screened by returning to the replica plaque that forms at 34°C.

8.3. POSITIVE AND NEGATIVE STRANDED RNA VIRAL GENOMES

The fragility of RNA and the propensity of its long strands to break impact the final size of the RNA virus genomes. Furthermore, because RNA genomes are copied less accurately than DNA genomes, they typically have higher mutation rates. RNA viruses may have tended to have smaller genomes as a result of this tendency. A restricted set of proteins are encoded by the genomes of RNA viruses. Double-stranded RNA, single-stranded RNAs with positive and negative strands, and monopartite and multipartite RNA viruses are the main categories of RNA virus genomes. RNA dependent RNA polymerase is one of the main proteins that all of these RNA genomes encode and is necessary for their replication. The ability of the RNA polymerase to be translated from ss(+) RNA while it is contained in ss(−) RNA distinguishes the + and − strand ssRNA viruses. The genome of monopartite ssRNA viruses encodes a single polyprotein, which is subsequently processed into several smaller molecules essential to the viral life cycle. Every segment in multipartite ssRNA genomes encodes for a single gene (Holmes et al., 2005).

8.3.1. Positive Stranded RNA Viral Genomes

Among the viruses possessing RNA genomes, single-stranded plus sense (+) RNA genomes represent an important subgroup, including many pathogenic plant, animal, and human viruses. These genomes contain cis-acting RNA elements that direct different viral processes, such as protein translation, genome replication, and subgenomic mRNA transcription mRNAs. Single stranded RNA genomes vary in size from coronaviruses (⊠30 kb) to those of phages such as MS2 and Qb(⊠3.5 kb). Although members of distinct families, most (+) sense RNA viruses share common features in terms of their genomes. Importantly, purified (+) sense virus RNA is capable of infecting the host cells in the absence of any viral proteins.

8.3.1.1. Picornavirus Genome

Picornaviruses are the causative agents of many diseases of medical and veterinary significance, including hepatitis, foot-and-mouth disease, poliomyelitis, the common cold, and the flu. These viruses have a positive-polarity single-stranded RNA genome that ranges in size from 7200 nucleotides for human rhinoviruses to 8500 nucleotides for the foot and mouth disease virus. It has several characteristics that are shared by all picornaviruses. There is a shorter 3′ UTR of 50–100 nucleotides, which is required for the (−) strand synthesis during replication, and a longer 5′ UTR of 600–1200 nucleotides, which is significant for translation, virulence, and potentially encapsidation. The internal ribosomal entry site (IRES), a "clover-leaf" secondary structure, is present in the 5′ UTR. Between 2100 and 2400 amino acids, a single polyprotein is encoded by the remaining portions of the genome. A small, basic VPg protein with a length of 23 amino acids is covalently attached to the 5′ end of the genome, while the poly(rA) tail modifies the 3′ end. The process of genome replication involves the use of the (+) strand as a template for the synthesis of the (–) strand, which in turn serves as a template for the synthesis of an excess of (+) strands. A uridylylated form of VPg, VPg-pUpU, is thought to prime both (+) and (–) strand RNA synthesis.

8.3.1.2. Toga Virus Genome

Comprises two genera: rubiviruses and alphaviruses. There are 27 members in the alphavirus genus, and many of them can be spread by insects. The only known member of the rubivirus family is the rubella virus. The togavirus genome comprises a single-stranded, non-segmented, + sense RNA that is roughly ~11.7 kb in length. These viruses' capsids are made up of 240 copies of a single, roughly ~264 amino acid capsid protein. Two virus-encoded glycoproteins, E1 and E2, are present in the envelope. With polyadenylation at the 3′ end and a cap at the 5′ end, the genome is similar to an mRNA. The nonstructural proteins needed for transcription and replication are encoded in the 5′ of the genome, while the structural proteins (capsid, E1, and E2) are encoded in the 3′. Genomic RNA will be used for the translation of non-structural proteins, producing a polyprotein that will eventually cleave into mature proteins. An internal promoter on the (–) strand is used to synthesize a subgenomic mRNA from the 3′ end of the genome. Three structural proteins are produced by translating the subgenomic mRNA: capsid, E1, and E2, in that order (Figure 8.4). Synthesis of the capsid protein occurs on free cytoplasmic ribosomes. It is co-translationally cleaved off, revealing a signal sequence that guides the ribosome to the ER membrane, where the remaining E1 and E2 proteins finish translating. E1 and E2 are produced in close proximity to the rough ER membrane, go through the Golgi apparatus, and subsequently are transferred to the plasma membrane.

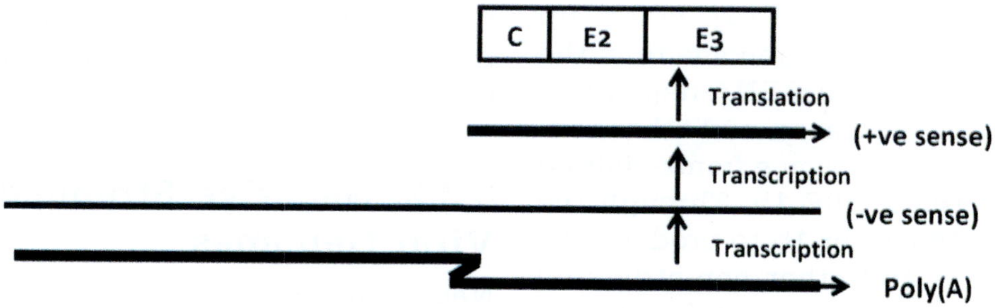

Figure 8.4. *Translation of togavirus subgenomic RNA.*

Source: Chaitanya KV. Structure and Organization of Virus Genomes. Genome and Genomics. 2019 Nov 18:1–30. doi: 10.1007/978-981-15-0702-1_1. PMCID: PMC7119911.

8.3.1.3. Flavivirus Genome

The three genera that make up the family Flaviviridae are: (1) Flaviviruses (also known as yellow fever viruses); (2) Pestiviruses (also known as bovine viral diarrhea viruses); and (3) Hepatitis C viruses. An abundance of Flaviviruses are spread by insects. The + strand genome of these enveloped viruses is 10p5 kb, capped at the 5′ end, and non-polyadenylated at the 3′ end. Structural proteins are encoded at the 5′ end of the genome, while non-structural proteins are encoded at the 3′ end. One polyprotein, which is cleaved into the mature proteins, is translated from the entire viral genome. Subgenomic RNA does not develop.

8.3.1.4. The Genome of Coronavirus

Two genera make up the family Coronavirides: coronaviruses and toroviruses. It is still not generally acknowledged that arteriviruses belong to this family. The large 27–30 kb + strand RNA genome that encases coronaviruses is polyadenylated at the 3′ end and capped at the 5′ end. A 60–80 nucleotide "leader" RNA and 200–500 untranslated nucleotides follow. These two overlapping open reading frames (ORF1a and ORF1b) make up approximately 60% of the genome from the 5′ end and encode the viral RNA-dependent RNA polymerase, proteases, and other non-structural proteins. Ribosome frame-shifting is used to translate ORF 1b. Through the translation of ORF 1a or 1b to produce the viral polymerase and other non-structural proteins, the transcription of (–) strand RNA using the viral polymerase, and the synthesis of both full-length viral RNA and subgenomic mRNAs using (–) strand RNA as a template, these viruses are able to replicate or express their genes (Holmes et al., 2005).

Each subgenomic mRNA is polyadenylated at the same location and has a similar leader sequence that is derived from the 5′ end of the genome. These transcripts are then followed by distinct regions of the genome to form a nested set. One translation unit, beginning at the first AUG following the leader, can be found in each of these transcripts, which are monocistronic. Though its exact mechanism is unknown, it does not involve RNA splicing because nested transcript synthesis occurs in the nucleus and coronaviruses replicate in the cytoplasm. Synthetic proteins and a few more non-structural proteins are produced by the translation of the subgenomic mRNAs. Either translation or viriopackaging of the new full-length + strands is an option. The electron microscope revealed that these viruses have a distinct crown-like structure. The envelopes comprise two viral glycoproteins: spike protein (SpP), which promotes receptor binding and cell fusion, and membrane protein (M protein), which attaches the viral nucleocapsid to the viral envelope during budding. The whole (+) strand is covered in a nucleocapsid protein and targeted at M protein-containing intracellular membranes. The Golgi smooth-walled vesicles carry the newly formed virions to the cell surface, where they bud through these membranes and merge with the plasma membrane to release the virus from the cell without lysis.

8.3.2. Negative Stranded RNA Viral Genomes

Negative-strand RNA viruses have a very different life cycle than other RNA viruses. To be more precise, infectious virus particles must also introduce their own RNA-dependent RNA polymerase into the infected cell in order to initiate the initial round of virus-specific mRNA synthesis. The genome of (–) RNA viruses is not contagious. The greater diversity observed in these viral genomes compared to (+) stranded viruses may be attributed to expression challenges. These organisms typically have big genomes that contain a greater amount of genetic information. This leads to the segmentation of the majority of these viral genomes. In its purified RNA form, none of these genomes are infectious. While some eukaryotic cells have been shown to harbor the RNA-dependent RNA polymerase gene, most uninfected cells lack sufficient RNA-dependent RNA polymerase activity to facilitate virus replication. These genomes are essentially inert because the (–) stranded RNA genome cannot be translated into mRNA without the help of the viral polymerase in every particle.

The linear, single-stranded (–) sense RNA found in the genetic material of non-segmented (–) stranded RNA viruses belongs to the order Mononegavirales. This order consists of four families: Rhabdoviridae, Paramyxoviridae, Filoviridae, and Bornaviridae. These families contain a wide range of pathogens that affect humans, animals, and plants, including the deadly Ebola and Marburg viruses, the recently discovered Nipah and Hendra viruses, the canine distemper virus, the rabies virus, and Rinderpest virus. One thing unites these four virus families: the transcription of a number of distinct monocistronic mRNAs expresses the genetic information found in their (–) sense RNA genomes. A single polymerase entry site close to the 3′ end of the genome is the starting point for transcription, which is required to proceed in a sequential fashion. However, attenuation happens specifically at each gene junction, progressively reducing the transcription of genes farther from the promoter. The order of the core genes in the Mononegavirales is supported by this straightforward regulatory mechanism, which is highly conserved across these four families. Genes with large product requirements are located close to the promoter, while genes with catalytically required products are located farther away. Nearly all genera in the four families of Mononegavirales have complete genome sequences known, ranging in size from ⊠8.9 kb for the Borna disease virus to ⊠18.9 kb for the Ebola virus, which is twice as large and has 5–10 genes. Four core genes that encode for a nucleocapsid protein gene (N), phosphoprotein gene (P), matrix protein gene (M), and an RNA-dependent RNA polymerase gene (L) are present in the genomes of all four virus families. The Rabies virus possesses an extra G gene that codes for the transmembrane attachment and glycoprotein entry. Three more genes that code for transmembrane glycoproteins have also been found in a few members of this order. Furthermore, it was discovered that the genus Rubulavirus contains a small hydrophobic gene that codes for a protein with an unclear function.

Numerous genes in this order encode nonstructural proteins, which can be encoded by several ORFs within a single gene or separately as distinct genes. Separate genes encoding two non-structural proteins involved in thwarting the host response are present in the pneumoviral genomes. The genomes of avian pneumoviruses do not contain these genes. Between the G and L genes on some rhabdoviruses is a gene that codes for a small nonstructural protein. The P gene's coding ability is expanded in the Paramyxovirinae to produce an unexpectedly large number of polypeptides through co-transcriptional editing or the use of multiple overlapping open reading frames on a single transcript. It has been reported that Pneumovirinae has an M2 gene that codes for the M2–1 protein, an additional transcription factor, and another overlapping

ORF that codes for the M2–2 protein. Certain viruses are ambisense, meaning they are not strictly (–) sense they have some (+) sense and some (–) sense. Animal viruses (the Phlebovirusgenus of bunyaviruses and arenaviruses) and plant viruses (the Tospovirusgenus of the bunyaviruses) both use ambisense coding techniques (Holmes et al., 2005).

8.3.2.1. Genome of Bunyavirus

Viruses in the family Bunyaviridae possess three distinct linear, single-stranded, negative or ambisense RNA segments in their genome named as small (S), medium (M) and large (L). RNA of the S segment is 0.9 kb, codes for the nucleocapsid protein and a non-structural protein NSs, which interferes with innate immunity. M segment RNA is 5.7 kb, encodes for a polyprotein, eventually giving rise to two glycol proteins Gn and Gc. The large segment (L) RNA is 8.5 kb, codes for the transcriptase, replicase p, and the large RNA-dependent RNA polymerase proteins. In addition to polymerase activity, Bunyaviral L proteins have an endonuclease activity which cleaves cellular messenger RNAs for the production of capped primers used to initiate transcription of viral messenger RNAs which is known as cap snatching. Exceptionally, in the members of the Phlebovirus and Tospovirus genera, the S segment is rather larger than that of M and L. All three genome segments have the same basic structure with the coding region flanked by untranslated regions (UTRs) at the 5′ and 3′ ends. In common with all (–) sense RNAs, the 5′ ends are not capped and the 3′ ends are not polyadenylated. Even though bunyaviruses are categorized into (–) strand viruses, some of the members have genome segments with an ambisense coding strategy, such as Phlebovirus and Tospovirus with 5′ end of each segment is (+) sense, but the 3′ end is (–) sense.

Bunyaviral genomes exhibit remarkable versatility. It is possible to interchange or significantly shorten the UTRs of the three segments, to switch around the ORFs within an ambisense segment, to extend the genomes by inserting extra ORFs and epitope tags, and to transform the tripartite genome into two- or four-segmented versions. The terminal sequences of the UTRs, which are mostly complementary and resemble panhandle structures, contain the promoters for transcription and replication of the genome segments (Horimoto et al., 2005).

8.3.2.2. Genome of Rhabdovirus

A member of a unique class of viruses, rhabdoviruses have a linear single-strand (–) RNA genome that is complementary to functional, virus-specific (+) sense messenger RNAs (mRNAs) and entirely non-infectious. These viruses are bullet-shaped, widely distributed, and have a very wide variety of host systems, including plants, invertebrates, and vertebrates. The human pathogen Rabies virus and the animal pathogen Vesicular Stomatitis Virus (VSV) are the two most well-known and extensively researched rhabdoviruses. Their genomes are up to 11 kb in size and are not segmented. The genome has a leader region at the 3′ end that is roughly 50 nucleotides long and 60 nucleotides UTR at the 5′ end. Nucleoprotein (N), phosphoprotein (P), matrix protein (M), glycoprotein (G), and polymerase (L) are its five genes. The five genes are separated by brief intergenic intervals, and each gene is terminated by a conserved polyadenylation signal. Rhabdoviral genomes comprise two structural elements: an envelope that surrounds the helical ribonucleoprotein core (RNP). The nucleoprotein in the ribonucleoprotein firmly encases the genomic RNA. The phosphoprotein and the large protein (L-protein), two viral proteins, are linked to the ribonucleoprotein. About 400 tightly packed trimeric spikes made of glycoprotein are present on the viral surface. The matrix protein may serve as the key protein in the rhabdovirus assembly process since it is connected to both the envelope and the ribonucleoprotein.

Warm-blooded animals can develop fatal encephalitis from the neurotropic rabies virus, which is a member of the genus Lyssavirus in the family Rhabdoviridae. The same five genes—nucleoprotein (N), phosphoprotein (P), matrix protein (M), glycoprotein (G), and polymerase (L)—make up the roughly 12-kb non-segmented single-stranded negative-sense RNA genome of this virus producing five different proteins. The G protein-encoding gene is known to be important for the pathophysiology of the rabies virus. This virus has a viral nucleoprotein that tightly encapsulates its genome. The viral-specific RNA-dependent RNA polymerase and its cofactor, phosphoprotein, will use the RNA-N complex as a template for transcription and replication. Along with other animal species like cats, ferrets, foxes, pigs, cattle, and donkeys, domestic dogs are the main reservoirs and vectors of rabies transmission. In Sihui, in the Guangdong Province of southern China, an outbreak of pig rabies in March 2011 claimed the lives of fourteen pigs. Recently, the entire genomic nucleotide sequence of the virulent wild RABV strain GD-SH-01, which was isolated from rabid pig brain tissue, was discovered.

8.4. SEGMENTATION IN VIRAL GENOMES

Segmented virus genomes are composed of two or more physically distinct nucleic acid molecules that are subsequently bundled into a single virus particle. The (–) sense RNA genome of the three viral families Arenaviridae, Bunyaviridae, and Orthomyxoviridae is not a single, continuous molecule but rather is composed of multiple segments. For these segmented genomes to produce and replicate mRNA, which enters the cell during infection along with other viral components, RNA-dependent RNA polymerase enzyme must be present.

Both many benefits and drawbacks can arise from viral genome segmentation. The physical characteristics of nucleic acids, in particular the shearing tendency of long molecules and the maximum length of each individual virus's nucleic acid chain that can fit inside the capsid, limit the size of a non-segmented virus genome. Since single-stranded RNA is far more brittle than double-stranded DNA, the shearing problem is especially important. The genome becomes biologically inactive when it is physically broken because it is unable to be fully translated, transcribed, or replicated. Genome segmentation increases the potential coding capacity of the entire genome, which increases the risk of shearing. The virus can produce reassortants because of genome segmentation. RNA molecules from various viral strains are rearranged in doubly infected cells during the course of this process, which involves morphogenesis and replication. Consequently, new RNA segment combinations can be acquired by progeny viruses, giving them unique characteristics. With influenza-A viruses, this mechanism—known as an antigenic shift—is more prevalent and thoroughly researched. The primary drawback of the segmentation strategy is that each virus particle needs to contain every single genome segment; otherwise, the virus will become malformed due to the loss of genetic material. Generally speaking, it is unclear how this packaging control is accomplished (Horimoto et al., 2005).

8.4.1. Influenza Virus

The most frequently spread virus that infects humans' respiratory systems is the influenza virus. Compared to other respiratory infections like the common cold, the flu frequently results in serious illness. The influenza viruses were classified in February 2014 by the World Health Organization into three categories: pandemic influenza, seasonal influenza, and zoonotic or variant influenza. Every year, particularly in the winter, seasonal influenza viruses of the A, B, and C types infect humans and cause the flu. These viruses are spread from person to person through sneezing, coughing, or touching contaminated surfaces. These seasonal viruses can cause mild to severe infections, and in certain high-risk individuals, they can even be fatal. The subtypes of type-A influenza viruses are further classified based on the particular variants and combinations of two proteins found on the virus surface: neuraminidase (N protein) and hemagglutinin (H protein). At the moment, the two most common seasonal influenza A virus subtypes are A(H1N1) and A(H3N2). Named after the regions where they were initially discovered, the two type B viruses are known as the Yamagata lineage and the Victoria lineage. Unlike influenza A and B, which cause mild infections linked to occasional cases and small localized outbreaks, type C influenza carries a far lower disease burden.

When a virus has not previously spread among humans and infects individuals who lack sufficient immunity, pandemic influenza strikes. It is possible for these viruses to appear, spread, and trigger significant outbreaks outside of the typical influenza season. A sizable section of the populace is becoming infected because most people lack immunity to these viruses. The 2009 H1N1 pandemic was caused by the A(H1N1) virus, which had not previously been identified. The virus spread throughout the world in 2009. Since then, as was previously mentioned, this virus has become established in human populations as a seasonal influenza virus. As of right now, no pandemic virus is known to be active worldwide.

Animal-borne influenza viruses, such as avian influenza virus subtypes A(H5N1) and A(H9N2) and swine influenza virus subtypes A(H1N1) and A(H3N2), are known as zoonotic or variant influenza. Dogs and horses are domestic animals with unique influenza virus types. These animal viruses are different from human influenza viruses and do not readily spread to humans, despite sharing the same subtype name as human viruses. Sometimes, some may infect people, resulting in anything from a serious case of pneumonia to a fatal case of mild conjunctivitis. Zoonotic influenza infections, which do not spread quickly among humans, are typically contracted by direct contact with infected animals or contaminated environments. Such a virus, like A(H3N2), would be classified as an epidemic or pandemic if it obtained the ability to spread easily among humans either by adaptation or by acquiring specific genes from human viruses. "swine flu" refers to the influenza viruses infecting swine and is never used when such viruses infect humans. In 2011, influenza A(H3N2) circulating in swine also started to infect people in the United States. These viruses were isolated and labeled as variant (v) in order to distinguish them from human viruses of the same subtype. Avian influenza viruses are other animal viruses, like A(H5N1), A(H7N7), A(H7N9), and A(H9N2) that infect humans. Animal influenza viruses are called after the natural animal host they infect. Examples of these include avian, swine, and equine influenza viruses.

8.4.2. The Genome of Influenza Virus

The influenza viruses A, B, and C each have eight genome segments, while influenza C has seven. Every segment is (–) sense RNA that is single-stranded. Each genomic segment's corresponding protein was identified through mutant analysis and electrophoretic mobility measurements. Table 8.1 lists each segment's size and the list of proteins it encodes. Common nucleotide sequences at the 5′ and 3′ ends of each of these eight segments are required for genome replication. These sequences complement each other, and base pairing holds the ends of the genome segments together to create a structure resembling a panhandle that may be involved in replication. The genome segments are not packed as naked nucleic acid, but rather in conjunction with the nucleoprotein, which appears as helical structures in electron micrographs. This suggests some interaction between the genome segments and may account for the influenza virus particles' low error rate when packing the genome segments within each viral particle. The precise process by which proteins and genome sequences interact to produce this delicate mechanism remains entirely unknown (Nelson et al., 2008).

Table 8.1. *Size of Influenza Virus Segments and the Proteins They Encode*

RNA Segment	No. of Nucleotides	Encoding Protein	No. of Amino Acids
1	2341	Polymerase PB2	759
2	2341	Polymerase PB1	757
3	2233	Polymerase PA	716
4	1778	Haemagglutinin HA	566
5	1565	Nucleoprotein NP	498
6	1413	Neuraminidase NA	454
7	1027	Matrix protein M1	252
		Matrix protein M2	97
8	890	Non-structural protein NS1	230
		Non-structural protein NS2	121

Source: Chaitanya KV. Structure and Organization of Virus Genomes. Genome and Genomics. 2019 Nov 18:1–30. doi: 10.1007/978-981-15-0702-1_1. PMCID: PMC7119911.

Human respiratory tract infections that are acute and highly contagious are caused by the influenza A virus. The World Health Organization (WHO) estimates that infection with this seasonal virus results in between 250,000 and 300,000 fatalities as well as five million cases of severe illness. Reassortant influenza: A strain of influenza virus that develops when genes from one variant of the virus merge with another when influenza virus variants occur simultaneously in an animal population, such as swine or bird reservoirs. Over the past ten years, there have been significant human pandemics brought on by reassortant virus transmission and zoonotic transmission of swine or avian influenza. The eight single-stranded (–) sense RNA molecules that make up the genome of the influenza A virus (family Orthomyxoviridae) span roughly 13.5 kilobases (kb). The segments encode up to 11 proteins and vary in length from 890 to 2341 nucleotides (Table 8.1). The genes encoding the viral envelope proteins that attach to cells and spread the infection are called hemagglutinin and neuraminidase. The internal viral proteins needed for viral assembly and replication are encoded by the remaining six viral genes, PB2, PB1, PA, NS, M, and NP. The replicative complex of influenza A virus proteins, which includes nucleoprotein and polymerases PB2, PB1, and PA, is essential for the virus's ability to infect hosts in a variety of settings. It has been discovered that the NS gene has a dual role in controlling the host cell response and enhancing viral replication within the host. It has been demonstrated that the NS protein's C-terminus is essential for the suppression of interferon expression (Nelson et al., 2007).

8.4.3. Reassortment of Influenza Viral Genomes

Viral reassortment is a genetic recombination process unique to segmented RNA viruses, where multiple viruses infecting the same host cell mix gene segments to create new virus combinations. This phenomenon occurs in all segmented viral families but is most studied in influenza viruses, where it plays a key role in interspecies transmission and the development of pandemic strains. The emergence of new influenza virus strains in humans, causing pandemics, is often linked to reassortment events between novel and existing viruses. Reassortment occurs when multiple virus particles enter a host cell, producing new genome segments. Two proposed models explain how this reassortment process occurs (Tautenberger et al., 1997).

Random packaging model posits the incorporation of the viral RNA into the virions without any discrimination, permitting the likelihood of forming viable reassortants with an entire genome set occurs by chance.

Selective packaging model states the packaging of a virus particle into eight unique viral RNA segments through specific packaging signals.

An epistatic interaction of packaging signals between viral segments has been identified by experimental evaluation of RNA interactions during viral segment assembly. This interaction may be crucial in controlling the reassortment. By manipulating the packaging signals of distinct influenza viruses experimentally, Essere et al. successfully overcame the bias seen toward particular genotypes. Baker, along with others, has demonstrated the crucial role that packaging signals play in reassortment by demonstrating how switching the packaging signals of two distinct influenza species has allowed the reassortment of viable particles that have not been seen in nature. These observational studies show some significant implications for viral evolution through epistatic interaction between gene segments and the emergence of novel reassortants, despite the lack of a mechanistic understanding of the role of packaging signals.

Mixed infections with multiple influenza subtypes are common in avian hosts, accounting for up to 25% of all infections. Genomic analysis has revealed various restrictions on reassortment between co-circulating influenza viruses, influenced by factors such as host species, virus subtype, and genetic compatibility. Wild aquatic birds, the natural reservoir of influenza, exhibit the highest rates of reassortment, leading to the creation of new genetic combinations. This process results in amino acid mutations, particularly in the hemagglutinin glycoprotein, causing antigenic changes. Selection pressure from host immunity may also drive evolutionary change through reassortment. The emergence of drug-resistant mutations, such as amantadine-resistant H3N2 and oseltamivir-resistant H1N1 viruses, highlights the challenges of controlling influenza through traditional methods (Nelson et al., 2008).

8.5. MULTIPARTITE VIRAL GENOMES

Genome segmentation is a common trait found in various viruses, with multipartite viruses being particularly unique as their genome segments operate independently, potentially reducing their infectivity. Multipartite viruses have fragmented genomes consisting of up to eight separate segments, each containing essential genes for the virus to replicate. These viruses rely on complementation, requiring all segments to come together to produce new viral particles. This complementation process greatly impacts how multipartite viruses are transmitted, as multiple viral particles must enter a cell to ensure the representation of each segment. The multiplicity of infection (MOI) plays a critical role in the biology of multipartite viruses. Additionally, multipartite viruses display asymmetry in host distribution, as all known viruses of this type exclusively infect plants, with no instances of animal infection reported. This distribution pattern may be due to the complementation requirement acting as a limiting factor, potentially leading to viral extinction.

8.5.1. The Genome of Gemini Virus

Gemini viruses are small, twinned, isometric particles measuring 20 × 30 nm. They contain circular, single-stranded DNA genomes of about 2.7 kb in size. These viruses are divided into two main subgroups based on the host plants they infect, the insect vectors involved, and the number of DNA components they possess. One subgroup includes viruses with a single DNA component that infect monocotyledonous plants and are spread by leafhoppers. The other subgroup consists of viruses with a bipartite genome, labeled as DNA A and B, that infect dicotyledonous plants and are transmitted by whiteflies. Some viruses, like Beet curly top virus (BCTV), share characteristics of both subgroups. The Mastrevirus and Curtovirus genera have single-stranded (+) sense DNA genomes of 2.7 kb inside their virions. However, both (+) and (–) sense strands are present in infected cells, containing protein-coding sequences. The genus Begmovirus has a bipartite genome with two 2.7 kb circular, single-stranded DNA molecules. These strands differ significantly in sequence, except for a common 200 nucleotide non-coding sequence necessary for DNA replication. Each strand is packaged in separate capsids. For a successful infection, both parts of the bipartite genome are required, with at least one copy of each segment found in a minimum of two viral particles. Both DNA strands in infected cells contain coding information, with overlapping open reading frames (Vijaykrishna et al., 2011). The multipartition of the genomes may be necessary to regulate the expression of this dense genetic information.

CASE STUDY

GENOME ANALYSIS OF SARS-COV-2 CASE STUDY: AN UNDERGRADUATE ONLINE LEARNING ACTIVITY TO INTRODUCE BIOINFORMATICS, BLAST, AND THE POWER OF GENOME DATABASES

The COVID–19 pandemic has significantly impacted scientific research and higher education. With the chaos, confusion, and concern caused by the SARSCoV2/COVID19 pandemic, there are lots of questions and general curiosity about coronaviruses and infectious disease. This global health crisis provides an optimal platform for teaching and discussing scientific research. COVID19-specific learning activities can help guide students and the public to accurately find scientific answers to their questions and help educate and teach true scientific principles to those eagerly wanting answers about COVID19. A SARSCoV2/COVID19 case study has been developed that allows students to align the genomes of 4 different coronaviruses to identify potential origins of the virus. Additional questions direct students to think critically about how an RNA viral genome may have different implications than a DNA viral genome, or why mutations in the surface glycoprotein may have big impacts on viral transmission and immunity. This case study was developed and implemented in Spring 2020 and was well received by students. It provides an introduction to genomics and bioinformatics through a highly relevant and applicable topic that everyone is interested in: COVID19.

Introduction

The 2020 pandemic of COVID-19 has had a devastating impact on human health, economies, cultural practices, and higher education. However, it has also provided a unique opportunity to teach about microbes in a highly relevant context. Modern technology has allowed a robust and rapid research response to this virus that has not been seen in viral outbreaks in the past. Data have been made publicly available at a record rate, allowing open access to the latest results and helping to direct public health decisions. The genome for SARSCoV2 (causative agent for COVID-19) was sequenced and made publicly available before most of the general public even knew it existed. Professors and instructors teaching microbiology and other biology courses can capture the enthusiasm and true curiosity surrounding COVID-19 by providing engagement with scientific literature and helping students find answers for themselves. Free online databases allow students to access cutting-edge genomic data for the virus causing COVID-19, SARS-CoV-2. With a basic introduction to genomics and bioinformatics, students can generate alignments, search for related ancestors, and discover unique mutations and sequences that make COVID-19 different from other coronaviruses. This is a COVID-19 case study that was generated and used during the COVID-19 pandemic and will allow students to compare genomic sequences of different coronaviruses to understand the origin of SARS-CoV-2 better.

Procedure

SARS-CoV-2 Genome Analysis Case Study

Case studies have been used in active learning classrooms for decades to help the learner connect course objectives with real-world applications. Some courses are designed exclusively around case studies and have shown impressive learning and beneficial retention of course concepts. The emergence of COVID-19 surprised most of the public population and there was a lot of confusion regarding where the virus came from. This case study was developed with the following objective: to introduce bioinformatics and genomics to the students so they can compare genomic sequences of different coronaviruses to better understand the origin of SARS-CoV-2.

Students enrolled in a 300-level Microbiology course with little or no prior bioinformatics knowledge were given this case study as an online learning activity during unit 3 (viruses and infectious disease). They were given a week to work on it and were instructed to complete it individually or with a partner. The case study directed students to the National Center for Biotechnology Information (NCBI) website and the Basic Local Alignment Search Tool (BLAST) for finding genomes for four different coronaviruses: SARS-CoV-2 (COVID-19 GenBank #MN908947.3), a bat coronavirus (RatG13; GenBank #MN996532.1), the SARS coronavirus (GenBank #NC_004718.3), and the coronavirus that causes the common cold (J0304/229E; GenBank #JX503061.1). Students perform paired alignments to look at genetic similarities and differences as the genomes align. As students complete the case study handout, they identify that the SARS-CoV-2 (COVID-19) genome is most similar to the bat coronavirus genome.

Students then have to scroll through the alignment to identify the largest difference between the bat coronavirus and the novel SARS-CoV-2 and determine which gene is affected. Upon discovery that the surface glycoprotein is the most mutated between the two genomes, several questions and learning opportunities are presented for students to think critically regarding why mutating a surface glycoprotein will affect viral transmission. Depending on the course and the learning objectives, the case study can provide great discussion points for molecular biology, immunology, virology, serology, etc. In the molecular biology course where this case study was implemented, the focus was on the viral genome, and further analysis discussed why an RNA genome would have different implications than a virus with a DNA genome.

Student Results

Case studies have been implemented in this course (300-level Microbiology) for a few years. Each of the three units has a corresponding case study for the students to work through. In 2020, the usual Unit 3 case study was replaced with a case study of COVID-19 and genome alignment. Interestingly, students were curious and much more engaged with the COVID-19 case study. Compared with the prior two case studies, the COVID-19 genome alignment learning activity had significantly more questions and student inquiries

(Figure 1A). Most of the questions were conceptual, with students inquiring deeper about the viral genome and asking what certain results mean. Additionally, students spent nearly double the amount of time on the assignment than they did on the other case studies (Figure 1B). Although the data do not reveal whether the extra time was due to a more technical case study or due to interest and curiosity, feedback from the students indicated the extra time was not because it was harder but because they were genuinely interested and continued their search for knowledge as they worked through the questions. Student comments stated that they appreciated the COVID-19 course material and case study because it helped them understand the virus outbreak and the scientific reasoning for the pandemic and global response (Table 1).

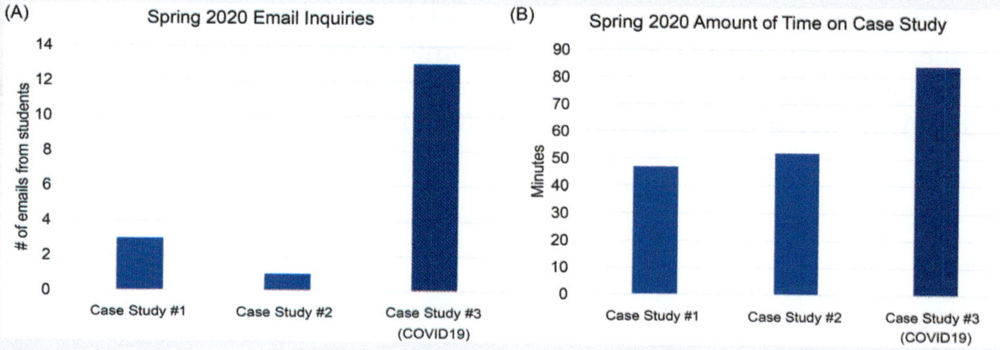

Figure 1 *Analysis of student involvement in the COVID-19 Case Study. (A) The total number of e-mail inquiries sent to the professor regarding the assigned case study ($P<0.05$). (B) The average amount of time (minutes) spent on each case study. Data was determined through email analysis, blackboard analytics, and student feedback ($P<0.005$). Spring 2020 enrollment was 49 students.*

Source: J. Jordan Steel, "Genome Analysis of SARS-CoV-2 Case Study: An Undergraduate Online Learning Activity to Introduce Bioinformatics, BLAST, and the Power of Genome Databases," 2021, DOI: https://doi.org/10.1128/jmbe.v22i1.2245.

Table 1. *Student Comments*

"I think that it was very relevant and informative to learn about the COVID-19 outbreak in the class and outside of class. Trying to educate people with their limited knowledge of COVID-19 and being aware of its capabilities was really helpful."
"I like how it offered perspective for the entry mechanism of the virus into the cell. Combined with Biochem and Genetics from last semester I feel like I knew what was happening and was very educated on attempts and efforts to help and reasons why masks, social distancing, and research are effective."
"I enjoy direct applications. The COVID-19 material helped bring a sense of purpose to the material."

Conclusion

This COVID-19 genome analysis case study provides relevant and very applicable hands-on opportunities for students to test out bioinformatics, genome alignments, and BLAST tools to discover real-world answers for where the COVID-19 virus originated. This curriculum activity can be implemented through online learning formats or in-person classrooms to teach critical molecular and genomic aspects of virology, microbiology, immunology, or any course that is relevant to COVID-19. The case study can be modified and extended to include elements of transmission, prevention, treatment, etc. Responses from students indicate that the COVID-19-specific course material was greatly appreciated and helped them understand the course objectives by providing a real-world context.

As educators and scientists prepare for future semesters, it is important to consider the COVID-19 world that we are living in and understand the risks but also the great opportunities that this provides for teaching and educating students and the general public about microbiology and the power of genomic data. The world is seeking answers for COVID-19 and relevant, data-backed, and science-driven information must be provided to understand SARS-CoV-2 and the resulting COVID-19.

CLASS ACTIVITY

MAPPING GENE THERAPY CONCEPTS

Abstract

- **Activity 1:** In this exercise, students are introduced to concept mapping and are encouraged to consider and apply the technique
- **Activity 2:** This exercise can be used to gauge students' comprehension of the subject at the start and finish of the Gene Therapy: Molecular Bandage? There are word lists available for beginners, intermediate users, and advanced users.

Learning Objectives

With the aid of concept maps, students will be able to arrange information.

Estimated Time

- Class time 45 minutes
- Prep time 30 minutes

Materials

- Copies of student pages
- Optional – Inspiration® software and computers.

Instructions

Activity 1: Teaching Concept Mapping

- As a class, discuss why it would be useful to organize facts, thoughts, ideas, etc.
- Option 1: Give students the example concept map (page 1) and go through it as a class.
  - Have students answer the questions on their own or with a partner.
  - Lead a class discussion of the questions.
  - Discuss why items are placed in one location and not another.
  - Remind students that sometimes things can be placed in multiple places.
- Option 2: Have students create the sample concept map using Inspiration® software.
  - Students can label the Main Idea, Large Topics, and Subtopics using the program and then
  - print out the concept map.

  - Proceed with a class discussion as in Option 1, above.
- To assess understanding, have students make their own concept maps utilizing the provided template. This can be done as homework or as an in-class exercise.

Activity 2: Concept Maps on Gene Therapy

Utilize concept maps as both pre- and post-assessments for the Gene Therapy: Molecular Bandage? module to evaluate students' comprehension of the subject. The initial assessment can help guide the direction of the learning activities for this module and aid students in pinpointing areas for further investigation. By comparing the results of the pre- and post-tests, you and the students can gauge the level of understanding gained throughout the module.

Pre-Test:

- Distribute the Beginning Word List (page 3) and instruct students to organize the words on the list into a concept map.
- Optional: Have students create their concept map using Inspiration® software.
- Allow students time to complete their maps without rushing them.
- Collect the maps and file them for reference at the end of the module.
- **Post-Test:** Three-word lists, Beginning, Intermediate, and Advanced, are provided.
- **Beginning:** Covers basic concepts from the module.
- **Intermediate:** Requires that students go into more detail.
- **Advanced:** Requires even more details.
- Make use of the word list that is suitable for the students' level and/or the material you have covered.
- Give the students the relevant word list and instruct them to arrange the words into a concept map.
- Optional: Have students create their concept map using Inspiration® software.
- Allow students time to complete their maps without rushing them.
- Return both the pre- and post-test to students.
- Have students compare the two maps, identifying what items they have learned or still need to learn.

Adaptations

- Students can arrange the words and phrases in the Word Lists into potential groups by cutting them apart. The paper slips can be adhered to or taped together to create a concept map once they have chosen an organization
- Rather than utilizing the word lists found in the Student Pages, have students create their own (either individually or as a class). This can be especially useful

when you first assess the students' prior understanding of gene therapy at the beginning of the module

- Have students add to and rearrange the pre-test concept map as they learn about gene therapy, rather than saving it until the end of the module. Students may choose to use a different color for each set of additions if they are working on paper. Each map they make can be printed out if they are using Inspiration® software. You and the students can track how the students' understanding of gene therapy has changed with this progressive set of concept maps (Figure 2).

Building a Concept Map

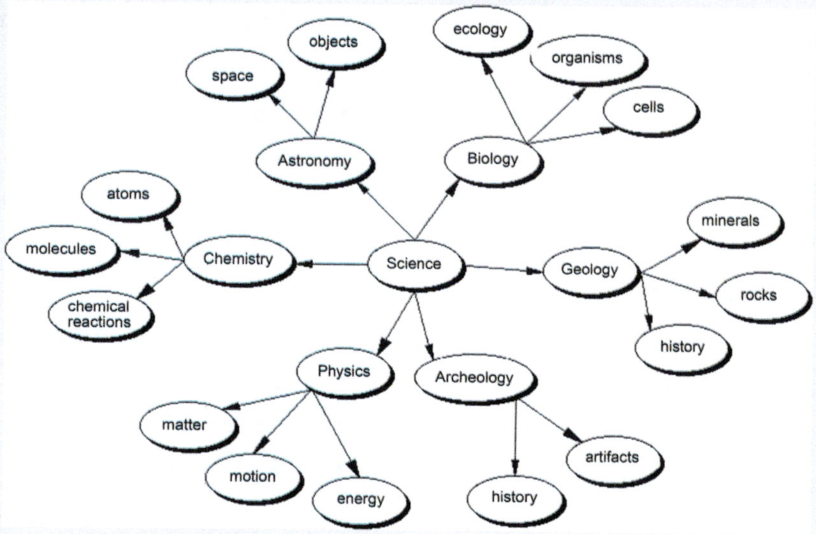

Figure 2. *Concept map.*

Source: https://teach.genetics.utah.edu/content/genetherapy/Mapping-Gene-Therapy-Concepts.pdf.

- Why is Science in the center circle?
- Why is history in more than one place?
- Do all Large Topics need to have the same number of Subtopics? Why or why not?
- What is the minimum number of Subtopics a Large Topic should have?
- Label the following on the sample concept map above: Main Idea, Large Topics, and Subtopics.
- Now, design your own concept map using the template on the next page.

Building a Concept Map

- You may add more stars where you need them.

- Make sure you have at least four Large Topics.
- Make sure you have at least two Subtopics for each of these.

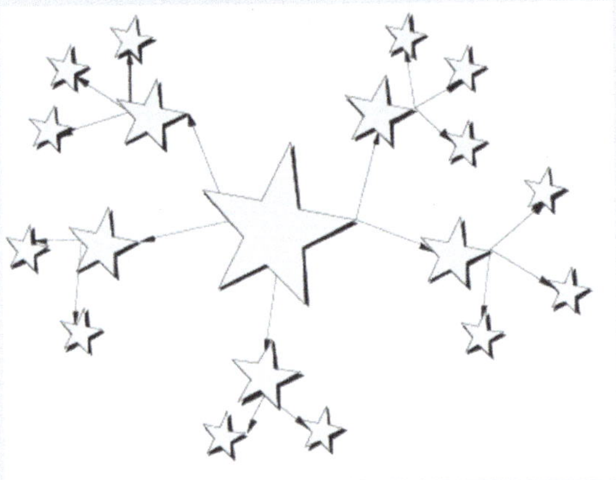

Beginning Concept Map Word List

- Targeting the right tissue
- Immune response
- Vectors
- Single gene disorders
- Naked DNA
- Activating the gene
- Disruption of existing genes
- in vivo
- Integrating the gene
- Fit inside vectors
- Keys to success
- Viruses
- Candidate diseases
- Liposomes
- Procedure
- Integration into the germline
- Risks/Challenges
- ex vivo

Intermediate Concept Map Word List:

- Targeting the right tissue
- Immune response
- Vectors
- Single gene disorders
- Naked DNA
- Activating the gene
- Disruption of existing genes
- in vivo
- Integrating the gene
- Fit inside vectors
- Keys to success
- Viruses
- Candidate diseases
- Liposomes
- Procedure
- Integration into the germline
- Risks/Challenges
- ex vivo
- Enhancement
- Viral ligands must match cell receptors
- Gene integrates into the middle of a functioning
- gene
- Herpes Simplex Virus
- Vector injected into the body
- No dominant-negative mutations
- Preventative medicine
- Transcription and translation must take place
- The introduced gene is passed on to offspring
- Retroviruses
- Cosmetic Purposes
- Adeno-associated viruses

- Vector injected into cells removed from the
- body; these cells are then returned to the body
- Adenovirus

Advanced Concept Map Word List:

- Targeting the right tissue
- Immune response
- Vectors
- Single gene disorders
- Naked DNA
- Activating the gene
- Disruption of existing genes
- *in vivo*
- Integrating the gene
- Fit inside vectors
- Keys to success
- Viruses
- Candidate diseases
- Liposomes
- Procedure
- Integration into the germline
- Risks/Challenges
- *ex vivo*
- Enhancement
- Viral ligands must match cell receptors
- Cosmetic Purposes
- Adeno-associated viruses
- Vector injected into cells removed from the
- body; these cells are then returned to the body
- Adenovirus
- Gene integrates into the middle of a functioning
- gene
- Herpes Simplex Virus
- Vector injected into the body
- No dominant-negative mutations
- Preventative medicine
- Transcription and translation must take place
- The introduced gene is passed on to offspring
- Retroviruses
- Triple-helix forming oligonucleotides
- Repairing mutated proteins
- New approaches
- Antisense
- SMaRT
- Preventing the production of mutated proteins
- Ribozymes

SUMMARY

- Viruses are the most abundant biological entities on this planet. Virology studies show the occurrence of at least 1033 viruses on our planet, which is roughly 10 times higher than the bacterial number. They are found everywhere, including the oceans and soil, in plants and the human body.
- Currently, over 4000 viruses have been described, classified into 71 families. Even though viruses possess small genomes, they exhibit enormous diversity compared with plants, animals and even bacteria. With respect to the genome, viruses are broadly divided into DNA viruses and RNA viruses.
- Cellular life forms of all categories possess genomes with double-stranded DNA and utilize the same as a standard scheme for their replication and expression. In contrast, viruses and other pathogenic and selfish elements exploit all possible inter-conversions of nucleic acids, with their genomes that can be either DNA or RNA, which are single-stranded or double-stranded.
- Restriction mapping is a method used to map an unknown segment of DNA by breaking it into pieces and then identifying the locations of the breakpoints. This method relies upon the use of proteins called restriction enzymes, which can cut or digest DNA molecules at short, specific sequences called restriction sites.
- A mutation occurs when there is a change to the genetic material —either the DNA or RNA —from the original normal or 'wild type' version of the genome of a particular organism or biological entity. The mutation could be entirely novel, or it could have occurred before.
- The ability of the HSV TK- mutants to replicate in the presence of an antiviral drug is an example of a selective advantage for the virus under these conditions of growth. Further, a selection scheme allows for a selective advantage for the growth of TK+ organisms. The easy selection of cells that are TK+ or TK- is an important tool because the HSV TK gene can be removed from the viral genome and used essentially as a portable selectable marker.
- Genomes of RNA viruses encode for a limited number of proteins. RNA viral genomes are broadly divided into double-stranded RNA, positive and negative strand single-stranded RNAs, and monopartite and multipartite RNA viruses.
- Among the viruses possessing RNA genomes, single-stranded plus sense (+) RNA genomes represent an important subgroup, including many pathogenic plant, animal, and human viruses. These genomes contain cis-acting RNA elements that direct different viral processes, such as protein translation, genome replication, and subgenomic mRNA transcription mRNAs.
- The life cycle of negative-strand RNA viruses differs from that of the other RNA viruses in many ways. Specifically, the genome of (–) RNA viruses is not infectious, and infectious virus particles must also deliver their own RNA-dependent RNA polymerase into the infected cell to start the first round of virus-specific mRNA synthesis.

- Segmented virus genomes are those that are divided into two or more physically separate molecules of nucleic acid, all of which are then packaged into a single virus particle.
- Genome segmentation is one of the most common traits in a wide variety of viruses, which is particularly striking in the case of multipartite viruses as their genome segments achieve complete independence at the apparent cost of reducing their infectivity.

REVIEW QUESTIONS

1. What are viral genomes?
2. What is the cleavage site for the restriction endonuclease?
3. What is the mapping of restriction endonuclease sites?
4. What are the recognition sites of restriction endonuclease? Explain the types of sites.
5. How are viruses modified for gene therapy?
6. How does viral replication result in genetic variation?
7. What is the size of the influenza virus segments?

REFERENCES

1. Baker, S. F., Nogales, A., Finch, C., Tuffy, K. M., Domm, W., et al. (2014). Influenza A and B virus intertypic reassortment through compatible viral packaging signals. *Journal of Virology*, *88*, 10778–10791. https://doi.org/10.1128/JVI.01440-14 (accessed on 4th October 2024).
2. Bitner, R., & Kuempel, P. (1982). P1 transduction mapping of the trg locus in rac+ and rac strains of *Escherichia coli* K-12. *Journal of Bacteriology*, *149*(2), 529–533. https://doi.org/10.1128/JB.149.2.529-533.1982 (accessed on 4th October 2024).
3. Essere, B., Yver, M., Gavazzi, C., Terrier, O., Isel, C., et al. (2013). Critical role of segment-specific packaging signals in genetic reassortment of influenza A viruses. *Proceedings of the National Academy of Sciences of the United States of America*, *110*, E3840–E3848. https://doi.org/10.1073/pnas.1308649110 (accessed on 4th October 2024).
4. Holmes, E. C., Ghedin, E., Miller, N., Taylor, J., Bao, Y. M., et al. (2005). Whole-genome analysis of human influenza A virus reveals multiple persistent lineages and reassortment among recent H3N2 viruses. *PLoS Biology*, *3*, 1579–1589. https://doi.org/10.1371/journal.pbio.0030300 (accessed on 4th October 2024).
5. Horimoto, T., et al. (2005). Influenza: Lessons from past pandemics, warnings from current incidents. *Nature Reviews Microbiology*, *3*, 591–600. https://doi.org/10.1038/nrmicro1208 (accessed on 4th October 2024).
6. Nelson, M. I., et al. (2007). The evolution of epidemic influenza. *Nature Reviews Genetics*, *8*, 196–205. https://doi.org/10.1038/nrg2053 (accessed on 4th October 2024).

7. Nelson, M. I., Viboud, C., Simonsen, L., Bennett, R. T., Griesemer, S. B., et al. (2008). Multiple reassortment events in the evolutionary history of H1N1 influenza A virus since 1918. *PLoS Pathogens*, *4*(e1000012). https://doi.org/10.1371/journal.ppat.1000012 (accessed on 4th October 2024).

8. Tautenberger, J. K., et al. (1997). Initial genetic characterization of the 1918 "Spanish" influenza virus. *Science*, *275*, 1793–1796. https://doi.org/10.1126/science.275.5307.1793 (accessed on 4th October 2024).

9. Vijaykrishna, D., Smith, G. J. D., Pybus, O. G., Zhu, H., Bhatt, S., et al. (2011). Long-term evolution and transmission dynamics of swine influenza A virus. *Nature*, *473*, 519–522. https://doi.org/10.1038/nature10004 (accessed on 4th October 2024).

CHAPTER

9 Replication of Positive-Strand RNA Viruses

LEARNING OBJECTIVES

After studying this chapter, you will be able to:

- Discuss on positive-strand RNA virus;
- Evaluate the host factors in positive-strand RNA virus genome replication;
- Determine the reinventing positive-strand RNA virus reverse genetics.

KEY TERMS FROM THIS CHAPTER

Brome mosaic virus (BMV)
Cellular receptors
Endoplasmic reticulum
Gene expression
Golgi apparatus
Plasma membrane
Cell membranes
Coronaviridae
Eukaryotic viruses
Genetic model systems
Intracellular membranes
Polymerase

INTRODUCTION

The genomic structure and replication mechanism of positive-strand RNA viruses distinguish them from other types of viruses in their varied group. The existence of a single-stranded RNA molecule that functions as mRNA right upon entry into the host cell is what distinguishes them. The host cell machinery uses this mRNA to transcribe it into viral proteins, which speeds up the virus's replication.

Positive-strand RNA viruses have a directly infectious genome, which means that ribosomes can translate it into viral proteins as soon as it enters the cytoplasm of the host cell. On the other hand, negative-strand RNA viruses need to produce a complementary RNA strand beforehand to act as a template for mRNA synthesis. Positive-strand RNA viruses can start replicating and generating new viral particles inside the host cell very fast because of their direct translatability.

Positive-strand RNA viruses usually start their replication cycle when they connect to and enter the host cell, which is facilitated by interactions between viral surface proteins and cellular receptors. Viral RNA enters the cell and is released into the cytoplasm, where it functions as messenger RNA (mRNA) for the translation of viral proteins. These proteins include the structural proteins that make up the viral capsid and replicase proteins that aid in the replication of viral RNA (Lai, 1998).

Replication complexes, which are membrane-bound structures that aid in RNA synthesis, are the product of positive-strand RNA virus replication. These complexes help in the replication of viral RNA and protect it from host cell defenses; they are frequently linked to altered host cell membranes. The viral RNA is replicated through the use of RNA-dependent RNA polymerase (RdRp), which creates new RNA strands using the viral RNA as a template

New viral particles are assembled inside particular organelles or in conjunction with the membranes of host cells. Progeny virions are formed when newly synthesized RNA genomes and viral structural proteins come together and are expelled from the host cell. Different methods, such as the host cell lysing or budding from the plasma membrane, might cause release.

Positive-strand RNA viruses can be pathogenic and show a variety of clinical symptoms. They can cause diseases ranging from minor ailments like the common cold to serious conditions like hepatitis C and dengue fever, and they can infect a wide range of hosts, including plants, animals, and humans. Their high rates of mutation and recombination events enable them to develop quickly, which enhances their flexibility and leads to the creation of new strains with modified virulence or antigenicity.

9.1. POSITIVE-STRAND RNA VIRUS

A class of similar viruses, positive-strand RNA viruses (+ssRNA viruses), contain single-stranded, positive-sense ribonucleic acid genomes. The ribosomes of the host cell can directly convert the positive-sense genome into viral proteins by acting as messenger RNA (mRNA). An RNA-dependent RNA polymerase (RdRp), encoded by positive-strand RNA viruses, is utilized during genome replication to produce a negative-sense antigenome, which is subsequently employed as a template to construct a new positive-sense viral genome.

The phyla Kitrinoviricota, Lenarviricota, and Pisuviricota (more precisely, classes Pisoniviricetes and Stelpavirictes) comprise positive-strand RNA viruses. These viruses belong to the realm Riboviria and kingdom Orthornavirae. They share a common RNA viral progenitor and are monophyletic. Group IV of the Baltimore classification system includes +ssRNA viruses.

Pathogens including the West Nile virus, dengue virus, MERS, SARS, and SARS-CoV-2 coronaviruses, as well as less dangerous pathogens like the coronaviruses and rhinoviruses that cause the common cold, are examples of positive-sense RNA viruses.

9.1.1. Genome

RNA viruses with positive-strand genomes typically have a small number of genes, ranging from three to ten, which includes an RNA-dependent RNA polymerase (Lai, 1998). Coronaviruses have the largest RNA genomes known, with lengths between 27 and 32 kilobases, and are thought to have mechanisms for proofreading during replication, such as an exoribonuclease found in the nonstructural protein nsp14.

9.1.2. Replication

Positive-strand RNA viruses contain genetic material that can act as messenger RNA and a genome, and their host cell's ribosomes can translate this genetic material straight into protein. Viral replication complexes established in conjunction with intracellular membranes are formed by the recruitment of the positive-strand viral genome by the first proteins to be expressed upon infection. These complexes can be linked to the membranes of various organelles, such as the rough endoplasmic reticulum, but also those from mitochondria, vacuoles, the Golgi apparatus, chloroplasts, peroxisomes, plasma membranes, autophagosomal membranes, and novel cytoplasmic compartments. They contain proteins from both the host cell and the virus (Figure 9.1).

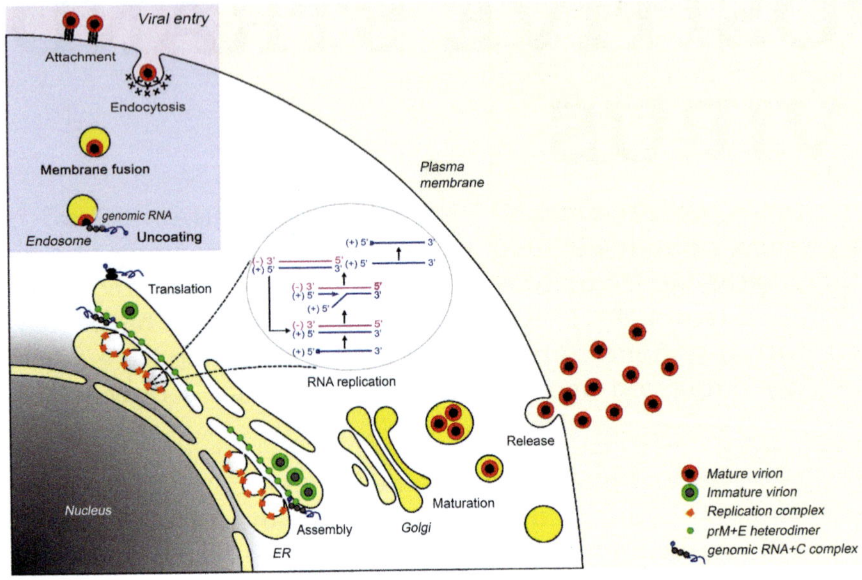

Figure 9.1. *The life cycle of Japanese encephalitis virus a +ssRNA virus: attachment, endocytosis, membrane fusion, uncoating, translation, RNA replication, assembly, maturation, and release.*

Source: https://upload.wikimedia.org/wikipedia/commons/thumb/8/80/Pathogens-07-00068-g002.webp/220px-Pathogens-07-00068-g002.webp.png.

Through double-stranded RNA intermediates, the positive-sense RNA genome is replicated. The goal of replication in these membrane invasions may be to evade the cell's reaction to the presence of dsRNA. Subgenomic RNAs are frequently also produced during replication. The internal ribosome entry site (IRES) elements of the viral genome have a very high affinity for ribosomes, which can cause the entire translation machinery of the host cell to be diverted towards the production of viral proteins after infection. In certain viruses, such as poliovirus and rhinovirus, normal protein synthesis is further disrupted by viral proteases that break down components needed to start translating cellular mRNA.

The viral protein known as RNA-dependent RNA polymerase, which synthesizes RNA from an RNA template, is encoded in the genomes of all positive-strand RNA viruses. RNA-binding proteins, chaperone proteins, membrane remodeling, and lipid synthesis proteins are among the host cell proteins that +ssRNA viruses recruit during replication. These proteins work together to take advantage of the cell's secretory route for viral replication (Lai, 1998).

9.1.3. Recombination

Genetic recombination is a common phenomenon for many positive-strand RNA viruses that occur when two or more viral genomes coexist in the same host cell. Recombination is a typical feature of human +ssRNA viral pathogens. RNA recombination seems to be a primary factor influencing the genome architecture and evolution of the Picornaviridae (poliovirus) family of viruses. Strand flipping, a type of recombination, appears to prevent genome damage during reverse transcription in the Retroviridae (e.g., HIV). The Coronaviridae family undergoes recombination (e.g., SARS). In RNA viruses, recombination seems to be an adaptation for handling damage to the genome. Occasionally, recombination between +ssRNA viruses belonging to different lineages within the same species can also happen. The recombinant viruses that result can occasionally lead to an outbreak of infection in humans, such as with SARS and MERS (Figure 9.2).

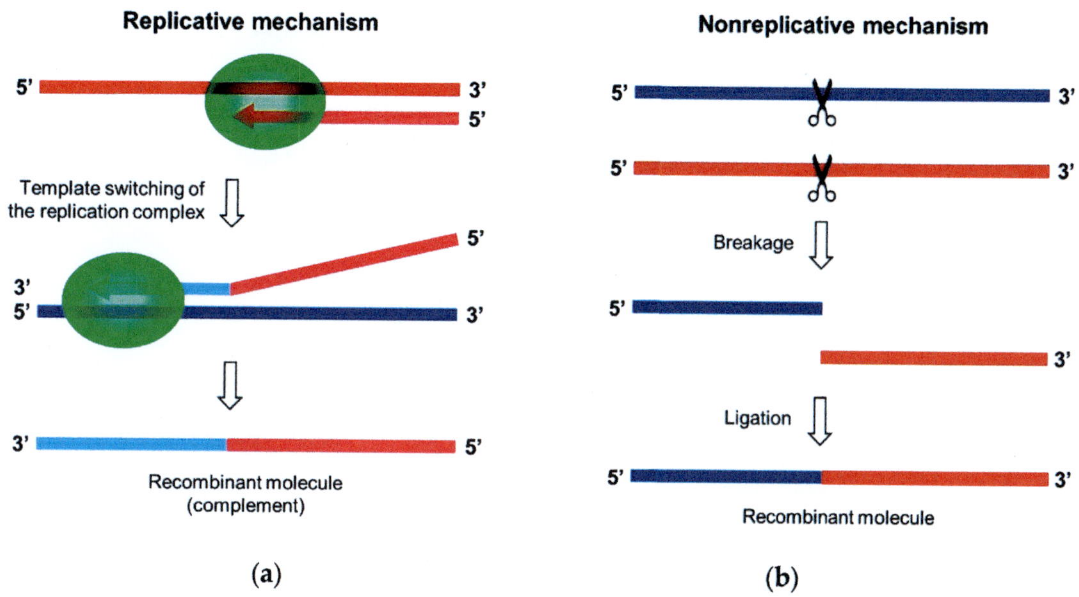

Figure 9.2. *Mechanisms of replicative and nonreplicative RNA recombination.*

Source: https://upload.wikimedia.org/wikipedia/commons/thumb/a/af/Viruses-11-00859-g004.png/220px-Viruses-11-00859-g004.png.

In plants, positive-strand RNA viruses are frequently found. During replication, RNA recombination happens often in tombusviruses and carmoviruses. These viruses' RNA-dependent RNA polymerase can switch RNA templates, which is consistent with the copy choice model of RNA recombination and may represent an adaptive method for the virus to deal with genomic damage. Recombination has also been observed to occur in several +ssRNA viruses of plants, including Sindbis virus and Brom mosaic bromovirus.

9.1.4. Classification

Three phyla—Kitrinoviricota, Lenarviricota, and Pisuviricota—assigned to the kingdom Orthornavirae in the realm Riboviria are home to positive-strand RNA viruses. +ssRNA viruses are categorized as group IV in the Baltimore classification system, which puts viruses in groups according to how they synthesize mRNA (van Regenmortel, 2000) (Figure 9.3).

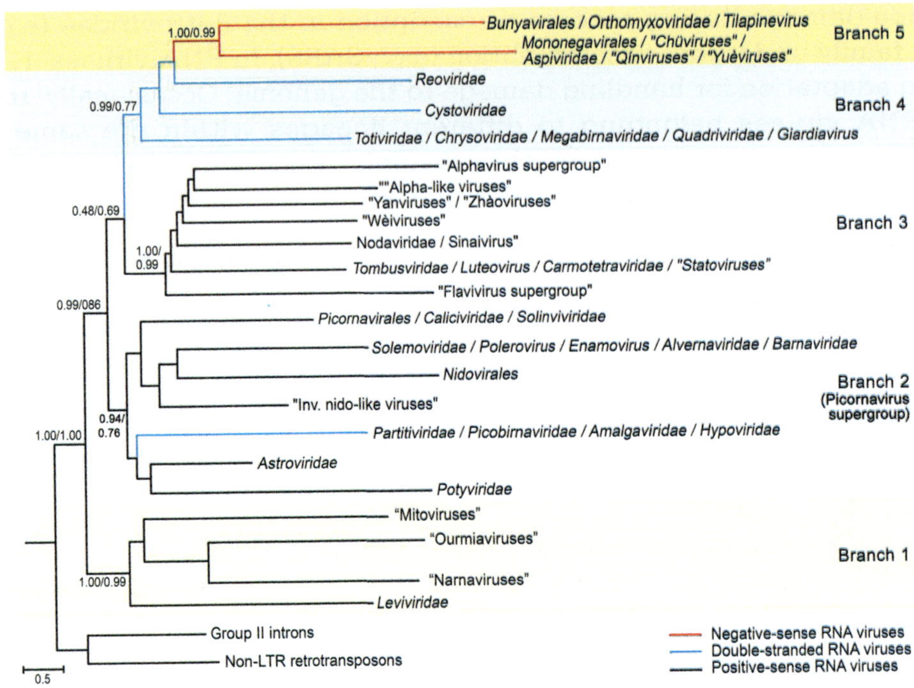

Figure 9.3. *Phylogenetic tree with phylum branches highlighted. Negarnaviricota (brown), Duplornaviricota (green), Kitrinoviricota (pink), Pisuviricota (blue), and Lenarviricota (yellow).*

Source: https://upload.wikimedia.org/wikipedia/commons/thumb/e/eb/MBio.02329-18.F1.large.jpg/220px-MBio.02329-18.F1.large.jpg.

- **Kitrinoviricota:** This is the first phylum of +ssRNA. Together with many other short-genome viruses, the phylum is home to the so-called "alphavirus supergroup" and "flavivirus supergroup." There are four recognized classes in the phylum: Tolucaviricetes, which mainly consists of plant viruses; Flasuviricetes, which contains flaviviruses; Magsaviricetes, which contains nodaviruses and sinhaliviruses; and Alsuviricetes, the alphavirus supergroup, which contains a large number of plant viruses and arthropod viruses.
- **Lenarviricota:** The second +ssRNA phylum is Lenarviricota. It includes the prokaryotic-infecting class Leviviricetes and the eukaryotic-infecting class Leviviruses, which appear to be descended from leviviruses. Four classes make up the phylum: Amabiliviricetes (which includes narnaviruses and their relatives), Howeltoviricetes (which includes mitoviruses and their relatives), and Miaviricetes (which includes botourmiaviruses and their relatives). Leviviricetes is the class that contains leviviruses and their relatives. Regarding Lenarviricota, all other RNA viruses are regarded as belonging to a sister group based on the phylogenetic study of RdRp (van Regenmortel, 2000).
- **Pisuviricota:** Also known as the "picornavirus supergroup" colloquially, Pisuviricota is the third phylum containing +ssRNA viruses." A vast variety of eukaryotic viruses that are known to infect protists, fungi, plants, and mammals are found within this phylum. Pisoniviricetes, which includes nidoviruses,

picornaviruses, and sobeliviruses, and Stelpaviricetes, which contains potyviruses and astroviruses, are the two groups in the phylum that solely include +ssRNA viruses. Double-stranded RNA viruses that are descended from +ssRNA viruses make up the third class, known as Duplopiviricetes (Figure 9.4).

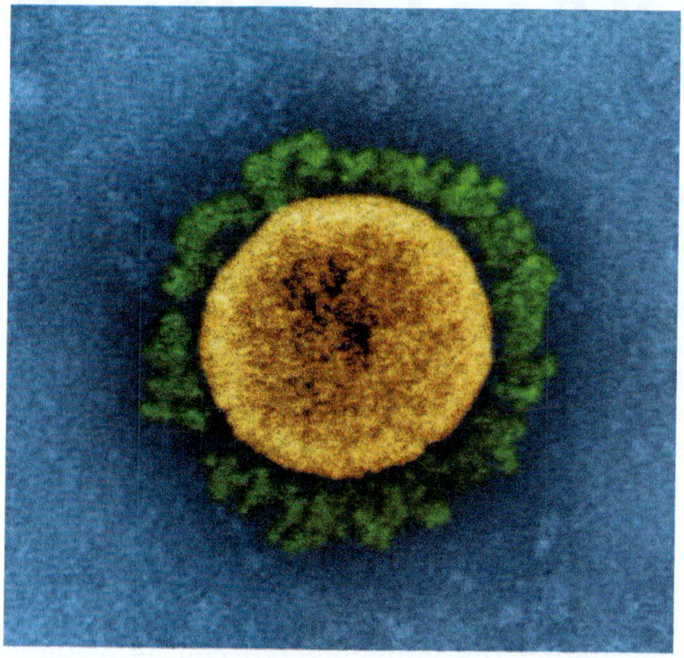

Figure 9.4. *False-color transmission electron micrograph of a SARS-CoV-2 virion. Coronaviruses like SARS-CoV-2 fall in the phylum Pisuviricota.*

Source: https://upload.wikimedia.org/wikipedia/commons/thumb/1/1d/Novel_Coronavirus_SARS-CoV-2_%2850960620707%29_%28cropped%29.jpg/800px-Novel_Coronavirus_SARS-CoV-2_%2850960620707%29_%28cropped%29.jpg.

9.2. HOST FACTORS IN POSITIVE-STRAND RNA VIRUS GENOME REPLICATION

Compared to their hosts, all viruses are gene sparse; even the biggest viral genomes only contain hundreds of genes, whereas host cell genomes often include tens of thousands of genes. As a result, the majority of viral infection processes include interactions between comparatively few distinct viral component types and far more intricate host factor pools. This ocean of host variables is a huge, manipulable resource for viruses with little gene content, as well as the vital environment to which viruses must adapt in order to survive. As a result, host variables are crucial to most viral infection stages, and pinpointing these host factors and their contributions has long been acknowledged as a key research area. The growing trend of close integration between viral and host functions in infection indicates the need to shift from viewing viruses and hosts as separate entities to considering the virus-infected cell as a unified functional unit in infection.

One area where host influences are becoming more and more important is in positive-strand RNA virus replication. More than one-third of all virus genera are classified as positive-strand RNA viruses, which are responsible for many pathogens, including hepatitis C virus (HCV), the severe acute respiratory syndrome coronavirus (SARS), and many of the viruses in the U.S. Health and Human Services Department's Select List of possible bioterrorism agents. Most of the positive-strand RNA virus infection process, if not all, involves host factors. These processes include entrance, viral gene expression, virion assembly, and release. Furthermore, positive-strand RNA viruses target host factors to alter host gene expression and defense mechanisms.

This review focuses on host factors that are involved in the genome replication of positive-strand RNA viruses. Numerous genetic and biochemical techniques have provided evidence for this host factor participation (Lai, 1998). These comprise, among other things, research on the various degrees of permissiveness of certain cell types and extracts for RNA replication; the discovery of numerous host proteins that interact with replication proteins or viral genomic RNAs and, in certain situations, have been functionally connected to replication; and mutational screens in genetic model systems, like the yeast Saccharomyces cerevisiae and the plant Arabidopsis thaliana.

According to recent studies, the assembly of the viral RNA replication complex, the recruitment and selection of viral RNA replication templates, the activation of the complex for RNA synthesis, and other processes are all facilitated by host factors. Each of these interactions between the virus and the host might influence the pathophysiology, tissue specificity, or host specificity of an infection. Every interaction between a virus and its host has the potential to be a target for viral control or optimization, which can enhance the advantageous applications of viruses and their constituent parts.

According to distinct RNA replication genes and characteristics, positive-strand RNA viruses may be categorized into many superfamilies (van Regenmortel, 2000). However, there are enough commonalities in these viruses' RNA replication methods to warrant a class discussion about their replication and an attempt to extrapolate general lessons from the comparison of instances from other superfamilies. The need to coordinate the use of the infecting viral genomic RNA as a template for translation and replication, the assembly of replication complexes on intracellular membranes, and the production of 10- to 100-fold excesses of positive- over negative-strand RNA are some of the common characteristics shared by positive-strand RNA viruses, as will be covered in more detail below.

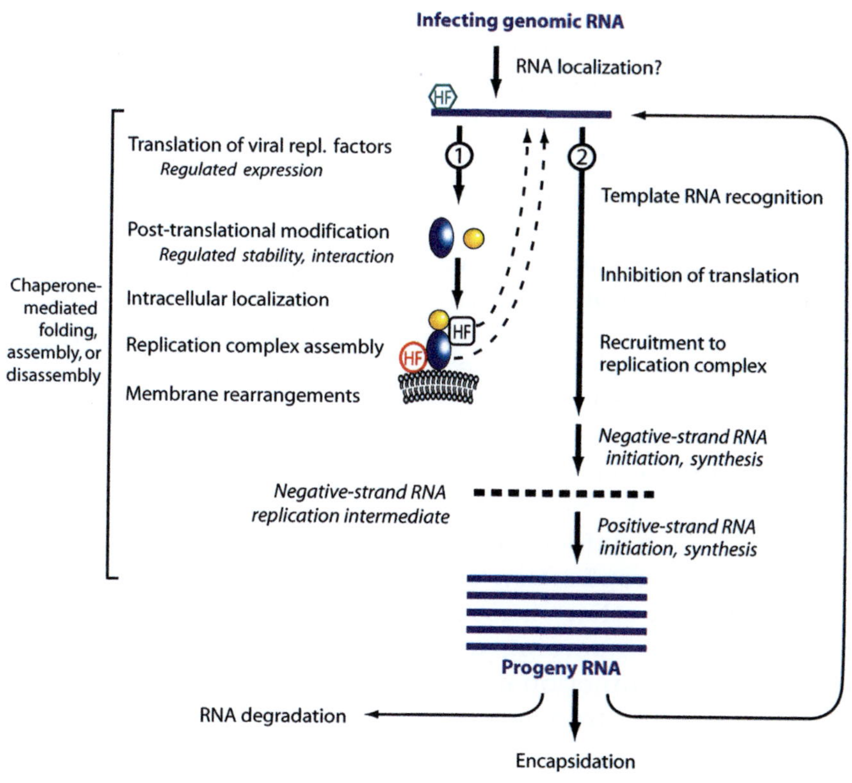

Figure 9.5. *Highly simplified schematic of major steps in positive-strand RNA virus genome replication. Arrows 1 and 2 depict the sequential use of the infecting, positive-strand genomic RNA as a template first for translation and then for RNA replication. As discussed in the text, precedents exist for host factor involvement in most steps shown. Based on examples cited in this review, sample host factors (HF) are shown to interact with viral RNA replication proteins and host membranes. Some host factors may be associated with viral components only transiently during particular steps of the replication process.*

Source: https://www.ncbi.nlm.nih.gov/pmc/articles/PMC165243/bin/jv1530381001.jpg.

Figure 9.5 depicts a generalized simplified strategy for positive-strand RNA virus RNA replication. Using examples from several viruses, we will go over how host components are involved in each stage of this RNA replication process in the following sections. We are sorry that we are unable to include all of the research being done in this fascinating field due to space constraints.

9.2.1. Translation and RNA Replication

Since the genomes of positive-strand RNA viruses serve as templates for both translation and replication, host translation factors and RNA replication interact with each other on a number of levels. The genes for an RNA-dependent RNA polymerase (RdRp), which is involved in genome replication, are present in all known positive-strand RNA viruses. In contrast to other RNA viruses, positive-strand RNA viruses do not encapsidate this polymerase. Therefore, until the genomic RNA is translated to make polymerase and, for the majority of positive-strand RNA viruses, other replication components involved in membrane targeting, template recruitment, RNA capping, and other tasks, viral RNA replication cannot start upon infection of a new cell. Additionally, the replication of RNA by poliovirus and certain other viruses necessitates translation in cis, serving as a limited control on genome functionality. The host factors crucial in virus translation and the viral alteration of translation have been discussed in other publications.

In methods crucial for RNA replication, host factors can either selectively regulate the translational processes of certain viral components or influence viral translation in general. The alphavirus-like superfamily of positive-strand RNA viruses, which includes the brome mosaic virus (BMV), controls RNA replication, subgenomic mRNA transcription, and virion assembly in yeast. Many host genes necessary for BMV RNA replication have been found using yeast genetic techniques. DED1, one of these host genes, encodes an RNA helicase that is necessary for the commencement of translation on all yeast mRNAs. Despite being a necessary gene, DED1 mutations may prevent BMV RNA replication without impairing cell proliferation.

This block is closely connected to 5′-proximal, cis-acting regions that downregulate polymerase translation in comparison to a second, helicase-like RNA replication factor translated from a different BMV genomic RNA. This block is linked to the selective loss of BMV polymerase translation. For many other positive-strand RNA viruses and retroviruses to replicate efficiently, downregulating polymerase levels in comparison to other replication factors is crucial. However, this is typically accomplished through translational frameshift or readthrough mechanisms rather than at translation initiation. Below are some other ways that positive-strand RNA viruses and retroviruses are related. Certain positive-strand RNA viruses employ host turnover processes in addition to translational regulation to control the relative buildup of viral RNA replication components.

3′-to-5′ polymerase copying of viral RNA is prevented by 5′-to-3′ ribosome trafficking despite the requirement for the translation of the viral genomic RNA first. Therefore, positive-strand RNA viruses have to change the incoming genomic RNA's template role from translation to replication when the first replication factor is synthesized. A variety of host and viral components, including a multifunctional 5′ cloverleaf structure, interact with both ends of the viral RNA to coordinate poliovirus translation and negative-strand RNA production. This complex consists of the viral polymerase precursor 3CD and the host poly(C) and poly(A) binding proteins PCBP and PABP, respectively. Translation is aided early in infection by the interaction of PCBP and PABP with the 5′ and 3′ RNA ends, respectively, as well as with one another.

The interaction of 3CD with the 5′ cloverleaf and related PCBP-PABP-RNA complex later in the infection, when it accumulates, slows translation and encourages the start of negative-strand RNA synthesis at the 3′ end of the RNA. Another host RNA binding protein, nucleolin, is likewise bound by the 3′ noncoding region of poliovirus RNA. Immunodepletion of nucleolin prevents the early yields of poliovirus virions in a cell-free replication system. Therefore, when viral RNA or protein levels are low during early infection, nucleolin may promote one or more interactions of poliovirus RNA. Nucleolin and a few other host proteins linked to viral replication experience substantial relocalization from the nucleus to the cytoplasm during poliovirus infection (Ahlquist, 2002).

Viral protein 1a is a helicase-like replication component that recruits BMV genomic RNAs from translation to replication through cis-acting recognition elements (REs) on each viral RNA. This process may be similar to the poliovirus 3CD-RNA interaction. BMV genomic RNAs 1 and 2 have REs at the 5′ end of the RNA, similar to the 5′ poliovirus RNA cloverleaf, whereas BMV RNA 3 has an internal RE that seems to interact with its 5′ end. A host gene called LSM1 is also necessary for the effective 1a recruitment of viral RNAs (Diez et al., 2000). A conserved Sm motif-containing protein that is encoded by LSM1 joins forces with other conserved Sm motif proteins to form complexes that are involved in mRNA turnover and other functions.

Recent findings demonstrate that LSM1 and the Sm complex, which are involved in cellular mRNA turnover, are also necessary for the translation of BMV genomic RNAs intended for replication, tying BMV RNA translation and replication in a further parallel to the coordination of poliovirus translation and replication. It's interesting to note that host Sm protein Hfq, which controls the turnover of several bacterial mRNAs, is also involved in the beginning of RNA synthesis by positive-strand RNA bacteriophage Qβ. Since the REs include the conserved sequence and structural features of the TΨC stem-loops of host tRNAs, and these functions are necessary to the 1a responsiveness of the REs, BMV RNA recruitment may involve host factors that interact with tRNAs in addition to 1a and LSM1.

The relationship between translation factors and active viral RNA replication complexes reveals further links with translation. Host translation initiation factor eIF3 subunits are linked to virus-specific RdRp-complexes that are extracted from tobacco mosaic virus (TMV) or BMV-infected cells. BMV RdRp copurified a 41-kDa eIF3 subunit, which interacted with BMV 2a polymerase. The addition of either pure eIF3 or the 41-kDa subunit prompted BMV RdRp to synthesize negative strands. TMV RNA replication proteins exhibited interactions and coimmunoprecipitation with the host GCD10 component of eIF3, whereas TMV RdRp negative-strand synthesis was suppressed by anti-GCD10 antibodies. These host factors may play roles in the functions associated with the aminoacylatable 3′ tRNA-like elements found in TMV and BMV genomic RNAs, where negative-strand RNA synthesis begins, as well as in the tRNA TΨC stem-loops of the BMV RE template-recruitment signals, or perhaps in both (Diez et al., 2000).

Viral RNA needs defense against cellular RNA breakdown processes, such as strong 5′ and 3′ exonucleases, both before and after replication. Cellular 5′ cap-binding proteins and 3′ PABPs shield cellular mRNAs and many virus RNAs from these turnover mechanisms. Because poliovirus RNA does not have a 5′ cap, host PCBP binding to the 5′ cloverleaf increases the stability of the virus. On the other hand, many viral RNAs feature 3′-terminal secondary structures that bind host proteins in place of a 3′ poly(A), which may help shield the RNA from 3′ exonucleases. As mentioned previously, 3′ tRNA-like structures are multifunctional features seen in the RNAs of BMV and several other positive-strand RNA viruses. These tRNA-like structures, measuring 3′ in length, interact with translation initiation factors and are processed by the host tRNA nucleotidyltransferase. This enzyme adds untemplated nucleotides to BMV RNAs, mimicking the function of telomerase to maintain the viral RNA 3′ ends. The resulting CCAOH 3′ ends of the viral RNAs resemble tRNA molecules and serve as substrates for host aminoacyl tRNA synthetases *in vivo*. These synthetases attach a specific amino acid (such as tyrosine for BMV) to the 3′ hydroxyl group, similar to the process of charging a tRNA.

9.2.2. Assembly of Membrane-Associated RNA Replication Complexes

The RNA replication complexes of all known positive-strand RNA viruses are assembled on internal membranes, typically in conjunction with the development of membrane vesicles or other membrane rearrangements. For instance, several viruses within and outside of the alphavirus-like superfamily replicate their RNA in correlation with spherical internal membrane invasions that range in size from 50 to 70 nm. The structure, construction, and function of such replication complexes are similar to those of retroviruses and double-stranded RNA (dsRNA) viruses, according to recent studies using BMV, one of these viruses. In particular, the helicase-like BMV 1a protein specifically packages the viral genomic RNAs and induces the development of spherical membrane invaginations, imitating the functions of the Gag protein in retrovirus capsid construction.

Just like a budding capsid surrounded by a membrane, each new spherule created invaginates or partially buds into the endoplasmic reticulum (ER) lumen and contains hundreds of 1a proteins. The BMV REs act as specific RNA packaging signals for the virus, similar to retroviral RNA packaging signals. With the coexpression of BMV 2a polymerase, a few copies of 2a are included in each capsid-like spherule, which then serve as locations for viral negative- and positive-strand RNA synthesis. The final replication complex has a 1a/2a polymerase ratio around 25, comparable to the Gag/GagPol ratio in retrovirus capsids. The host membrane is an essential host component with several functions for these replication complexes. Initially, the membrane offers a surface for the localization and concentration of replication components in preparation for assembly. Like a growing virion, the membrane envelops and shields the virus-induced replication compartment, which isolates the genomic RNAs and RNA replication factors from rival RNA templates and opposing activities like translation. Additionally, according to Ahlquist (2002), this organization aids in shielding any dsRNA replication intermediates from host defense reactions triggered by dsRNA, such as RNA interference or interferon-induced responses. Thus, the RNA replication complex may be thought of as a virus-specific organelle for RNA replication, forming a novel, membrane-bounded compartment with distinct components and activities. While some positive-strand RNA viruses create spherical invaginations, others produce different membrane structures, such as vesicles or appressed membranes. These alternate membrane rearrangements may have similar roles in concentrating and isolating viral RNA replication components. Although their specific organization is not fully understood, they may be related in structure to the spherical replication complexes. Different viruses put together their RNA replication complexes on distinct membranes or membrane subsets that are typically specialized. Whereas nodaviruses use mitochondrial membranes, alphaviruses use endosomal and lysosomal membranes, BMV uses ER membranes, and poliovirus replication complexes form on tightly packed vesicles that are thought to be produced from the cellular COPII transport vesicle budding route. Although interaction with host components is necessary for such precise membrane localization, many aspects of such targeting are yet unclear.

The localization of several viral replication proteins to these membranes is contingent upon their interaction with another viral protein. The viral NS3 protease-helicase is localized to the perinuclear ER by HCV NS4A, whereas alphavirus nsP1 guides nsP2, -3, and -4 to endosomal membranes. In a similar manner, the 2a polymerase is guided to ER membranes by interaction with the BMV 1a protein. The similar cucumber mosaic virus experiences a disruption of the 1a-2a connection due to the phosphorylation of 2a by a host kinase. Such 2a phosphorylation takes place *in vivo* and may aid in maintaining the about 25:1 ratio of 1a to 2a in replication complexes, or it may prevent the production of additional replication complexes later in infection. In addition to performing other tasks, host membrane proteins may play a role in the loading or retention of viral replication factors on membranes for at least certain viruses. These proteins may also be transient or permanent members of the RNA replication complex. Interestingly, dual mutations of the host integral membrane proteins TOM1 and TOM3, homologs, prevent TMV replication in Arabidopsis. TOM1 cofractionates with TMV replication proteins and RdRp activity and interacts with the common helicase-like domain of the overlapping TMV 130- and 180-kDa RNA replication factors. In order for TMV replication to be effective, TOM1 also interacts with TOM2A, another host integral membrane protein (Barton et al., 1995).

9.2.3. Viral RNA Synthesis

Some viruses require extra contributions from host components to commence RNA synthesis once viral replication factors and templates are formed into a complex on host membranes. In order for preformed, sedimentable poliovirus RNA replication complexes to become competent for viral RNA synthesis, soluble cytoplasmic components from uninfected cells must be added after they are recovered prior to RNA start by reversible inhibition with guanidine (Barton et al., 1995). In a similar vein, coinjection of human cell extracts is necessary for a post-translation stage in the replication of poliovirus RNA in Xenopus laevis oocytes; however, which posttranslational step is impacted remains unclear. Similar to this, mutations in the yeast gene YDJ1 prevent BMV RNA replication once the first replication complex is assembled. Many early stages of RNA replication, such as the viral 1a protein's interaction with membranes, the recruitment and stabilization of viral RNA replication templates by 1a, and the recruitment of the viral 2a polymerase by 1a, are all supported at normal levels in cells with such YDJ1 mutations. However, the commencement of negative-strand RNA synthesis is not supported by YDJ1 mutant cells. A chaperone that is partly localized to ER membranes, the location of BMV RNA production, and linked to Escherichia coli DnaJ is encoded by YDJ1. Furthermore, a significant portion of 2a polymerase was discovered in a membrane-free, quickly sedimenting state in YDJ1 mutant yeast but not in wild-type yeast, indicating aggregation. Thus, the *YDJ1* chaperone function may direct a 2a polymerase folding or interaction step that is required for RNA synthesis. Host chaperones are also involved in the proteolytic processing of HCV nonstructural proteins (Tellinghuisen and Rice, 2002).

Mutations in OLE1, a different host gene, also prevent BMV RNA replication following the membrane connection of viral RNA, 1a, and 2a polymerase, but before to the start of negative-strand RNA synthesis. Encoding Δ9 fatty acid desaturase, the crucial enzyme responsible for converting saturated fatty acids into unsaturated ones, is encoded by the vital gene OLE1. Medium supplementation experiments reveal that BMV RNA synthesis is much more sensitive than cell growth to reductions in unsaturated fatty acid levels, even though this enzyme is an integral ER membrane protein and is therefore localized to the site of BMV RNA replication. The functional significance of membrane connection for RNA replication is clearly demonstrated by this remarkable sensitivity to the lipid composition of the membrane. The need for high unsaturated fatty acid levels also suggests that the replication complex may require significant membrane fluidity or plasticity and thus may be in a dynamic relationship with the host membrane. Thus, host factors can play a variety of roles in the commencement of negative-strand RNA synthesis, such as blocking the translation of viral RNA, turning on certain replication complex activities, and identifying and using the 3′ end of positive-strand RNA as an initiation site. The start of positive-strand RNA synthesis from negative-strand RNA templates may also be mediated by host factors. Base pairing of the 5′ cloverleaf structure in the nontemplate, positive-strand RNA is necessary for poliovirus positive-strand RNA production. One theory has been put forth, the possibility that the negative-strand template RNA is bonded in a replicative form of dsRNA. To start the new positive-strand RNA, the 3′ end of the negative-strand RNA template needs to be separated from the dsRNA structure. This process may be aided by the 5′ end of the positive-strand RNA forming a cloverleaf secondary structure, which is then supported by the attachment of host PCBP and viral proteins.

A 3′-terminal stem-loop on negative-strand RNA of the flavivirus West Nile virus interacts with a number of host proteins, such as TIAR and its homolog TIA-1, which are host RNA binding proteins involved in RNA splicing and translation (Anderson et al., 2002). Murine cell lines missing TIAR showed a substantial reduction in West Nile virus growth. TIAR could aid in the production of positive-strand RNA because this negative-strand RNA's sole known function is to serve as a replication intermediary. Although the function of this protein *in vivo* has been disputed, coronaviruses may employ host RNA binding factor hnRNP A1 to attach to specific locations in viral negative- and positive-strand RNA for subgenomic mRNA transcription and RNA replication.

9.3. REINVENTING POSITIVE-STRAND RNA VIRUS REVERSE GENETICS

The International Committee on the Taxonomy of Viruses has identified approximately 2300 viral species across 60 virus families, making positive-strand RNA viruses the most varied virus genome type (Tellinghuisen and Rice, 2002). Alphaviruses, flaviviruses, and picornaviruses are examples of positive-strand RNA viruses that are lipid enveloped; nodaviruses, on the other hand, have small genomes (~4.5 kb), whereas coronaviruses have massive genomes (~ 30-kb) (Figure 9.6). Most of these viruses have only one genome segment, but some, like bromoviruses and nodaviruses, have multiple genome segments. Though their genome sizes and replication strategies vary, all positive-strand RNA viruses have characteristics in common that are important to comprehending their genetic makeup.

9.3.1. General Principles of Positive-Strand RNA Virus Replication

Positive-strand RNA viruses enter cells, and their genomes are translated directly into the viral proteins required for viral replication by cellular ribosomes. An RNA-dependent RNA polymerase (RdRP) is encoded by all positive-strand RNA viruses, and the phylogenetic categorization of RNA viruses is largely based on the evolution of this enzyme. Generally, viruses larger than 5 kilobases (kb) also encode an RNA helicase. Viruses with 5′ caps on their genomes need to encode capping enzymes, and those containing a 5′ genome-linked viral protein need to encode a matching VPg protein.

Flaviviruses produce a single long polyprotein through genome translation, which is then processed by a viral serine protease (NS2B-NS3) and cellular proteases (ER-resident signal peptidases). This results in the production of three viral structural proteins, namely capsid, prM, and E, as well as seven nonstructural proteins, namely NS1, NS2A, NS2B, NS3, NS4A, NS4B, and NS5 (Lindenbach et al., 2020) (Figure 9.6). When it comes to coronaviruses, the main translation result is the open reading frame (orf) orf1a, a polyprotein that is split into ten nonstructural proteins, nsp1–nsp10, by two virally encoded cysteine proteases (nsp3 and nsp5) (Figure 9.6). These proteins block host defenses, create ER-derived replication organelles, and cleave viral polyproteins to prepare the cell for viral reproduction.

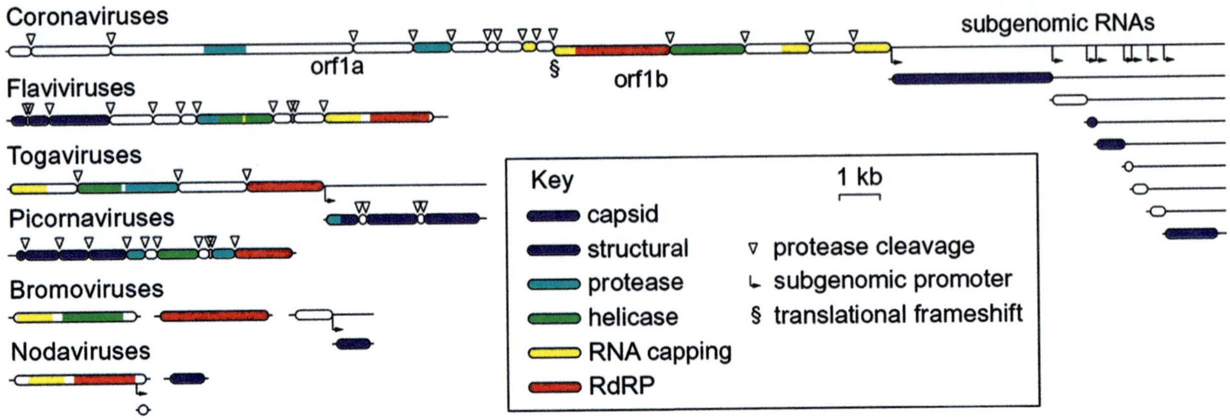

Figure 9.6. *Representative positive-strand RNA virus genomes. Genomes are drawn to scale, showing viral protein expression and polyprotein cleavage strategies, subgenomic RNA transcripts, a programmed ribosomal frameshift, and the location of conserved viral enzymes and structural proteins, as per the Key.*

Source: https://www.ncbi.nlm.nih.gov/pmc/articles/PMC9273853/bin/f01-01-9780323989909_lrg.jpg.

Furthermore, a bigger polyprotein, orf1ab, that is similarly cleaved to yield the viral RNA replication enzymes, nsp12-nsp16, is consistently produced by a ribosomal frameshift that is encoded by the virus. A portion of the basic set of viral replication proteins must pull the genome from translation and into a viral replication compartment after they have been created. It is assumed that a single RNA molecule cannot be concurrently translated in the 5′ to 3′ direction and copied in the 3′ to 5′ direction; hence this shift from genome translation to replication is considered a crucial step. In fact, this shift is a rate-limiting stage in establishing productive replication, as verified by live cell imaging of incoming Coxsackie virus B3 genome translation events.

The production of a complementary negative-strand RNA, which acts as a template to create new viral genomes, frequently in large excess, is the process by which positive-strand RNA viruses replicate. As a result, the semi-conservative replication mechanism employed by positive-strand RNA viruses results in a comparatively limited number of negative strands, each of which gives birth to many daughter genomes. In order to enable late gene expression, certain viruses, including coronaviruses, additionally create or "transcribe" one or more sgRNAs that are colinear with the 3′ end of the viral genome. These comprise distinct transcripts for coronaviruses that encode the structural proteins Spike, Envelope, Membrane, and Nucleocapsid, in addition to auxiliary proteins like SARS-CoV-2 orf3a that regulate the interface between the virus and host.

Within the cytosol of infected cells, membrane-bound replication organelles (ROs) with two main morphologies are involved in positive-strand RNA viral genome replication and transcription processes. Certain viruses, notably flaviviruses and nodaviruses, create "spherule"-like RNA-containing radial oligomeric structures (ROs), which are sac-like invaginations into bigger membranes with a neck that are contiguous with the cytosol. For some viruses, ROs are double-membrane vesicles (DMVs) with unknown topologies, such as hepaciviruses and coronaviruses. There are several additional membrane virus-induced changes that have been noted, and this article is not intended to cover the specifics of RO structure and function.

However, I would be remiss if I did not also mention that recent developments in cryo-electron microscopy and electron tomography have demonstrated that the interior of these compartments is connected to the cytosol by virus-encoded proteinaceous pores with 6- or 12-fold symmetry found in both spherules (e.g., nodavirus and alphavirus ROs) and DMVs (coronavirus ROs). What is most relevant to this discussion is not only the amazing structural conservation of these structures over enormous evolutionary distances but also the fact that viral RNAs are infectious because the viral proteins required for RO formation and RNA synthesis are all produced through translation of a positive-strand RNA virus genome. An important exception to this pattern is the Nucleocapsid protein of the coronavirus, which is essential for viral RNA synthesis. This protein is not produced until after RNA replication and transcription have begun, entering cells alongside the viral genome.

The fact that RdRP enzymes have high mistake rates—typically one misincorporation for every 10–3 to 10–6 nucleotides synthesized—is another significant aspect of positive-strand RNA virus genetics. Consequently, almost all of the developing daughter genomes of several viruses include one or more nucleotide changes. As a result, populations of RNA viruses have a swarm-like appearance of related genotypes and might not have a distinct "wild-type" sequence. This leads to the theory that RNA viruses act like "quasi-species," a term that Manfred Eigen and Peter Schuster used to characterize a hypothetical primordial replicator with high mutation rates. It should be noted that the mutant swarm seen in populations of RNA viruses is sometimes mislabeled as a "quasi-species."

Rather than any one individual genotype, this phrase explicitly refers to the idea that a population of linked genotypes is a unit of natural selection (i.e., a purely population genetics model of evolution). Although Holmes and Moya (2002) maintain that this evolutionary model is still debatable, the quasispecies idea has offered a helpful framework for comprehending the dynamics of RNA virus populations. One implication of the quasispecies idea, for example, is that RNA viruses reproduce extremely close to their mutational limit and that little increases in mutation rate can wipe out a viral population—a process known as the "error catastrophe."

It has been demonstrated that the nucleoside analog ribavirin mediates its antiviral activities against the poliovirus by fatal mutagenesis; similarly, molnupiravir, an antiviral nucleoside used to treat SARS-CoV-2, acts through a similar mechanism. Further support for population-based natural selection comes from findings indicating variety plays a crucial role in positive-strand RNA viruses' *in vivo* fitness. In particular, it was demonstrated that in mouse models of poliomyelitis, mutant polioviruses with higher replication fidelity were attenuated. Their lethality was, however, restored by synthetically increasing genetic diversity within the input, high-fidelity virus population using a chemical mutagen. The increased variety in the input population presumably gave the virus a way to adapt *in vivo* to various tissues and host defense responses. The key points for our discussion are that: (i) genetic diversity is generally high, even within small positive-strand RNA virus populations, and (ii) this diversity can be important for viral phenotypes.

9.3.2. The Origins of Positive-Strand RNA Virus Reverse Genetics

Infectious viruses can be saved by transfecting genomic RNA into receptive host cells, as all the viral proteins required to start RNA replication are generated by translation of a positive-strand RNA virus genome. These pioneering findings made it possible to directly modify the viral genomes and track the resulting phenotypic changes in order to do reverse genetics on positive-strand RNA viruses. RNase H and poly(dT) were used in a series of groundbreaking studies to enzymatically remove the 3′ poly(A) tail from pure poliovirus genomes. This process was shown to eliminate RNA infectivity. Additionally, using a purified replicase, mutant nucleotide triphosphates were selectively incorporated at specific places in the RNA genome of bacteriophage Qβ during *in vitro* synthesis, resulting in some of the earliest site-directed mutagenesis studies ever conducted. Some of these defined mutants had altered replication fitness in bacterial cells.

These early attempts did yield valuable ways to assess and compare RNA-specific infectivity (i.e., the number of infectious units recovered normalized to the amount of input RNA), even though direct manipulation of viral RNA is technically difficult and limited in scope. To initiate plaque development, for example, in an infectious center assay, cells are transfected with viral RNA, serially diluted, and then plated with untransfected cells. Every plaque, also known as an "infectious center," is assumed to represent a single productively transfected cell. This can be utilized to calculate the transfection efficiency (%) and RNA-specific infectivity (e.g., PFU/μg RNA), which is commonly understood to indicate the efficiency with which viral replication is started. The saturation of RNA levels and the potential for cells to be transfected with multiple RNAs, depending on the method used, necessitates that RNA-specific infectivity measurements show a correlation between the number of infectious units and the input RNA concentration. This is especially crucial when creating reverse genetics experiments with intricate mutant libraries (Li et al., 2002).

9.3.3. Construction and Application of Functional cDNA Clones

The first positive-strand RNA virus functional cDNA clones, or "infectious clones," were produced in 1975 at the Asilomar Conference on Recombinant DNA Technology, thanks to the discovery of reverse transcriptase (RT) in 1970 and the creation of biosafety and ethical criteria. In particular, Taniguchi and associates discovered that Escherichia coli, using this plasmid, spontaneously produced infectious viruses by inserting the whole cDNA of the Qβ genome into a bacterial plasmid. Later, after transfecting this DNA into monkey cell lines, Racaniello and Baltimore cloned a full-length poliovirus cDNA into the bacterial plasmid pBR322. This resulted in the formation of an infectious virus.

Surprisingly, neither of these cloned cDNAs included a particular promoter or mechanism to stop the transcription of the viral genome; it is likely that RNAs capable of replication somehow emerged by RNA degradation, termination, or cryptic transcription. The ineffective start of viral replication from these initial viral cDNA clones is consistent with the stochastic genesis of these transcripts. The next significant advancement was to produce viral RNAs with precisely defined ends and high specific infectivity by run-off *in vitro* transcription using purified *E. coli* RNA polymerase. This was made possible by flanking viral cDNAs with a distinct restriction enzyme site at the 3′ end and a well-defined bacterial promoter at the 5′ end. The use of the multi-component *E. coli* RNA polymerase was quickly replaced by single-chain bacteriophage SP6, T3, and T7 RNA polymerases. These became widely available in recombinant form, allowing for precise initiation of RNA transcripts over 1 kilobase in length from small promoters of 20 base pairs or less.

With the development of widely accessible reverse genetics techniques, infectious clones for a variety of positive-strand RNA viruses were created. This expanded era of clone building also yielded two additional principles: (1) the initiation of RNA replication for many viruses is contingent upon the precise generation of the correct 5′ and 3′ ends, and (2) given the genetic diversity within RNA virus populations, identifying and reconstructing a functional genome sequence is of utmost importance. On the latter issue, it was not unusual to laboriously build full-length cDNA clones only to discover that the resulting RNA transcripts were either non-functional or had extremely poor specific infectivity. Most notably, the first successful hepatitis C virus infectious clones were developed only after identifying cDNAs that matched the consensus sequence of the input virus population. Presumably, the nonfunctional clones contained deleterious mutations that prevented viral replication (Rovner et al., 2015).

An infectious clone should, by definition, be clonal, meaning that before the virus is saved, it should be feasible to isolate a single genotype. Therefore, an infectious clone only represents one sequence, even while quasispecies-like swarms of positive-strand RNA viruses exist. In actuality, though, the error-prone nature of viral replication means that once a virus has been released from an infectious clone, it may rapidly restore genetic diversity among the population of rescued viruses. Reverse genetic systems can be population-based or clonal, as will be covered in more detail below.

Utilizing an infectious clone for viral rescue with a well-defined genotype is the simplest application. By acting as a genetic repository, an infectious clone can reduce the amount of viral variation that would result from the serial transmission of virus stocks. This genetic stability has proven particularly appealing to preserve live, attenuated vaccination strains that can return to virulent phenotypes, such as the oral poliovirus vaccine. It is also possible for researchers to use DNA synthesis to recreate and save a virus using an infectious clone, even if the virus has already been cultivated or is extinct.

Replication-dependent reporter viruses that produce readily assayable proteins, including fluorescent or luciferase enzymes, are frequently generated from infectious clones. Live viral vaccines containing foreign antigens can also be administered via infectious clones. It has been possible to transmit tiny antigenic peptides or even big protein antigens like SARS-CoV-2 Spike using the yellow fever virus (YFV) vaccine strain 17D, for example. Chimeric live attenuated vaccines that express the structural glycoproteins of other flaviviruses instead of the YFV structural glycoproteins have also been made using the YFV-17D backbone.

Many times, RNA replication can be separated from virus assembly, leading to the deletion of viral structural genes to form RNA replicons. These replicons replicate within cells without producing infectious virus particles. Replicons, especially those expressing reporter enzymes, are valuable for performing high throughput screening assays with less stringent biosafety protocols. Of course, an infectious clone also enables scientists to make hypothesis-driven modifications to a viral genome, such as deletions, insertions, or point mutations, and then evaluate the resulting phenotypic outcomes. Even the most basic methods may easily produce single mutants or tiny mutant panels (Figure 9.7A). Nevertheless, robust and effective reverse genetics systems are needed for complex mutant libraries (Figure 9.7B), such as those produced by genetic bar-coding a few residues, saturation mutagenesis of individual residues, deep mutational scanning of many residues (Fowler and Fields, 2014), or genetic footprinting via transposon insertion mutagenesis, in order to reduce genetic bottlenecks and guarantee that library complexity can be representationally oversampled at every step of library construction, amplification, and launch.

Launch efficiency is a crucial metric to optimize for complex library sampling efficiency, for example, by measuring the complexity of the resulting library, as RNA specific infectivity provides the most accurate approximation of launch efficiency. For instance, the Ebel lab created barcoded libraries with 6.55 × 104 possible genotypes using a non-clonal Zika virus (ZIKV) reverse genetic technique (*see below*); nevertheless, only 18–146 legitimate barcodes were originally found among rescued ZIKV stocks. Sampling efficiency increased dramatically once viral launch circumstances were optimized, producing up to 1.40 x 104 distinct ZIKV genotypes.

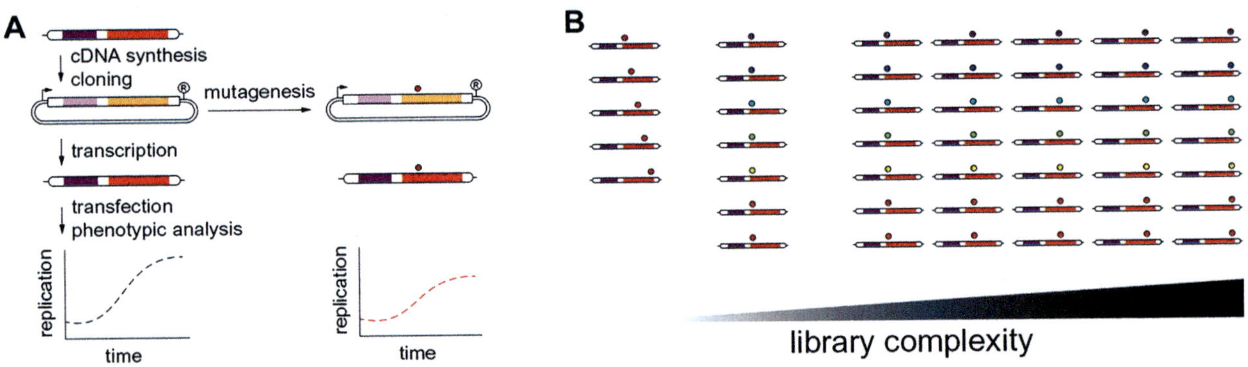

Figure 9.7. *Standard reverse genetics workflows. (A) Construction of a viral cDNA clone, rescue of the wild-type virus, and the construction and testing of a mutant virus. Here, ® represents a unique restriction site just downstream of the viral genome, allowing for full-length, run-off transcripts corresponding to the viral genome. (B) Illustration of mutant virus library complexity. To the left is a theoretical mutational scanning analysis (low complexity); in the middle is a theoretical deep mutational analysis (randomization) of a single codon; on the right is a deep mutational scanning analysis, where multiple residues are targeted for randomization.*

Source: https://www.ncbi.nlm.nih.gov/pmc/articles/PMC9273853/bin/f01-02-9780323989909_lrg.jpg.

Lastly, the ability to study the mechanics of RNA replication in model systems, such as cell-free responses and cells that are not normally conducive to infection, has been tremendously aided by the development of infectious clones. For example, *in vitro* processes, including cell-free extracts from HeLa cells, may translate, reproduce, and create infectious virus particles from the poliovirus genome. Certain positive-strand RNA viruses can replicate their RNA when their viral RNAs are expressed in Saccharomyces cerevisiae, brewer's yeast. These viruses include the carnation Italian ringspot virus, the brome mosaic virus, the tomato bushy stunt virus, the Flock House virus, and the Nodamura virus, which affects both plants and mammals. Then, the capacity of yeast genetics made it possible to do the first thorough genome-wide screens for host components necessary for positive-strand RNA virus replication.

9.3.3.1. Flavivirus Infectious Clones

Yellow fever virus (YFV) is the type species and namesake of a major clade of positive-strand RNA viruses (genus Flavivirus, from Latin flavus, "yellow"); family Flaviviridae; order Amarillovirales, from Spanish amarillo, "yellow"; and phylum Kitrinoviricota, from Greek kítrinos, "yellow"). It was the first human virus to be fully sequenced and described. Moreover, YFV was the first flavivirus for which an infectious clone could be successfully produced; this took a lot of work because the full-length cDNA of YFV was unstable in bacterial plasmid and ⊠ phage vectors (Fowler and Fields, 2014).

The strategy that worked in the end was to clone two distinct portions of the viral genome onto distinct plasmids that could be cut with restriction enzymes, ligated to create a full-length cDNA right before *in vitro* transcription, and kept stable in *E. coli* (Figure 9.8A). Later, a full-length YFV genome was reassembled in this vector and discovered to be stable in *E. coli* after unstable cDNAs for two pestiviruses, bovine viral diarrhea virus (BVDV) and classical swine fever virus (CSFV), was successfully cloned into the low-copy vector pACYC184 bearing a p15A origin of replication.

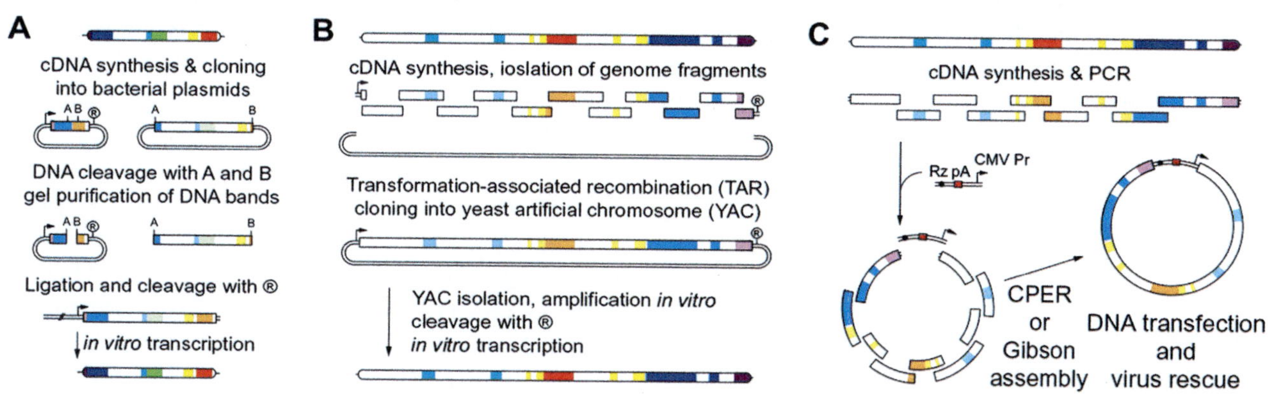

Figure 9.8. *Solutions to flavivirus and coronavirus cDNA instability. (A) Construction of a stable, two-piece YFV-17D infectious clone. Here, the arrow indicates an SP6 promoter, restriction enzyme A represents ApaI, enzyme (B) represents AatII, and enzyme ® represents XhoI. (B) Construction of a stable, full-length SARS-CoV-2 infectious clone within a YAC via transformation-associated homologous recombination (TAR) cloning. Here, the arrow indicates a T7 promoter, and enzyme ® represents EagI. (C) In vitro assembly of non-clonal SARS-CoV-2 cDNAs. Overlapping viral cDNA fragments are amplified, combined with a DNA adaptor encoding an HDV Rz, polyadenylation (pA) signal, and CMV promoter, and circularized via CPER or Gibson assembly to produce DNAs that can be transfected into mammalian cells.*

Source: https://www.ncbi.nlm.nih.gov/pmc/articles/PMC9273853/bin/f01-03-9780323989909_lrg.jpg.

Many more flavivirus infectious clones were created in the 1990s and 2000s. While many cDNA clones, including DENV1 strain WP74, DENV2 strain NGC, DENV4, Japanese encephalitis virus (JEV) strain JaOArS982, JEV strain K87P39, and tick-borne encephalitis virus (TBEV) strain Hypr, showed instability and/or a strong preference for low copy number vectors, some, like dengue 2 virus (DENV2) strain 16,681, Murray Valley encephalitis virus (MVEV) strains 1–51, TBEV strain Neudoerfl, West Nile virus (WNV) subtype Kunjin, and WNV lineage II, were fortunately stable in standard bacterial vectors.

Some studies applied the *in vitro* ligation approach, first used for YFV, to get over these issues, while others used bacterial artificial chromosomes (BACs), which are kept at 1-2 copies/bacterial cell. Others created techniques to use homologous recombination in yeast to build full-length DENV clones; nevertheless, in order to produce enough material for RNA transcription, the resulting yeast shuttle plasmids still needed to be amplified in *E. coli*.

A few teams embarked on a project to determine the molecular causes of flavivirus cDNA instability in bacteria at this time. Large bacterial colonies were seen by Yamshchikov and colleagues when spontaneous mutations developed inside their JEV infectious clone. These mutations mapped to the prM and E domains, indicating that these areas encoded components that were toxic to bacteria. The researchers used a 5′ cytomegalovirus (CMV) immediate early gene 1 promoter to drive a JEV clone that disrupted the unstable sections with two synthetic introns and ended with a 3′ mammalian polyadenylation signal in order to overcome this instability. As a consequence of RNA Pol II transcription, 5′ capping, proper splicing, nuclear export, translation, and selective replication of mRNAs with correct 3′ ends, the resultant clone was stable in *E. coli* and generated infectious virus upon transfection of DNA into human cells. Zheng et al. discovered cryptic, transcriptionally active bacterial promoters in the 5′ one-third of the JEV cDNA. The promoters showed weak transcription during bacterial growth at 25 °C, which also improved plasmid stability. In line with these results, two groups demonstrated that full-length DENV-2, JEV, and ZIKV cDNA clones were stabilized by inserting silent mutations into the viral cDNA to lower bacterial promoter activity.

It is important to acknowledge that utilizing mammalian transcription and RNA processing machinery in the nucleus to create viral transcripts carries certain risks. Cellular mRNAs are typically only exported from the nucleus once they have been capped, spliced, and polyadenylated. Additionally, positive-strand RNA viruses are designed to replicate in the cytosol, which may result in the presence of unintended splice sites in their genomes. This can produce nonfunctional edited transcripts and nuclear retention signals that hinder export to the cytosol. Another potential issue is ensuring the correct formation of the 3′ ends of the transcripts. Flavivirus genomes do not have 3′ polyadenylated tails, so most Pol II-driven flavivirus infectious clones use a 3′ self-cleaving hepatitis delta virus (HDV) genomic or antigenomic ribozyme (Rz) to create a genuine 3′ end before the polyadenylated tail. However, this process removes a key element for mRNA nuclear export (Stewart, 2019). Some argue that splicing efficiencies can vary, and some non-polyadenylated RNAs are still able to leave the nucleus, so ultimately, only one intact viral transcript is necessary to start a viral infection.

Roughly half of the sequenced transcripts retained the intron without RNA replication, but viral replication selectively increased the spliced, intron-less transcripts. Notably, in order to reach the same viral titers as viruses produced by transfection of RNAs produced by *in vitro* transcription, many Pol II-driven flavivirus infectious clones require longer rescue times (sometimes as long as a week) and/or additional passages. This suggests that the launch efficiency of these viruses is lower than that of RNA transfection-based techniques. I urge other researchers using DNA transfection-based launch systems to critically examine the efficiency of RNA processing in the absence of replication and the overall efficiency of virus rescue by quantifying library complexity for experiments where efficiency is crucial (e.g., complex library sampling).

9.3.3.2. Coronavirus Infectious Clones

At first, coronavirus cDNAs were very difficult to work with because of their enormous genomes, which were around 30-kb long—nearly three times longer than flaviviruses (Figure 9.7)—and because they had several troublesome regions for bacterial cloning. Consequently, the first coronavirus infectious clones were not produced until 2000, by two different labs working independently. The transmissible gastroenteritis virus (TGEV) genome was successfully stably cloned by Yount et al. as six cDNA segments in different bacterial plasmids. Full-length cDNAs could be assembled *in vitro* by restriction digestion, gel purification of individual fragments, and DNA ligation; ligation products were then used directly to transcribe viral genomic RNA, as was the case with the first YFV infectious clone (Figure 9.8A), because each fragment was flanked by restriction sites with uniquely compatible ends.

Crucially, this group found that co-transfection with a synthetic mRNA expressing the TGEV nucleocapsid (N) gene significantly enhanced viral rescue. We now understand that this is because the coronavirus N protein is crucial for RNA synthesis. Although this multi-piece cloning and reassembly method was initially ineffective, it has since been improved upon by the Baric lab and others by forcing directional ligation of the cDNA fragments using type IIS or IIG restriction sites; this method has been successful for several coronavirus infectious clones, such as SARS-CoV, MERS-CoV, and SARS-CoV-2.

As an alternative to the previously mentioned multi-piece cloning method, Almazón et al. were able to clone the whole TGEV cDNA into a single copy/cell bacterial artificial chromosome (BAC), which is flanked by a 3′ HDV Rz and a 5′ CMV promoter. However, the efficacy of rescuing viruses was very low and required long serial passage to multiply viral titers; in the meantime, the rescued virus stock acquired mutations. The researchers discovered that although some spliced viral transcripts were generated, they were functional and selectively reproduced when left unspliced. Furthermore, the TGEV Spike gene is a significant predictor of viral tropism and pathogenicity *in vivo*, as the investigators showed with this approach. Later on, the team demonstrated that inserting an intron into the TGEV-containing BAC at two different locations greatly stabilized the full-length viral cDNA. Despite ongoing stability issues and low success rates with DNA-based launch systems, similar full-length cDNA BAC systems have been created for SARS-CoV and SARS-CoV-2 as well.

The aforementioned achievements and failures sparked fresh ideas for full-length coronavirus cDNA cloning. Thiel et al. conducted an impressive investigation whereby they discovered that a recombinant vaccinia virus DNA genome sustainably housed a full-length human coronavirus 229E cDNA. This cDNA could be sufficiently purified to create high-quality infectious transcripts *in vitro*. After that, a second recombinant poxvirus expressing T7 RNAP was co-infected with vaccinia to produce a full-length avian infectious bronchitis virus (IBV) cDNA, which could then be produced by plaque purification of the resulting IBV. The ability to deliver genetic modifications *in vivo* by homologous recombination—a very effective method in poxviruses—is a significant benefit of the vaccinia-based approach. On the other hand, a significant drawback is that utilizing vaccinia virus vectors can heighten biosafety and biocontainment concerns, requiring the resulting coronavirus to be isolated from the poxvirus(es) through plaque purification before further analysis (Stewart, 2019).

Thiel's team has been at the forefront of utilizing yeast artificial chromosomes (YACs) to successfully maintain full-length coronavirus and other viral cDNAs in S. cerevisiae. Homologous recombination in yeast enables efficient construction and adjustment of the viral genome, particularly with vaccinia virus. Although YAC yields can be limited for *in vitro* transcription, this challenge can be addressed by amplifying the isolated YAC through rolling circle amplification (RCA) with phage ø29 DNA polymerase.

9.3.4. Non-Clonal Reverse Genetic Systems

Many teams have created techniques to construct and amplify viral cDNAs only *in vitro* in order to get around the problems of cDNA instability and toxicity in bacterial hosts. Gritsun and Gould amplified the tick-borne encephalitis genome as two overlapping cDNAs bordered by an upstream SP6 promoter, providing the first example of this. The researchers were subsequently able to create functioning, full-length transcripts in a matter of days by assembling these products into full-length cDNA by cross-primed PCR amplification. To simplify their strategy, Aubry and colleagues attempted to skip the assembly PCR step and instead transfect mammalian cells with three overlapping flavivirus cDNA constructs surrounded by CMV and HDV Rz elements. They believed that full-length cDNAs would be reassembled through homologous recombination and then transiently transcribed. Although this approach was successful, it was inefficient and required significant time and serial passage to rescue the virus, potentially resulting in the accumulation of additional mutations.

Several groups have used circular polymerase extension reactions (CPER) to construct functional full-length cDNAs of flaviviruses and coronaviruses, with a strong 5′ RNA Pol II promoter and 3′ HDV Rz. This PCR-based technique involves cross-priming target amplicons and a circularizable DNA vector to create full-length, nicked circles that can be transfected directly into mammalian or insect cells for virus rescue. Like other DNA transfection methods, this process may require substantial rescue times and/or passage. However, the CPER method has been successful in partially representing the genetic diversity of input virus populations, allowing for the examination of complex mutant libraries, such as a large number of theoretical codon variants in the ZIKV E glycoprotein, to identify mutations that impact host cell tropism.

James Weger-Lucarelli and the Ebel lab introduced a new method for assembling viral cDNA amplicons *in vitro*. This method involves using Gibson assembly reactions to combine cDNA fragments with a 5′ CMV promoter and 3′ HDV Rz, creating circular DNAs. These circular DNAs can then be amplified using RCA with phage ø29 DNA polymerase. Unlike the CPER method, this technique produces covalently closed circular DNAs, allowing for more efficient amplification via RCA. This method has the potential to generate enough cDNA products to create infectious transcripts *in vitro*, facilitating a more accurate determination of infectivity.

These non-clonal reverse genetics systems may have the benefit of properly representing viral populations; in fact, they may even be able to be used as a tool to investigate the quasispecies character of RNA virus evolution. All of these techniques, meanwhile, rely on DNA-based rescue in its current incarnations, which might be vulnerable to genetic bottleneck events, as previously mentioned. The fact that these population-based techniques make it difficult to clearly identify a "wild-type" for a thorough genetic study on an isogenic background raises further concerns. As a result, these techniques are not as good for examining how specific mutations affect phenotype.

Additionally, without the ability to purify clones by isolating a single colony, the continuous amplification of cDNA populations in a lab setting could result in the spread of errors induced during PCR or assembly processes. As a result, cell-free cloning techniques show potential for managing viral genomes in populations, but further strict measures are necessary to ensure that a particular phenotype is directly caused by a specific genotype (Almazan et al., 2004).

9.3.5. Novel Methods and Future Directions

Although the description above has concentrated on reverse genetic techniques that result in the creation of a functioning viral replicase, it should be highlighted that these techniques may also be used to examine other elements of the virus life cycle that are not dependent on RNA replication. For example, Drs. Jennifer Doudna and Melanie Ott's labs discovered a packaging signal in the SARS-CoV-2 genome, enabling them to co-express reporter gene mRNAs with the viral Spike, Envelope, Membrane, and Nucleocapsid genes and package them into virus-like particles. The Nucleocapsid gene alterations that improve viral RNA packaging and raise virus titers and are seen in SARS-CoV-2 variants of concern were then identified by the scientists using this technique.

Although the impressive advancements in cell-free reverse genetics techniques and their possible use in population-based genetics excite me, some issues need to be resolved with these approaches' current inefficiency and inability to operate with a genetically distinct clone on an isogenic background. Furthermore, every reverse genetic technique covered in this chapter requires substantial modifications to *in vitro* molecular biology. I can't help but think that perhaps we have things reversed; perhaps it would be simpler to keep viral cDNA clones within host cells that are receptive and edit them in situ utilising the *in vivo* genome editing techniques that have recently been created by the synthetic biology community.

For example, it is easy to use CRISPR/Cas9 and/or homologous recombination to replace specific sections within huge YACs. CRISPR/Cas9 is limited, though, since it relies on generating and fixing double-stranded DNA breaks, which can be hazardous if several are generated at once and frequently result in insertions and deletions (indels) at altered locations. I am especially excited about two technologies: eukaryotic MAGE (eMAGE), which allows exact editing of the yeast genome and likely YACs, and multiplex automated genome engineering (MAGE), which allows precise editing of bacterial genomes and, probably, BACs. These techniques are basically *in vivo* versions of site-directed mutagenesis, in which a locus is modified during DNA replication by using mutagenic oligonucleotides during the synthesis of the lagging strand (Figure 9.9A).

These methods are highly effective (with gene conversion rates over 10%), versatile (allowing for multiple changes simultaneously through multiplexing), and iterative (edited populations can be further edited). Other promising genome engineering techniques include retron-associated recombineering, which uses bacterial retron reverse transcriptase to continuously generate mutagenic oligos in living cells, as well as synthetic base editing enzymes and prime-editing enzymes (Almazan et al., 2006).

I am intrigued by the recent research, which demonstrated that yeast containing a Sindbis virus cDNA controlled by the potent, galactose-inducible Gal1 promoter can generate infectious transcripts and begin replication when the induced yeast is combined with mammalian cells. This innovative approach eliminates the need to isolate cDNA and transcribe RNA in a laboratory setting. I envision a future where much of standard molecular virology can be delegated to yeast: We can program yeast to generate desired mutants, like with eMAGE, and then produce infectious RNAs by supplying them with galactose (Figure 9.9A).

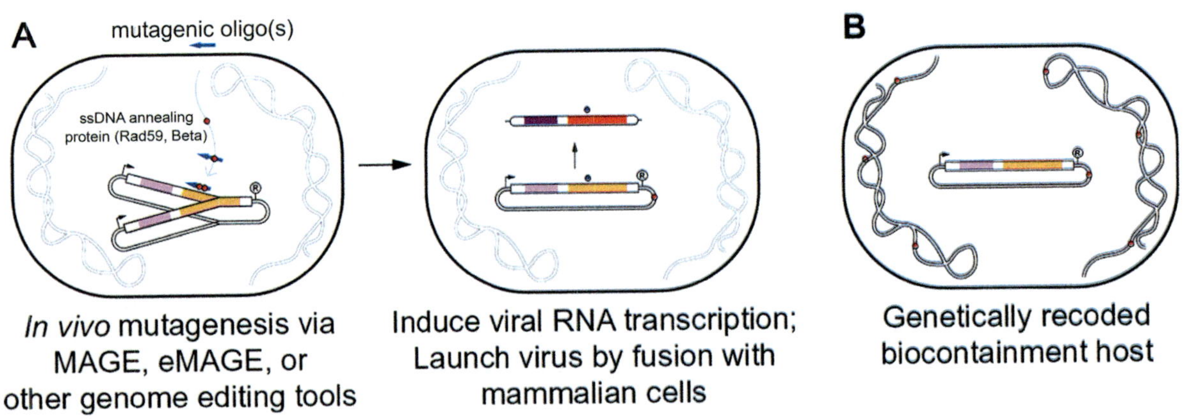

Figure 9.9. *Proposed synthetic biology approaches to reverse genetics. (A) In vivo editing of the viral cDNA via genome engineering methods, such as MAGE or eMAGE, and induction of infectious transcripts in vivo, such as in yeast. In this case, the arrow represents a Gal1 promoter, ® represents an HDV Rz and pA signal. (B) Engineered biocontainment of host cells stably harboring a viral infectious clone. Red bullets represent amber codons re-coded for a synthetic amino acid.*

Source: https://www.ncbi.nlm.nih.gov/pmc/articles/PMC9273853/bin/f01-04-9780323989909_lrg.jpg.

Of course, molecular virologists need to be aware of the biosafety and biosecurity issues associated with new technologies being developed to facilitate the creation, modification, and rescue of viral cDNAs. To solve these issues, synthetic biology may be employed to create cells containing viral cDNA impenetrable biocontainment systems. For example, all known amber (TAG) stop codons from 321 bacterial open reading frames were removed from the *E. coli* genome using MAGE genome editing. The TAG codon was then able to precisely code for an artificial amino acid when this codon was reassigned to a synthetic orthogonal tRNA-tRNA synthase pair. The team utilized MAGE to introduce amber codons into several crucial bacterial genes, making bacterial growth reliant on an unnatural amino acid (Rovner et al., 2015). Comparable techniques could prevent horizontal gene transfer of viral cDNAs or unauthorized use of a cell-based reverse genetic system (Figure 9.9B).

CLASS ACTIVITY

To investigate and understand the replication cycle and characteristics of positive-strand RNA viruses through a guided exploration and discussion.

Assessment: Divide the class into small groups (3-4 students per group). Assign each group a specific positive-strand RNA virus to research. Provide guidelines on what aspects to focus on, such as:

- Viral structure and genome organization;
- Mechanism of viral entry into host cells;
- Replication strategy (e.g., replication complexes, RNA-dependent RNA polymerase);
- Interaction with host cell machinery (e.g., translation, immune evasion);
- Clinical significance and disease implications.

SUMMARY

- Positive-strand RNA viruses are a diverse group characterized by their genome structure and replication strategy. Their defining feature is the presence of a single-stranded RNA molecule that serves directly as mRNA upon entry into the host cell
- Replication of positive-strand RNA viruses occurs through the creation of replication complexes, which are membranous structures that facilitate RNA synthesis. These complexes are often associated with rearranged host cell membranes, aiding in viral RNA replication and shielding viral RNA from host cell defenses
- Positive-strand RNA viruses (+ssRNA viruses) are a group of related viruses that have positive-sense, single-stranded genomes made of ribonucleic acid. The positive-sense genome can act as messenger RNA (mRNA) and be directly translated into viral proteins by the host cell's ribosomes.
- Positive-sense RNA viruses include pathogens such as the Hepatitis C virus, West Nile virus, dengue virus, and the MERS, SARS, and SARS-CoV-2 coronaviruses, as well as less clinically serious pathogens such as the coronaviruses and rhinoviruses that cause the common cold.
- Positive-strand RNA viruses have genetic material that can function both as a genome and as messenger RNA; it can be directly translated into protein in the host cell by host ribosomes
- The replication of the positive-sense RNA genome proceeds through double-stranded RNA intermediates, and the purpose of replication in these membranous

invaginations may be the avoidance of cellular response to the presence of dsRNA. In many cases, subgenomic RNAs are also created during replication

- Numerous positive-strand RNA viruses can undergo genetic recombination when at least two viral genomes are present in the same host cell. The capability for recombination among +ssRNA virus pathogens of humans is common
- Positive-strand RNA viruses are found in three phyla: Kitrinoviricota, Lenarviricota, and Pisuviricota, each of which are assigned to the kingdom Orthornavirae in Riboviria
- Positive-strand RNA virus genomes are templates for translation and replication, leading to interactions between host translation factors and RNA replication at multiple levels
- Host factors can mediate viral translation generally or be involved in selective translational regulation of specific viral factors in ways important for RNA replication. Brome mosaic virus (BMV), a member of the alphavirus-like superfamily of positive-strand RNA viruses, directs RNA replication, subgenomic mRNA transcription, and virion assembly in yeast
- Mutations in another host gene, OLE1, also block BMV RNA replication after membrane association of 1a, 2a polymerase, and viral RNA but before initiation of negative-strand RNA synthesis. OLE1 is an essential gene encoding Δ9 fatty acid desaturase, the key enzyme for converting saturated to unsaturated fatty acids

REVIEW QUESTIONS

1. How does translation and RNA replication occur in positive-strand RNA viruses?
2. What is the process involved in the assembly of membrane-associated RNA replication complexes?
3. What are the general principles underlying positive-strand RNA virus replication?
4. How are functional cDNA clones constructed and applied in virology?
5. What are non-clonal reverse genetic systems, and how are they used in studying RNA viruses?
6. What are some novel methods and future directions in the study and manipulation of positive-strand RNA viruses?

REFERENCES

1. Ahlquist, P. (2002). RNA-dependent RNA polymerases, viruses, and RNA silencing. *Science*, *296*, 1270–1273.
2. Almazán, F., Dediego, M. L., Galán, C., Escors, D., Alvarez, E., Ortego, J., Sola, I., Zuñiga, S., Alonso, S., Moreno, J. L., Nogales, A., Capiscol, C., & Enjuanes, L. (2006). Construction of a severe acute respiratory syndrome coronavirus infectious cDNA clone and a replicon to study coronavirus RNA synthesis. *Journal of Virology*, *80*, 10900–10906.

3. Almazán, F., Galán, C., & Enjuanes, L. (2004). The nucleoprotein is required for efficient coronavirus genome replication. *Journal of Virology*, *78*, 12683–12685.

4. Barton, D. J., Black, E. P., & Flanegan, J. B. (1995). Complete replication of poliovirus in vitro: Preinitiation RNA replication complexes require soluble cellular factors for the synthesis of VPg-linked RNA. *Journal of Virology*, *69*, 5516–5527.

5. Diez, J., Ishikawa, M., Kaido, M., & Ahlquist, P. (2000). Identification and characterization of a host protein required for efficient template selection in viral RNA replication. *Proceedings of the National Academy of Sciences of the United States of America*, *97*, 3913–3918.

6. Fowler, D. M., & Fields, S. (2014). Deep mutational scanning: A new style of protein science. *Nature Methods*, *11*, 801–807.

7. Holmes, E. C., & Moya, A. (2002). Is the quasispecies concept relevant to RNA viruses? *Journal of Virology*, *76*, 460–465.

8. Lai, M. M. (1998). Cellular factors in the transcription and replication of viral RNA genomes: A parallel to DNA-dependent RNA transcription. *Virology*, *244*, 1–12.

9. Li, W., Li, Y., Kedersha, N., Anderson, P., Emara, M., Swiderek, K. M., Moreno, G. T., & Brinton, M. A. (2002). Cell proteins TIA-1 and TIAR interact with the 3′ stem-loop of the West Nile virus complementary minus-strand RNA and facilitate virus replication. *Journal of Virology*, *76*, 11989–12000.

10. Lindenbach, B. D., Randall, G., Bartenschlager, R., & Rice, C. M. (2020). In D. M. Knipe & P. M. Howley (Eds.), *Fields virology: Flaviviridae* (6th ed.). Lippincott Williams & Wilkins.

11. Rovner, A. J., Haimovich, A. D., Katz, S. R., Li, Z., Grome, M. W., et al. (2015). Recoded organisms engineered to depend on synthetic amino acids. *Nature*, *518*, 89–93.

12. Stewart, M. (2019). Polyadenylation and nuclear export of mRNAs. *Journal of Biological Chemistry*, *294*, 2977–2981.

13. Tellinghuisen, T. L., & Rice, C. M. (2002). Interaction between hepatitis C virus proteins and host cell factors. *Current Opinion in Microbiology*, *5*, 419–427.

14. van Regenmortel, M. H. V. (Ed.). (2000). *Virus taxonomy*. Academic Press.

CHAPTER

10 Replication Strategies of Small and Medium-Sized DNA Viruses

LEARNING OBJECTIVES

After studying this chapter, you will be able to:

- Define DNA viruses express genetic information and replicate their genomes in similar, yet distinct, ways;
- Understand the papovavirus replication;
- Discuss replication of adenoviruses;
- Explain the replication of some single-stranded DNA viruses.

KEY TERMS FROM THIS CHAPTER

- Agnoprotein
- DNA synthesis
- Herpesviruses
- Replication fork
- Replicative intermediate
- Virus-host interaction
- DNA genomes
- Encapsidation
- Polyomaviruses
- Replication origin
- SV40 virus

INTRODUCTION

Small and medium-sized DNA viruses exhibit a variety of replication strategies tailored to their specific needs and host environments. These viruses, typically characterized by their compact genomes, have evolved efficient mechanisms to hijack host cellular machinery to replicate their DNA and produce progeny virions. Despite their size, they deploy sophisticated tactics to ensure their propagation and survival within the host.

One common strategy involves the use of host DNA polymerases for replication. Many small DNA viruses, such as the parvoviruses, rely heavily on the host cell's replication machinery. They generally infect cells already in the S phase of the cell cycle, where the host DNA polymerases are active, ensuring that the viral DNA can be efficiently replicated. This dependence on host enzymes reduces the genetic burden on the virus, allowing it to maintain a smaller genome. However, this strategy also means that the virus must synchronize its life cycle with the cell cycle of the host, often manipulating host cell regulatory pathways to enter the S phase if necessary (Breitburd et al., 1995).

In contrast, some medium-sized DNA viruses, such as adenoviruses and polyomaviruses, encode their own replication proteins, including DNA polymerases, primases, and helicases. This allows these viruses more control over their replication process and enables them to infect a broader range of cell types, not just those in the S phase. For instance, adenoviruses initiate replication by encoding early proteins that alter the host cell environment to favor viral DNA synthesis, including inhibiting cellular DNA synthesis and activating viral replication origins. By encoding their own replication machinery, these viruses can circumvent some host cell cycle restrictions, leading to more efficient and versatile replication.

Furthermore, these DNA viruses often employ mechanisms to evade host immune responses. For instance, they can produce proteins that inhibit apoptosis, ensuring the host cell remains alive long enough to produce viral progeny. They may also encode proteins that interfere with host antiviral pathways, such as the interferon response, allowing them to replicate undetected by the host immune system. This immune evasion is critical for the survival and persistence of the virus within the host.

Overall, the replication strategies of small and medium-sized DNA viruses highlight a balance between utilizing host resources and maintaining the flexibility to adapt to different cellular environments. These strategies reflect their evolutionary adaptations to maximize replication efficiency while overcoming host defenses, ensuring continued survival and propagation.

10.1. DNA VIRUSES EXPRESS GENETIC INFORMATION AND REPLICATE THEIR GENOMES IN SIMILAR, YET DISTINCT, WAYS

Since DNA is the universal genetic material of all cells, it should be no surprise that a sizable fraction of all known viruses are viruses with DNA as their genome. The fact that these viruses frequently use a sizable portion of the cellular machinery involved in decoding and replicating genetic information encoded in double-stranded (ds) DNA—which is, after all, the substance that makes up the cellular genome—is also not especially surprising.

This is not the case, despite the possibility that all viruses with DNA genomes would replicate in a manner that is broadly similar. Indeed, just as RNA viruses with DNA genomes employ numerous variants on a common replication strategy. Some viruses contain DNA that are both naked and enveloped, and some DNA viruses only encapsulate one strand of DNA. Certain DNA viruses that infect plants have multipartite genomes and one class of animal viruses that use DNA as genetic material replicates in the cytoplasm of eukaryotic cells. A significant and crucial group transforms RNA into DNA, and a related group, as the virus matures, transforms RNA contained in the virion into DNA.

Although it is possible to draw helpful generalizations about how DNA viruses replicate (in fact, one must if the subject is to be easily mastered), it is best to view these generalizations as merely basic guidelines. Therefore, RNA is expressed by viruses of eukaryotic cells that replicate through the nucleus using the transcription machinery of the cell; however, novel RNA polymerases or specificity factors are encoded by bacterial DNA viruses and at least one group of insect DNA viruses (the baculoviruses) to guarantee that only viral mRNA is expressed after infection. Similarly, many enzymes involved in transcription and mRNA modification are encoded by the cytoplasmic-replicating DNA genome-containing poxviruses of eukaryotes. When it comes to DNA replication, most viruses use enzymes and mechanisms that are similar to those found in uninfected cells. However, there is one significant complexity that must be taken into account. This is because not all viral groups use RNA primers, even though primers are necessary for all viral DNA replication (Schiffman et al., 1993).

A significant issue that also impacts the replication of cellular chromosomal DNA affects viruses with linear genomes. This "end problem" arises because the primer used in DNA replication needs to identify brief regions of the viral genome through base pairing or particular DNA–protein interactions. DNA replication can move on to the next fragment 5′ to 3′ down when the primer anneals to the template's exact 3′ end sequences. There is no space for a new primer to anneal upstream of this final gap to be filled, so how is the primer to be removed and replaced with DNA? This would require the viral genome to shrink with each replication, which would cause it to vanish quickly.

Eukaryotes have discovered a solution to the final issue in the replication of their linear chromosomal DNA: the replacement of lost sequences with the enzyme telomerase. DNA viruses with linear genomes have evolved different strategies to get around this final obstacle. The repeated sequences at the ends of the genomes of herpesviruses and many bacterial DNA viruses allow the virus to recombine after infection to form a circular genome. Consequently, replicating viral DNA in the cell is either circular or joined end to form long concatamers, despite the fact that virion DNA is linear. The resolution of these structures to linear ones occurs upon encapsidation of viral DNA.

Adenovirus, on the other hand, has found a solution by attaching a primer to a viral protein that binds to the end of the viral DNA through covalent bonding. Furthermore, there is no irregular strand synthesis in the adenovirus DNA; it only proceeds continuously.

The solution is found in small single-stranded (ss) DNA viruses such as parvoviruses, which encode a complementary repeat sequence at the end that permits the genome to form a "hairpin loop" at the end, preventing the end of the molecule from being free. The poxvirus genome structure demonstrates a comparable solution. This linear DNA genome has ends that are covalently closed, just like chromosomal DNA. Replication, therefore, essentially just moves "around the corner" onto the complementary strand.

Another significant "general" method of nuclear-replicating eukaryotic viruses and many bacteriophage replication is the creation of infections in which the viral genome coexists with the host for the entirety of the infection. Any pathogen can benefit greatly from such a process in terms of evolution, but the specifics of the process vary greatly amongst the groups once more.

Considering how much cellular function and cellular transcriptional machinery are required for productive replication is perhaps the best way to conceptualize the fundamental processes of DNA virus replication in light of these variations. This and the total size of the viral genome are roughly correlated. The benefit of this kind of grouping is that the replication strategies of the viruses in each group are somewhat similar. Crucially, they are comparable in that they can modify cells while they are replicating. These changes may have a significant and long-lasting impact on the host's health.

10.2. PAPOVAVIRUS REPLICATION

The acronym for "papilloma, polyoma, vacuolating" viruses is papovavirus. Members of the group actually belong to two different families: polyomaviruses and papillomaviruses. These two groups share characteristics such as circular genomes, icosahedral capsids, the capacity to bind to the host for extended periods of time, and the need to specifically change cell growth in the host cell's reaction to nearby cells in order to replicate the virus. Their genome sizes and various aspects of host cell specificity are different.

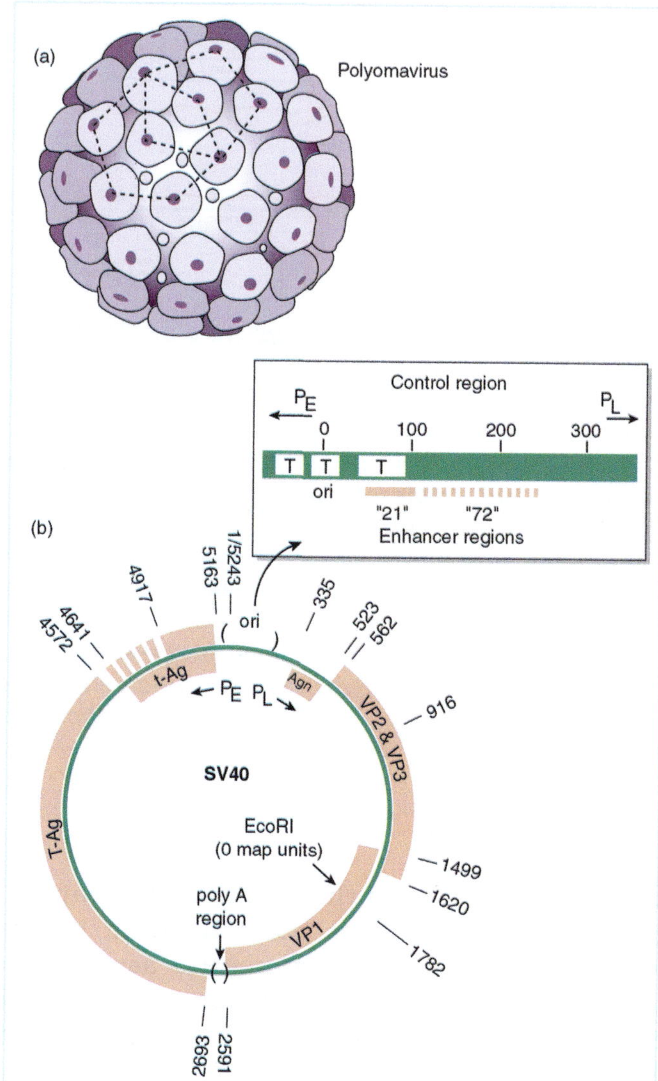

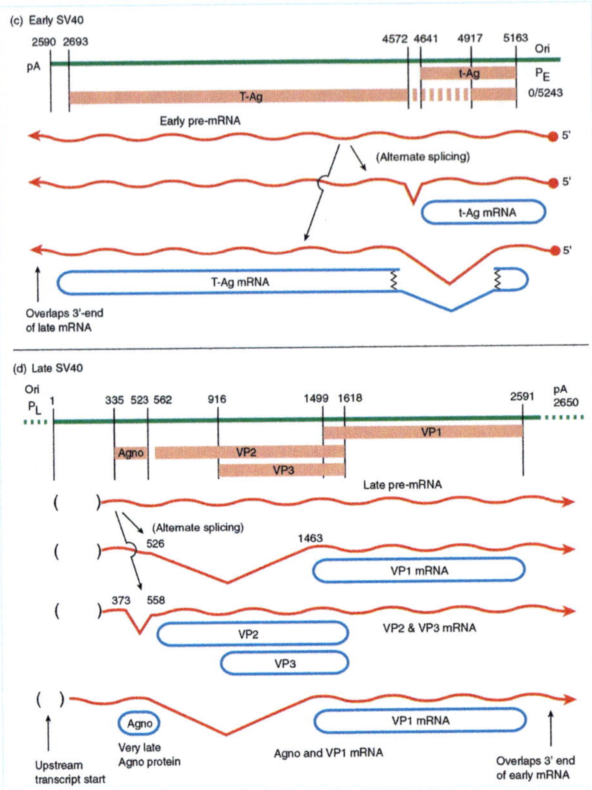

Figure 10.1. *Polyomavirus and the genetic and transcript map of SV40 virus. (a) The 60 pentameric subunits of the capsid proteins are arranged in a manner that the packaging of individual capsomers is not equivalent in all directions. The 5243-base pair dsDNA genome is condensed with host cell histones and packaged into the 45-nm-diameter icosahedral capsid. (b) The early and late promoters, origin of replication, and bidirectional cleavage/ polyadenylation signals are shown along with the introns and exons of the early and late transcripts. A high-resolution schematic of the approximately 500-base pair control region with the early and late promoters is also provided. Two early promoter enhancers, one containing the 21-base pair repeats, and the other containing the 72-base pair repeats, are shown. The origin of replication (ori) is situated between the enhancers and the early promoter, and the three binding sites for large T antigen (T) are indicated. (c) A higher resolution schematic of the processing of early viral mRNAs. Splice sites, translational reading frames, and other features are indicated by sequence number. (d) A higher-resolution schematic of the processing of late viral mRNAs. Splice sites, translational reading frames, and other features are indicated by sequence number.*

Source: Edward K. Wagner, Martinez J. Hewlett, David C. Bloom, David Camerini; Basic Virology Third Edition; ISBN-13: 978-1-4051-4715-6.

One peculiar structural characteristic of the polyomavirus capsid is that, despite being an icosahedron, the capsid subunits do not normally form hexon and penton arrays. Rather, the capsid can be formed by the asymmetric assembly of all 60 pentameric subunits, equivalent to one another. This can be seen in Figure 10.1 (a). It is unclear if polyomaviruses have any genetic or functional restrictions that would call for this peculiar structure (Dupuy et al., 1997).

10.2.1. Replication of SV40 Virus: The Model Polyomavirus

The genomes of the polyomaviruses consist of roughly 5000 base pairs. Three proteins—usually referred to as VP1, VP2, and VP3—make up capsids. Animal tumors can be caused by polyomaviruses, which can also alter the characteristics of primary cell growth in culture, particularly in animal cells that differ from the virus's natural host. Moreover, polyomaviruses continue to be persistently linked to their host, frequently in the absence of clear signs of severe pathology or illness. These viruses replicate slowly, even though the cells they infect die. There is no widespread virus-induced shutdown of host function, which is consistent with the need for substantial cellular activity during replication (Howley and Livingston, 2009).

Murine polyomavirus (Py) is one polyomavirus that has been extensively studied. It was first isolated from wild mice and given its name because it can cause a variety of small tumors in certain strains of newborn mice. The SV40 virus, also known as simian vacuolating agent 40, is another polyomavirus that has been extensively studied. Although there is no proof that this virus causes tumors in its natural host, it can easily cause cells grown in culture to undergo growth transformation. Originally discovered as a contamination of African Green monkey kidney cells (AGMK), poliovirus was being grown in these cells in order to produce a vaccine. Although early Salk polio vaccine recipients received a healthy dose of the virus, no pathology has been linked to this to date.

Although the Rous sarcoma virus, a retrovirus, has been known to cause tumors in chickens since the early part of this century, molecular biologists working in the 1950s and 1960s could not understand its mechanism of oncogenesis because its genome is RNA. Indeed, significant advancements were to come until Howard Temin and David Baltimore discovered reverse transcriptase in 1970. However, the ability of the DNA-containing SV40 and mouse polyomaviruses to induce growth transformation and tumors in the lab offered a model for the process that could be used to study it using then-current techniques. Tumor suppressor genes and their crucial function in controlling cell growth and division were eventually discovered due to the molecular biological study of carcinogenesis, which was largely initiated by examining these viruses.

The SV40 virus has been studied more than any other DNA virus, possibly due to its convenient genome size, ease of manipulation in the lab, and significance in basic research oncogenesis. Although there are some notable differences between Py and SV40 replication, the general approach remains the same. There are three human polyomaviruses known to science: BK, JC, and possibly another that has not yet been thoroughly identified. Similar to SV40, the BK and JC viruses are believed to be transmitted through the respiratory system. Primary infection happens in kids who don't have much visible disease. The majority of children in the United States contract the BK virus by the time they are 5 or 6 years old, and the only symptoms of the infection may be mild respiratory illness. The JC virus infections happen a little later; most children get infected between 10 and 14 (Howley and Livingston, 2009)

In children with normally functioning cell-mediated immunity, the infection has fully resolved. The kidney is the main site of persistence for the BK virus, which can occasionally be shed from it, even after resolution. The virus remains in the individual for the duration of their life. Additionally, brain biopsy specimens can yield the JC virus. Immunosuppression caused by HIV infection or before organ transplantation can have serious consequences, even though this persistence has no known clinical manifestations in healthy individuals and is thought to be the result of viral genomes persisting in an inactive state in nondividing, terminally differentiated cells. Progressive multifocal leukoencephalopathy (PML), a rare progressive CNS neural tissue destruction, is linked to the JC virus in immune-compromised individuals. The transcription of JC virus causes this neuropathology RNAs in oligodendrocytes in the adult brain (but not in other cells). It is unclear, however, exactly what kind of immune suppression, such as that caused by HIV infection, reactivates this dormant virus. Although the link between BK virus infections and urinary tract pathologies in immunocompromised individuals is not as clear-cut, it is highly likely.

The precise sequence of the human JC virus, which has been isolated from people across the globe, differs enough to be useful as a genetic marker of population distribution. Numerous studies conducted on natural isolates have demonstrated that particular virus variants are closely linked to particular racial and ethnic population groups and that the occurrence of particular virus variants can be used to track the movements of these groups over time. This indicates that the virus has been linked to people for a long time and that variations have emerged due to population divergence.

When laboratory strains of mice are infected with murine polyomavirus, the pattern of infection of young animals is followed by virus persistence and shedding. The pathology of this virus differs noticeably from that of SV40, JC, and BK viruses, though, in that infections in nursing mice can result in tumor formation—hence the term polyomavirus. While there is some suggestive evidence that human polyomaviruses can be associated with tumors, conclusive evidence of causation is lacking. Genetic studies suggest that the presence of specific endogenous retroviruses in the laboratory mouse strains is a major factor in the ability of the murine virus to cause tumors.

10.2.1.1. The SV40 Genome and Genetic Map

The genome of the SV40 virus has 5243 base pairs, and Figure 10.1 (b) shows a map of its key characteristics. Four functional regions comprise the genome, each covered in detail below.

10.2.1.1.1. The Control Region

This region includes the early promoter/enhancer, the late promoter, and the replication origin. It spans approximately 500 bases. Although there is a significant amount of sequence element overlap in this region, the bases directly linked to each function can be precisely located on the genome. To achieve this, specific sequence mutations have been made, and their effects on the replication of the viral genome and the expression of early and late genes have been examined.

The early promoter region has a TATA box and enhancer regions (indicated by 72-base and 21-base repeats). Remarkably, the late promoter lacks a TATA box, and the late mRNA starts at multiple locations within a 60-to 80-base region. One of the first indications that the TATA box serves to assemble transcription complexes at a particular location in relation to mRNA initiation was the multiple start sites for late mRNA expressed from this "TATA-less" promoter. Though the exact replacement for the TATA box in the late promoter is unknown, transcription complexes are thought to be able to form relatively easily throughout the region (Lai and Nathans, 1974).

The replication origin, or ori, spans approximately 150 base pairs and consists of multiple elements whose sequences are crucially connected by "spacers" whose length—rather than a specific sequence—is crucial to the origin's functionality. The far left region's sequence is repeated in the far right region in the opposite direction, exhibiting some dyad symmetry in the ori elements. It is believed that this symmetry plays a part in enabling the DNA helix to "melt" at the origin, which makes it easier for replication enzymes to enter and start new rounds of DNA replication.

10.2.1.1.2. The Early Transcription Unit

The early region of the SV40 genome is displayed in high resolution in Figure 10.1 (C). It is translated into a single open reading frame (ORF)-containing mRNA precursor that stretches roughly halfway across the genome. There are two distinct sites at which the single early pre-mRNA transcript can be spliced. An mRNA encoding a comparatively small protein (t (small t) antigen; about 20,000 daltons) is produced if a short intron is removed. This protein plays a part in enabling the virus to replicate in specific cells.

A larger intron in the pre-mRNA is spliced to produce a slightly smaller (and more abundant) mRNA. The small-t-antigen ORF is terminated by removing this translation terminator. The T (large T) antigen (about 80,000 daltons) is encoded by the smaller (!) mRNA. The large T antigen has several functions, including the following:

- Activation of cellular DNA and RNA synthesis by binding to the cellular growth control gene products named Rb and p53. This binding stops these control proteins from keeping the cell contact inhibited. This function causes the infected cell to begin a round of DNA replication
- Blockage of apoptosis normally induced in cells where p53 is inactivated at inappropriate times in the cell cycle.
- Binding to the SV40 ori to initiate viral DNA replication.
- Shutting off early viral transcription by binding to regions in and near the early promoter.
- Activating late transcription.
- Playing a role in virion assembly.

10.2.1.1.3. The Late Transcription Unit

Late mRNA is expressed from a region extending around the other half of the genome from the late promoter; this is shown in Figure 10.1(d). The late region contains two large ORFs that encode the three capsid proteins. Part of the expression of late proteins requires alternate splicing patterns, just as is seen with the generation of early mRNA. Splicing of a large intron from the primary late pre-mRNA transcript generates an mRNA that encodes the 36,000-dalton major capsid protein (VP1). A small amount of mRNA is generated by splicing a small intron near the 5′ end of the mRNA, allowing the first ORF to be translated into the 35,000-dalton VP2 protein.

By using a different translation initiation site, the same mRNA that codes for VP2 is also used to express the third capsid protein, or VP3. Sometimes, the first AUG of the 5′ ORF in the mRNA expressing VP2 is "missed" by ribosomes. In this scenario, the ribosome starts translation at an AUG downstream of the first one but in phase with it, resulting in the production of the 23,000-dalton VP3 protein. As a result, depending on where the ribosome begins translation, one mRNA can encode both VP2 and VP3. Since the first AUG is not visible and is, therefore, in the operational leader sequence of the mRNA, this "skipping" does not break the general rule that a eukaryotic ribosome can only initiate a protein at the 5′ ORF.

Only very late in infection is a fourth late protein expressed from the late region visible. The "agnoprotein," a fundamental protein, is encoded in a brief ORF upstream of the one that codes for VP2. This protein can be translated from certain mRNAs that are produced very late in the infection by transcription starting more upstream than it did earlier. Although the exact function of this product is unknown, it is possible that it contributes to the virus's ability to replicate in some cells that are normally inhospitable to viral replication.

10.2.1.1.4. The Polyadenylation Region

A second cis-acting control region is located approximately 180° from the ori/promoter region around the circular SV40 genome. Both the early and late regions' transcripts end in this region because it has polyadenylation signals on both DNA strands. Notably, there is a region of 3′ overlap between the early and late transcripts due to the positioning of the polyadenylation signals for the mRNAs. As a result, during the replication cycle, dsRNA may be produced, which may cause the infected cells to produce interferon.

10.2.1.2. Productive Infection by SV40

It is simple to study productive infection by SV40 in its normal host in cell culture using monkey kidney cells. It frequently takes 72 hours or longer for the replication cycle to finish, resulting in cell lysis and the release of fresh virus. The fact that the virus depends heavily on ongoing cellular activity for the majority of its replication process may be one explanation for this "leisurely" pace. Because the cultured cells have not yet reached confluence or are growth-transformed and not susceptible to contact inhibition of growth, the virus replicates effectively in these actively dividing cells.

The virus can multiply effectively in replicating cells, but it can also do so in cells in a growth arrest state. This is a result of early infection-related T-antigen expression. Expressions of this capacity led to the identification of the tumor suppressor genes p53 and Rb, which are covered in the section that follows. They also offer many helpful insights into the nature of the cell's capacity to regulate and control its own DNA replication.

10.2.1.2.1. Virus Attachment and Entry

The replication cycle of SV40 is outlined

in Figure 10.2. Virions interact with a specific cellular receptor. This leads to receptor-mediated endocytosis, and the partially uncoated virion is transported in the endocytotic vesicle to the nucleus, where viral DNA is released (Lederberg and Meselson, 1964).

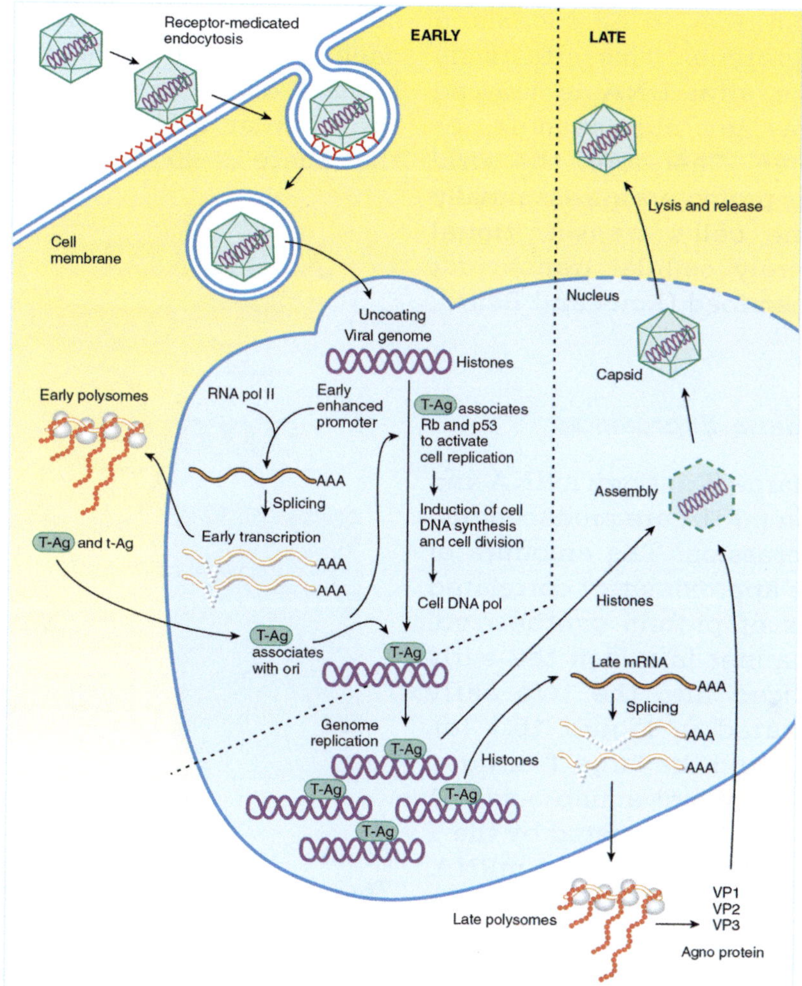

Figure 10.2. *The replication cycle of SV40 virus in a permissive cell. The replication is divided into two phases, early and late. During the early stages of infection, the virus attaches, and viral genomes with accompanying cellular histones are transported to the nucleus via receptor-mediated endocytosis. RNA polymerase II (pol II) recognizes the enhanced early promoter, leading to transcription of early pre-mRNA, which is processed into mRNAs encoding small t (t-Ag) and large T antigen (T-Ag). These mRNAs are translated into their corresponding proteins. Large T antigen migrates to the nucleus where it carries out a number of functions, including inactivation of the cellular growth control proteins p53 and Rb and binding of the SV40 origin of DNA replication (ori). Viral DNA replication takes place by the action of cellular DNA replication enzymes, and each round of DNA replication requires large T antigen to bind to the ori.*

Source: Edward K. Wagner, Martinez J. Hewlett, David C. Bloom, David Camerini; Basic Virology Third Edition; ISBN-13: 978-1-4051-4715-6.

When the viral DNA is enclosed in the virion of SV40 and other papovaviruses, it is linked to chromosomal proteins such as histones. Following its entrance into the nucleus, it continues to be connected to chromosomal proteins. This effectively indicates that the viral DNA is viewed by the cell as a tiny chromosome, or "mini" chromosome. Therefore, the viral chromosome and its promoters are essentially recognized by the cell's transcriptional machinery as merely cellular genes just waiting to be transcribed (Rous and Beard, 1935).

10.2.1.2.2. Early Gene Expression

Large amounts of large-T antigen mRNA and less small-t-antigen mRNA are produced due to early gene expression. The amounts of mRNA present are approximately correlated with the amounts of protein synthesized. Because of the manner in which the early pre-mRNA is spliced into the two early mRNAs, as illustrated in Figure 10.1 (c), the small t antigen and the large T antigen share the same N-terminal amino acids. A translation stop signal is eliminated by the T antigen-encoding splice-generating mRNA. On the other hand, protein termination is unaffected by the T antigen mRNA splice because it occurs outside of the ORF. Many viruses have the ability to produce two proteins with similar amino acid sequences but major or minor functional differences. When it comes to the expression of adenovirus proteins, it is crucial (Schiffman et al., 1993).

10.2.1.2.3. The Role of T Antigen in Viral DNA Replication and the Early/Late Transcription Switch

T antigen modifies the host cell to enable viral DNA replication. In order to initiate DNA replication and stop the synthesis of early mRNA, the T antigen also attaches itself to the SV40 ori. T antigen is necessary for each round of DNA replication in order for it to attach to the replication origin and start a new round of DNA synthesis. After each replication cycle, DNA cleavage and ligation divide the two daughter circles of the circular SV40 genome, which means that there is no end problem.

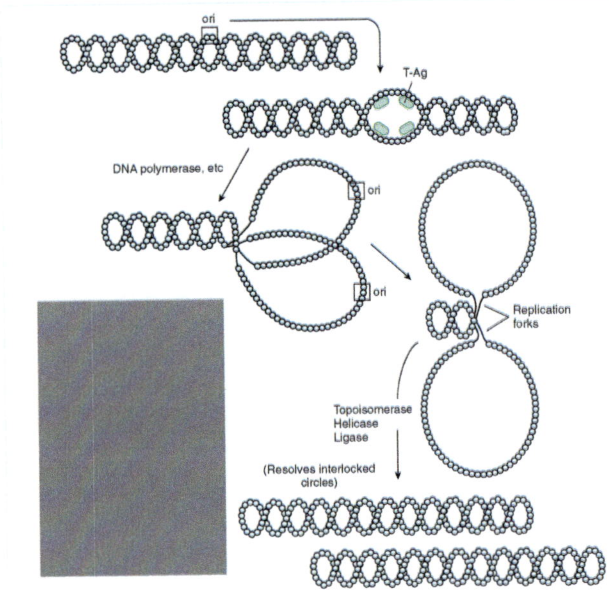

Figure 10.3. *The replication of SV40 DNA. The closed circular DNA has no end problem, unlike the replication of linear DNAs. Structures of the replication fork and growing points are identical to those in replicating cellular DNA, using cellular DNA replication enzymes and accessory proteins. Replication results in the formation of two covalently closed and interlinked daughter genomes that are nicked and religated into individual viral genomes by the action of cellular topoisomerase and other helix-modifying enzymes. T-Ag = large T antigen; ori = origin of replication.*

Source: https://www.blackwellpublishing.com/content/bpl_images/Content_Store/Middle_Sized/Wagner9781405103466/286to310.pdf.

Cellular enzymes, specifically topoisomerases and resolvase, mediate the resolution of the interconnected supercoiled DNA molecules into individual genomes. In Figure10.3, the procedure is depicted. The figure does not depict the association between the daughter DNA genomes and cellular histones, which is a crucial aspect to consider as it is required for the virus to be effectively encapsidated.

Due to the accumulation of large T antigen in the cell, which binds at the ori and early promoter to repress mRNA synthesis, the relative rate of early mRNA synthesis decreases while DNA replication continues. Even though the proportion of early mRNA in the cell decreases with time, early mRNA is always produced as long as a template is available for early mRNA expression that has not yet bound a large T antigen.

In addition to controlling and inhibiting its own synthesis, this adaptable protein also initiates the transcription of late pre-mRNA from replicating DNA templates. As previously mentioned, very late in infection, the start of late mRNA transcription shifts to a point upstream of that previously used, and the agnogene protein (the agnoprotein) can be encoded and translated from a novel subset of late mRNAs. Late transcripts also have heterogeneous 5′ ends.

10.2.1.3. Abortive Infection of cells Nonpermissive for SV40 Replication

In cells derived from a species other than the natural host of SV40, infection resulted in an abortive infection where no virus was produced relatively early in the study of polyomavirus replication. Nevertheless, it was demonstrated that viral infection promoted cellular DNA replication and cell division, and research on this phenomenon produced some of the first significant models for the study of carcinogenesis. Even though these unsuccessful infections might only occur in laboratories, the knowledge they yielded was crucial for comprehending both viral oncogenesis and the pathophysiology of papovaviruses in their natural hosts.

SV40 virus can infect and stimulate RNA and DNA synthesis in rodents and some non-primate cells by producing the large T antigen. This viral protein disables two cellular tumor suppressor or growth control genes, p53 and Rb.

There are two primary roles for the two proteins in question, p53 and Rb. Firstly, they bind to and inactivate the cellular proteins needed to start cell division, thereby mediating an active repression of it. Second, cells that break free from repression and start to divide undergo apoptosis when free protein levels rise above a certain threshold. Similar to the initial stages of productive infection, large T antigen binds to p53 with greater avidity in the early stages of nonpermissive cell infection, displacing active replication-initiation proteins bound to the protein. The proteins thus released can start the replication of cellular DNA, but no apoptosis is induced because there is no free p53.

These are the same steps that occur in the early stages of productive infection; however, viral DNA cannot be replicated in the nonpermissive cells. This failure is due to the inability of T antigen to interact effectively with one or more of its other cellular targets important in the early phases of infection. In this abortive infection, the cells in which T antigen is expressed do not die, but they replicate even while in contact with neighboring cells; this process is shown in Figure 10.4. The continued stimulation of cellular DNA replication by expression of viral T antigen can lead to continual cell replication (i.e., transformation). The stable transformation will require the viral genome to become stably associated with cellular DNA by the integration of viral DNA into the cellular genome. Such viral DNA replicates every time the cell replicates, and thus keeps the cell transformed (Sclair et al., 1973).

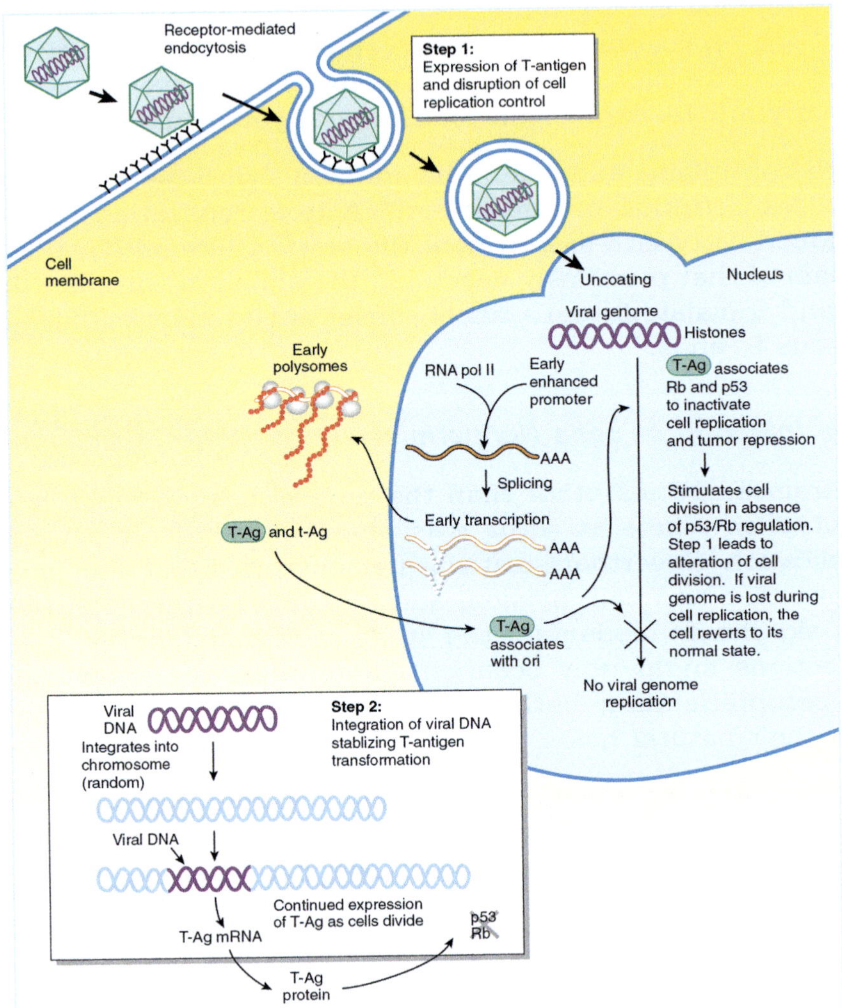

Figure 10.4. *Representation of the two steps in the transformation of a nonpermissive cell by SV40. The infection begins as described in Figure 10.2, and early mRNA is expressed into early proteins. The infection is abortive in that DNA replication and late gene expression cannot occur in the nonpermissive cell. Still, the large T antigen (T-Ag) can interfere with cellular growth control (tumor suppressor) proteins, leading to cell replication. Stable transformation requires a second step, the integration of the viral DNA. This is a random (stochastic) occurrence with SV40, and integration is random throughout the genome. A similar path is followed in the transformation of nonpermissive cells by other polyomaviruses. t-Ag = small t antigen.*

Source: https://www.blackwellpublishing.com/content/bpl_images/Content_Store/Middle_Sized/Wagner9781405103466/286to310.pdf.

T antigen or any other viral product does not play a role in the integration of viral DNA into a host cell chromosome. The majority of abortively infected cells will indeed divide for about a round until the viral DNA is lost, at which point they will return to their typical growth patterns. This transformation is sometimes referred to as transitory, transient, or abortive. A completely random recombination event leads to the integration of viral DNA into the host cell at locations where a small number of circular viral DNA bases can anneal to a small number of chromosomal DNA bases. Breaking and re-ligating the chromosome with the integrated viral DNA must come next. Although this is not very common, a focus of transformation will form if a large number of cells are abortively infected with the relevant polyomavirus and one or more of those cells integrate the viral chromosome and continue to express T antigen. A clump of transformed cells growing on the surface of a culture dish containing contact-inhibited cells is an example of such a focus. Similar statistical analyzes apply to both these foci and plaques formed by productive infection (Suzich et al., 1995).

10.2.2. Replication of Papillomaviruses

SV40-induced cell transformation seems to be a phenomenon unique to laboratories, and many polyomavirus-caused tumors can be considered terminal symptoms of viral infection. The stability of histone-associated viral genomes in nonreplicating cells, characterized by sporadic episodes of low-level viral replication brought on by an immune crisis or other events that alter the transcriptional environment of the host cell, appears to be the cause of persistence in such infections. On the other hand, papillomaviruses, a related family of viruses, replicate naturally in their host by requiring the development of tumors, typically benign tumors, during their replication cycle. Persistence in this virus replication strategy results from the ongoing replication of cells harboring viral genomes!.

Replication of the papillomavirus incorporates elements of the previously discussed unsuccessful and successful strategies. There are numerous varieties of these viruses that cause warts and papillomas, and the majority of them do not exhibit antigenic cross-reactivity with one another. Most papillomavirus infections are completely benign despite occasionally causing pain or irritation. However, certain types of the virus can be transmitted through sexual contact and result in persistent genital infections, particularly in females. Even though very few people actually develop cervical carcinoma, statistical analyzes comparing the incidence of the disease with patterns of persistent infection by some of these papillomaviruses—including human papillomaviruses 16 and 18 (HPV-16 and HPV-18)—show a highly significant correlation. These viruses are, therefore, unquestionably human cancer viruses.

10.2.2.1. The HPV-16 Genome

The circular genome of HPV-16, depicted in Figure 10.5, is approximately 7900 base pairs in length and bears a resemblance to that of SV40, with numerous early ORFs. The "LCR" region corresponds to the promoter/origin region of SV40. Studying the replication of papillomaviruses in cultured cells is challenging, necessitating a detailed examination of splicing patterns and transcripts produced during infection. This task is laborious and involves generating DNA copies from viral RNA with retrovirus reverse transcriptase, followed by cloning of the cDNA copies (Suzich et al., 1995).

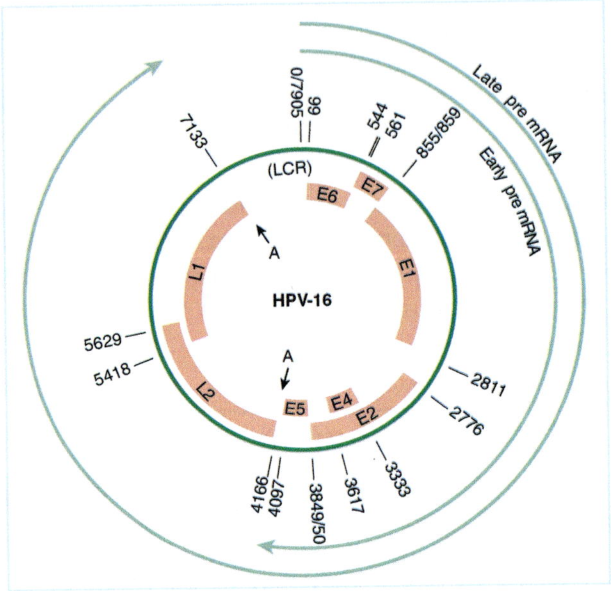

Figure 10.5. *The human papillomavirus (HPV)-16 genome. The 7-kbp circular genome contains a number of translational reading frames that are expressed from spliced mRNAs. Unlike the related polyomaviruses, papillomaviruses encode all proteins on the same DNA strand. The actual details of mRNA expression also appear to differ among different papillomaviruses. Analysis of the bovine papillomavirus genome and expressed transcripts shows that both early and late transcripts are produced from a limited number of promoters as pre-mRNAs. Unlike polyomaviruses, papillomaviruses exhibit extensive splicing of pre-mRNAs, with early and late promoters present in multiple regions of the genome.*

Source: https://www.blackwellpublishing.com/content/bpl_images/Content_Store/Middle_Sized/Wagner9781405103466/286to310.pdf, Edward K. Wagner, Martinez J. Hewlett, David C. Bloom, David Camerini; Basic Virology Third Edition; ISBN-13: 978-1-4051-4715-6.

10.2.2.2. Virus Replication and Cytopathology

A papillomavirus infection forms a wart, as shown in Figure 10.6. The virus enters the basal cells of the epithelium, or the skin in the case of warts. Early genes expressed by the virus cause cells to replicate their DNA at a higher rate than that of an uninfected epithelial cell. Consequently, the SV40 T antigen and one set of early functions are comparable. But unlike SV40 replication in permissive cells, which results in vegetative viral genome replication and cell death, papillomavirus DNA replicates when cell DNA replicates in the infected cell nucleus as an episome or "mini"-chromosome. It does not, however, replicate in the high numbers observed in viral DNA replication of a productive infection (Suzich et al., 1995).

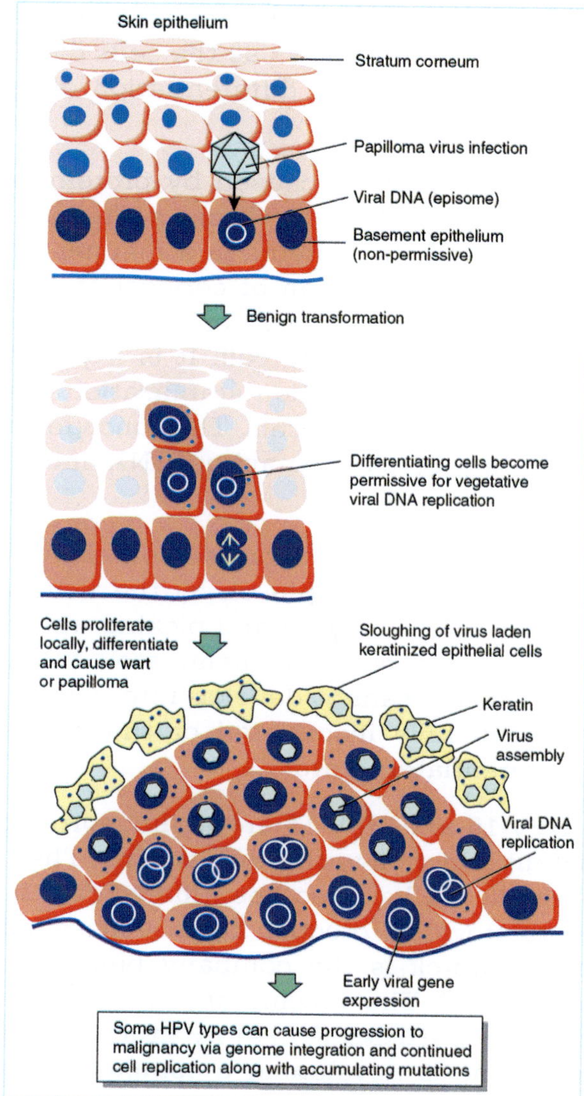

Figure 10.6. *The formation of a wart by cell proliferation caused by infection of basement epithelial cells with human papillomavirus (HPV). Early gene expression leads to stimulation of cell division and terminal differentiation. This results in late gene expression and virus replication in a terminally differentiated, dying cell, which produces large quantities of keratin.*

Source: Edward K. Wagner, Martinez J. Hewlett, David C. Bloom, David Camerini; Basic Virology Third Edition; ISBN-13: 978-1-4051-4715-6.

Cell-linked replication, also known as plasmid-like replication, occurs when cellular DNA replication proteins interact with the viral origin of replication. This origin acts like a cellular DNA replication origin during persistent infection and is regulated in a similar way. As cells divide and differentiate, they change their function and begin to produce proteins specific to their type of differentiation. For example, the synthesis of K5 and K14 keratins, common in basal cells, is stopped, and instead, keratins K1 and K10, found in suprabasal skin cells, are produced. Eventually, some of these differentiated cells become permissive for high levels of viral DNA replication and late gene expression, leading to the production of new virus particles as the cells die. This process is localized and simply accelerates the normal terminal differentiation of epithelial cells, resulting in the formation of a benign wart.

This enhanced growth is known to be caused by proteins encoded by the E5, E6, and E7 gene products for HPV-16 and HPV-18. These proteins interact with and inhibit the normal functions of the p53 and Rb proteins in a manner similar to large-T antigen activity in SV40. Likely, a persistent infection of cervical epithelium with one of these viruses may occasionally lead to the development of a true cancer cell through additional mutations in other control circuits within the cell. It is hypothesized that oncogenesis involves a process similar to the transformation stabilization observed in abortive SV40 infection of the appropriate nonpermissive cell. This oncogenic transformation occurs concurrently with the integration of papillomavirus DNA into cellular DNA (Dupuy et al., 1997).

Since no virus is generated in such a transformed cell, the development of the cancer can be understood as a random event brought on by the virus's ongoing stimulation of cell division. These altered cells keep dividing and picking up mutations that eventually let them infiltrate other tissues and create disseminated tumors (metastasis). Benign warts on the skin and other tissues cannot develop into cancer because either the p53 and Rb proteins have not been fully inactivated or the stimulated cells are so near death in their terminally differentiated state.

It is well known that HPV-16 and HPV-18 are highly correlated with cervical cancer in women who have persistent infection with these strains. They can also be transmitted through sexual contact. This illness has also been linked to other strains, most notably HPV-6 and HPV-11. In order to provide protection against these four strains, a quadrivalent vaccine has been developed. Merck is the company behind the recombinant vaccine known by its trade name, Gardasil. The vaccine uses recombinant DNA technologies to create self-assembling virus-like particles (VLPs) by utilizing the major capsid protein L from each virus. GlaxoSmithKline is developing a comparable product under the trade name Cervarix. June 2006 saw the Food and Drug Administration (FDA) approve Gardasil for general use.

The potential impact of this vaccine is significant. If given as a preventative measure to young women before they become sexually active, it has the potential to greatly reduce the global incidence of cervical cancer. Currently, cervical cancer affects 16 out of every 1000 women annually, with 9 out of every 1000 women dying from the disease, making it the third leading cause of death for women after breast and lung cancers. In the United States, the prevalence and mortality rates are lower, likely due to the widespread use of Pap smears for early detection. It is important to note that while the vaccine can help prevent many cases of cervical cancer, regular Pap smears are still necessary, as at least 10% of cases are not caused by the viruses targeted by the vaccine.

10.3. REPLICATION OF ADENOVIRUSES

A vast class of intricate, icosahedral, nonenveloped viruses that infect humans and other mammals is known as the adenoviruses. In humans, they are typically linked to respiratory illnesses that resemble the common cold or mild flu, but certain serotypes are also linked to upset stomachs. Adenoviruses and papovaviruses are not closely related, but they both have a lengthy replication cycle because virus replication requires the stimulation and utilization of numerous cellular processes. Additionally, they are both capable of inducing abortive infection in lab settings, which transforms cells. Adenovirus replication necessitates extensive splicing of a small number of pre-mRNAs, just like papovavirus replication does. When alternative splicing sites are used, many partially overlapping mRNAs encoding related proteins are expressed (Sweet BH, Hilleman, 1960).

The organization and replication of the viral genome, as well as the specifics of replication, differ significantly from papovaviruses despite these commonalities. Adenovirus infection has a relatively mild course and some useful properties that make it an appealing candidate for use as a therapeutic agent.

10.3.1. Physical Properties of Adenovirus

10.3.1.1. Capsid Structure

Adenoviruses have intricate icosahedral capsids with projecting spikes or fibers at each of the 12 vertices (pentons), and proteins that are not present in equimolar amounts. The core protein of the virus encapsidates the genome and functions somewhat like a histone to create a condensed chromatin-like structure inside the nucleocapsid.

10.3.1.2. The Adenovirus Genome

Adenoviruses have linear genomes with distinct viral proteins, or terminal proteins, at the 5′ ends. The genome consists of approximately 30,000 bp, with the sequence at one end being inversely repeated at the other end (100–150 base pairs, depending on the virus serotype). This is the viral DNA ori.

The genome map displayed in Figure 10.7 shows the locations of numerous transcripts that are expressed during infection. Since the genome is divided into 100 map units, there are 300 base pairs in each map unit. Complex splicing patterns and the presence of multiple promoters complicate the location of transcripts. E1 through E4 are the names of the four early transcription "units," each of which has a minimum of one promoter and one polyadenylation signal. A unique RNA known as "VA" is also produced by the action of host cell RNA polymerase III (pol III) and five "families" of late mRNAs can be generated from a single late promoter.

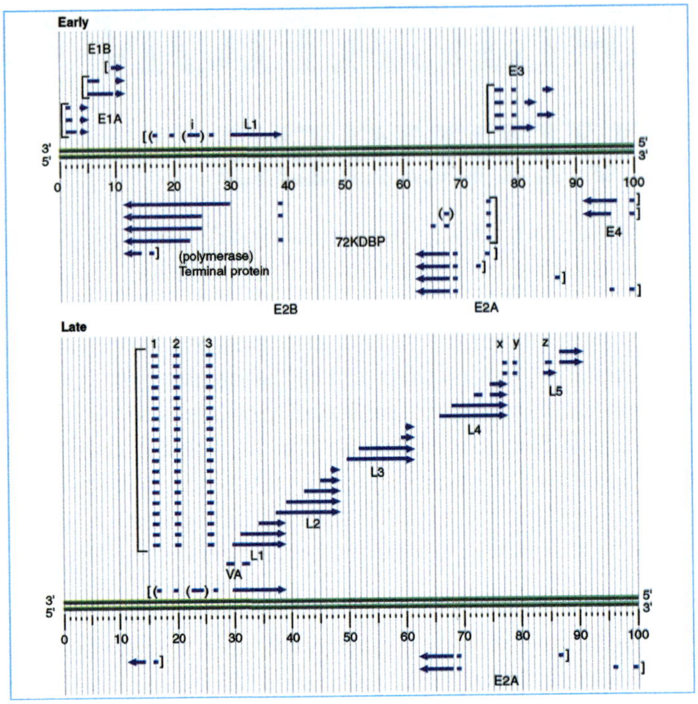

Figure 10.7. *The genetic and transcription map of the 36-kbp adenovirus genome. There are three kinetic classes of transcripts. The E1 transcripts are controlled by enhanced promoters and require no modification of the host cell because some functions of their expression are similar to those of T antigen in SV40 virus replication. These functions include stimulating cellular transcriptional activity and cell replication. Early in infection, only early transcripts are expressed. These include mRNAs encoding viral DNA polymerase and terminal proteins. There are a number of early promoters and transcription units. The E2 transcription unit also has a 72-kd DNA-binding protein (72KDBP) that shuts off early transcription. Two primary transcripts, E2A and E2B, are expressed from the same E2 promoter. The mRNA for the DNA-binding protein continues to be expressed late because there is a second promoter upstream of the E2 promoter that is not shut off by the DNA-binding protein. The major late promoter at map position 15 is always "on," but polyadenylation and splicing patterns change markedly as infection proceeds. Late in infection, the late transcription unit extends to one of five polyadenylation signals, and differential splicing results in the generation of a myriad of late mRNAs encoding structural proteins as well as proteins involved in host cell modification and virus maturation.*

Source: https://www.blackwellpublishing.com/content/bpl_images/Content_Store/Middle_Sized/Wagner9781405103466/286to310.pdf.

10.3.2. The Adenovirus Replication Cycle

10.3.2.1. Early Events

Adenoviruses enter cells similarly to papillomaviruses through receptor-mediated endocytosis. Virion fiber proteins interact with cellular receptors to start an infection. A particular terminal protein attached to each 5′ end of an adenovirus is released into the nucleus, where it binds to cellular histones. Adenoviruses need to encourage the infected cell to transcribe and replicate its genes in order to start gene expression. Expression of the spliced mRNAs encoding the immediate-early (or "pre-early") gene E1A and E1B protein "families" is how this is achieved. "Like the SV40 early promoter, these have enhanced promoters that can still function in the cell even without viral modification (Sweet BH, Hilleman, 1960). The products of the E1A gene prevent the growth suppressor genes Rb and p53 from suppressing cell division, and one or more E1B proteins prevent the stimulated cell from going through apoptosis. As a result, the way these two proteins cooperate is comparable to that of the polyomavirus large T antigen.

Activation of the infected cell's transcriptional machinery results in the production of four early pre-mRNAs, which are then spliced in different ways to create early proteins such as a DNA polymerase protein (140-kd pol), a terminal protein, and a 72-kd DNA-binding protein (DBP). DBP is responsible for turning off most early promoters, with the exception of the E2 region, where a second promoter becomes active when DBP levels are high. Surprisingly, the main late promoter is active early in the infection, but only the L1 region is transcribed as mRNA due to all transcripts being stopped at the polyadenylation signal at 40 map units. This cessation of transcription is caused by the prevention of splicing downstream of the L1 region through the binding of cellular splice factors. Additionally, late transcripts beyond L1 are not transported out of the nucleus.

10.3.2.2. Adenovirus DNA Replication

The process of adenovirus genome replication is illustrated in Figure 10.8 and is carried out by an unusual mechanism that involves the formation of ssDNA as intermediates. The replication of adenovirus DNA starts at either or both ends of the DNA and uses an 80,000-dalton precursor of the 50,000-dalton viral genome-bound terminal protein as a primer. During packaging, the larger priming terminal protein is proteolytically cleaved to the smaller terminal protein present in genomes associated with capsids. Initiation of DNA replication without the need for a short RNA primer has only been observed in this one case. Nonetheless, a covalently bound cytosine residue is present in the terminal protein, where DNA replication takes place. Replication is continuous and uses the adenovirus-encoded DNA polymerase; brief Okazaki fragments are not visible (Waga and Stillman, 1994). After the other strand is freed, replication can continue as ssDNA becomes circular due to the association of inverted repeat sequences at the end of the process. Thus, the two pathways depicted in Figure 10.8 can be used for adenovirus DNA replication. Type I replication takes place when DNA synthesis starts roughly at the same time at both ends of the genome. Type II replication is the result of DNA synthesis occurring from only one end of the genome.

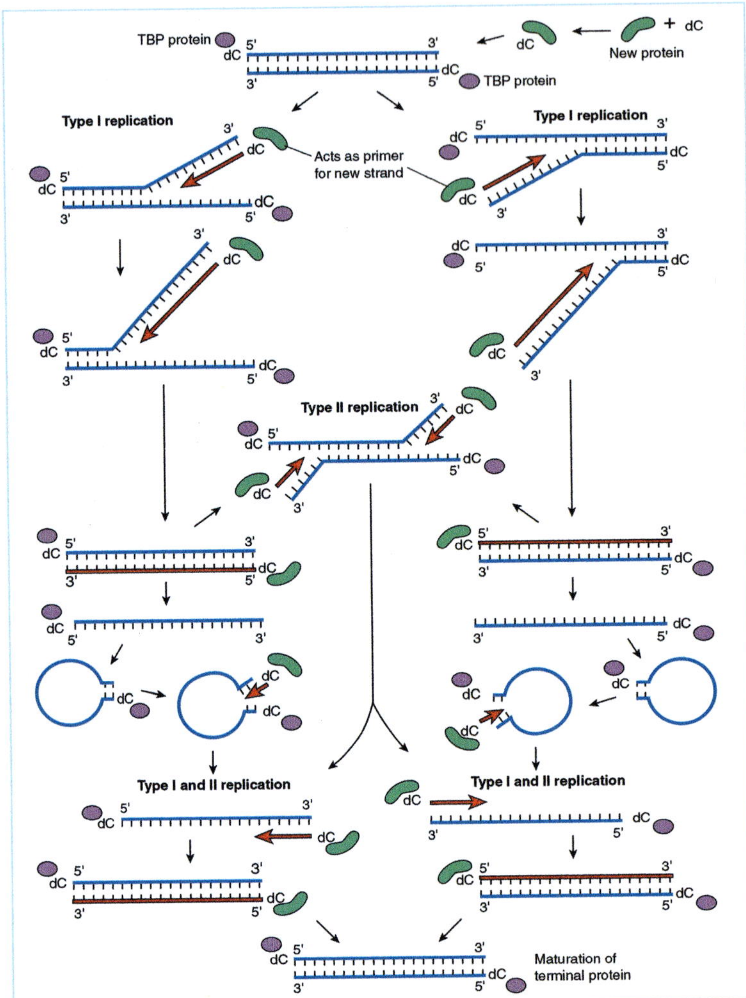

Figure 10.8. *Adenovirus DNA replication. The 5′ ends of the viral genome have 50,000-dalton terminal proteins bound to them. Of major interest is the fact that there is no discontinuous strand synthesis. The process is marked by the accumulation of a large amount of single-stranded DNA (unusual in eukaryotic DNA replication). Further, the initial priming event requires the first nucleotide of the new DNA strand being covalently bound to the 80,000-dalton precursor of the 50,000-dalton terminal protein. Following complete second-strand synthesis, the precursor end proteins are proteolytically cleaved to form the mature terminal proteins. TBP = precursor to terminal binding protein.*

Source: https://www.blackwellpublishing.com/content/bpl_images/Content_Store/Middle_Sized/Wagner9781405103466/286to310.pdf.

10.3.2.3. Late Gene Expression

As levels of early 72-kd DBP increase, early gene expression decreases. Simultaneously, the E4 protein disrupts splicing inhibition downstream of the L1 region, leading to changes in polyadenylation site usage. This results in the transcription of transcripts up to 24,000 bases long from the major late promoter. Differential polyadenylation and splicing create five families of late mRNAs, which are translated into structural proteins for capsids. Other late proteins modify cellular structure and metabolism to facilitate virus assembly and release. Some E4 protein species also play a role in transporting late mRNA from the nucleus to the cytoplasm, in addition to altering splicing patterns (Lai CJ, Nathans D. 1974).

10.3.2.4. VA Transcription and Cytopathology

The host cell must continue to function for a considerable amount of time after infection due to the intricate relationship between the human adenovirus infection and the cell. This inhibits the complete shutdown of host cell function caused by the virus; as a result, virus-induced cytopathology is sluggish and causes long-term cell death. In this time, the cell can produce interferons, which are cellular gene products that can make nearby cells resistant to viral infection, as a defense against viral gene expression. In order to get around this issue, the human virus synthesizes VA RNA, a brief but highly structured RNA molecule that prevents the cell from producing interferon and, most likely, other defense mechanisms. Indeed, siRNAs share many characteristics with this RNA molecule. Cellular RNA pol III, the same polymerase that transcribes cellular amino acid transfer RNAs (tRNAs), is the means by which VA RNA is expressed. Interestingly, several adenoviruses of domesticated animals do not express a homolog to VA RNA, despite the human Epstein-Barr herpesvirus expressing an analogous transcript, indicating that this is an important feature in virus-mediated immune evasion.

Adenovirus-host interaction shares a second feature that is similar to papillomavirus replication. Adenovirus persists as a persistent infection in the host for extended periods of time, particularly in the lung and adenoidal tissue epithelium. Although the virus can infect basement cells, it is only in terminally differentiated cells that it can begin DNA replication and viral assembly. In actuality, the virus speeds up these differentiated cells' apoptosis. Removing dying infected cells seems to have the benefit of creating more space for basement cell differentiation and growth. This gives the virus a constant and ready supply of cells to start fresh replication cycles in. The relative levels of E1A and E1B in critical cells differ from those in cells where the latter viral protein prevents apoptosis, which is most likely why apoptosis is stimulated in these cells (Waga and Stillman, 1994).

10.3.2.5. Transformation of Nonpermissive Cells by Adenovirus

Like SV40, some types of adenovirus can transform cells and lead to tumor formation if they infect cells that are not permissive. Although there is no current evidence linking adenovirus infection to human cancers, the transformation process appears to be similar to that of papovaviruses. In fact, under certain circumstances, adenovirus gene products can replace early papovavirus gene products in mixed infections.

10.4. REPLICATION OF SOME SINGLE-STRANDED DNA VIRUSES

Many plant viruses, as well as some animal and bacterial viruses, benefit from having a smaller capsid size. For plant viruses, this advantage is related to the restrictions of virus capsid size that can pass through the plant cell wall pores. The advantages of animal and bacterial viruses are not as well understood, but they must still exist.

10.4.1. Replication of Parvoviruses

The parvoviruses are tiny, icosahedral, nonenveloped viruses. Warm-blooded animals are infected by two of the three known groups, whereas insects are infected by members of the third group. The parvovirus genome is about 5 kb long, but its capsid diameter is only 26–30 nm, much smaller than that of polyomaviruses. Because it only encodes one DNA strand, the virus is able to pack the genome into such a tiny virion. Interestingly, a large number of parvoviruses are able to package the sense DNA strand in an equal or nearly equal amount, either opposite or equivalent to mRNA. This indicates that the packaging signals that the virus uses to encapsidate the genome must occur on both strands; the interaction between the distinct end structures of the two strands and the capsid proteins is most likely the mechanism for this.

Figure 10.9 displays the genome of the adeno-associated virus, a common parvovirus. It encodes two translational reading frames for proteins, which are expressed by different transcript types. The capsid protein is encoded in the second reading frame, while nonstructural proteins involved in replication are encoded in the first. In order for the sequences to form hairpin loops in solution and within the infected cell's nucleus, the ends of the genome contain 120–300 bases of inversely repeated sequences. Since these terminal hairpins are repeated at the ends of both (+) and (–) sense DNA strands, they can act as templates for DNA replication and act as primers for the start of DNA replication (Waga and Stillman, 1994).

DNA replication in the host cell is a prerequisite for parvovirus replication. The virus can, therefore, only spread among cells that are actively replicating. In spite of this, parvoviruses cannot promote cell division through the action of a protein encoded in the virus, in contrast to papovaviruses and adenoviruses. This incapacity leads to an extremely stringent limitation on virus replication in the dividing cells of the host, particularly immune system cells. Young animals may suffer greatly from this, and dog kennel infections due to the parvovirus are a serious issue. Infection with parvovirus can also be extremely harmful to an adult animal's actively growing cells. One serious pathogen that affects domestic cats is feline panleukopenia, a parvovirus disease that causes the immune system to collapse.

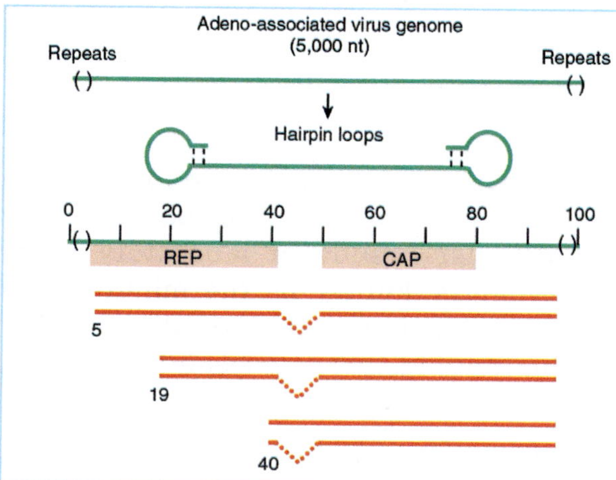

Figure 10.9. *The 5000-nucleotide (nt) linear genome of adenoassociated virus (AAV). This ssDNA has repeated sequences on both ends that allow it to form a "hairpin" structure. This serves as the template for cellular enzymes to convert into dsDNA. Cellular enzymes also mediate replication of the viral genome. Three families of coterminal mRNAs are expressed from the three AAV promoters; the genome encodes replication proteins and a capsid protein but depends on cell replication for its ability to replicate its genome. This cellular replication is induced by a helper virus such as adenovirus in the animal, but the virus can replicate in cultures of some actively replicating cells. Other groups of parvoviruses, such as the minute virus of mice (MVM), are able to replicate in some actively replicating cells of their natural host.*

Source: Edward K. Wagner, Martinez J. Hewlett, David C. Bloom, David Camerini; Basic Virology Third Edition; ISBN-13: 978-1-4051-4715-6.

After entering the nucleus during infection, cellular DNA repair enzymes transform the ssDNA into full dsDNA. One of the three viral promoters initiates transcription of the double-stranded viral DNA template into several 3′-coterminal transcripts only 5′ after the transcript begins. Each translational reading frame is translated into multiple proteins with related sequences because some of these transcripts are spliced. As was previously mentioned, the only cells in which viral genome replication is possible are those in which active cellular DNA replication (i.e., in the cell division process' S phase). The viral replication enzyme does not have polymerase activity; instead, it is involved in the cleavage of the covalently closed replicating viral DNA into single-stranded genomic DNA.

10.4.1.1. Dependovirus DNA Integrates in a Specific Site in the Host Cell Genome

Dependoviruses are a major subclass of parvoviruses, including adeno-associated viruses. It is typically linked to active adenovirus infections and sporadically to herpesvirus infections. Adeno-associated virus (AAV), which causes human parvovirus, is a well-known example. The dependoviruses rely on the adenovirus or herpesvirus helpers to stimulate the cell in a way that allows them to divide, even though they can be grown in culture in fetal cells or after appropriate chemical stimulation of some adult host cells. These viruses are, therefore, parasitic on other viruses, just like viroids.

The dependence on a helper virus might be expected to be a great impediment to virus replication for AAV, but this is overcome in part by its ability to integrate into chromosome 19 of the host when it infects a cell in the absence of the helper. The integrated viral DNA allows AAV to remain latent in host tissue for long periods of time but to "reactivate" if and when that cell is infected with a virus that can act as a helper

Short homologous sequence segments within a few hundred base regions of the host chromosome undergo integration. Although integrated viral DNA permits the viral genome to coexist with the host for extended periods, it functions as a biological "time bomb," ready to proliferate and destroy the cell upon infection with the right helper. Given that the replication of AAV impedes the helper virus's ability to replicate efficiently, it's possible that this process ultimately limits helper infection, which benefits the host.

10.4.1.2. Parvoviruses Have Potentially Exploitable Therapeutic Applications

AAV viruses may be used as antiviral or anticancer agents due to their stringent requirement for actively replicating cells and their competition with adenovirus and herpesvirus infections. Research in the lab demonstrated that this was possible. For instance, certain tumors are more common in particular breeds of laboratory mice. When young mice are infected with the murine parvovirus, known as the minute virus of mice (MVM), the animals live much longer and develop fewer tumors at an earlier age! It should be noted, though, that translating this finding into successful treatment of human cancers is a difficult task.

Another potential use for parvoviruses stems from their ability to integrate in a specific site in the chromosome. This integration is mediated by the hairpin loop ends of the viral genome and may be useful in designing viral vectors for delivering genes into cells (Howley and Livingston, 2009).

10.4.2. DNA Viruses Infecting Vascular Plants

While DNA viruses infecting vascular (i.e., "higher") plants might be expected to display genetic variability equivalent to that seen within animal and bacterial viruses, they do not appear to. The reason for this is that plant viruses must traverse a relatively thick and dense cell wall to approach and breach the plant cell's plasma membrane. Although at least one algal virus can insert its genome like bacterial viruses inject genomes, apparently, the dimensions of the vascular plant's cell wall preclude this accommodation. This results in the viruses of higher plants having a strict limitation on the size of their genomes, and although such viruses are not fully characterized, they may require a significant number of cellular functions for replication.

10.4.2.1. Geminiviruses

A particular class of plant-infecting viruses are encased in peculiar twinned capsids and have single-stranded, covalently closed circular DNA genomes. The genus name Geminivirus, derived from the Latin word geminae, which means "twins," is attributed to these "twin" capsid structures. The three main groups of these viruses are distinguished by the number of genes encoded and how they are arranged on the genome. Two of the groups have a monopartite genome because they encapsulate the same genome in both of the twinned capsids. In contrast, the third group's genome is bipartite, with the two distinct genomic segments enclosed in separate capsid halves. It is quite surprising to learn that a single geminivirus isolated from bananas has capsids with eight different genomic segments. It is unclear how the virus packs distinct genomic segments into various subcapsids, which is a pretty amazing feat.

Maize streak virus, which has a monopartite genome, and tomato golden mosaic virus, which has a bipartite genome, are examples of geminiviruses. Similar to papovaviruses, the geminiviruses have ORFs oriented in both directions around the circle in their genome (2.7–3.0 kb) organization. Because geminiviruses are single-stranded, after infection, the input genome strand needs to be transformed into dsDNA to provide the right template for mRNA transcription, whose translational reading frames are antisense to the virion DNA.

Whiteflies or leafhoppers are the insects that spread the geminiviruses from plant to plant. Geminiviruses do not replicate in their insect vectors, in contrast to RNA-containing arboviruses, although the virus can stay in the insect for extended periods of time. These viruses replicate and transcribe, utilizing a rolling circle mechanism inside the nuclei of infected plants. It is unknown what the precise roles of the gene products that sequence analysis predicted are. As a result, it is currently impossible to identify which particular viral protein may be engaged in this DNA replication.

10.4.3. Single-stranded DNA Bacteriophage ΦX174 Packages its Genes Very Compactly

The bacteriophage ΦX174's gene packaging indicates that genomic size compression provides clear benefits in the prokaryotic realm as well. With shorter fibers, this icosahedral virus resembles an adenovirus in structure. Its genome is a circular ssDNA that is roughly 3.4 kb long. A bacterial cell's ssDNA genome is changed into dsDNA upon infection. Although referred to as the replicative form (RF) or replicative intermediate, this is not the same as the ribonucleoprotein complex of the same name that is observed during the replication of ssRNA viruses (Sclair et al., 1973).

Examining the genetic map of this virus, which is displayed in Figure 10.10, provides a startling illustration of the lengths to which it has compressed its genome. The virus codes for nine different genes; however, only 36 bases (less than 1% of the genome) are free of translational reading frames, whereas one may anticipate that 200–300 bases of the DNA sequence would contain information unrelated to proteins. Because of this configuration, translational reading frames contain all transcriptional control sequences.

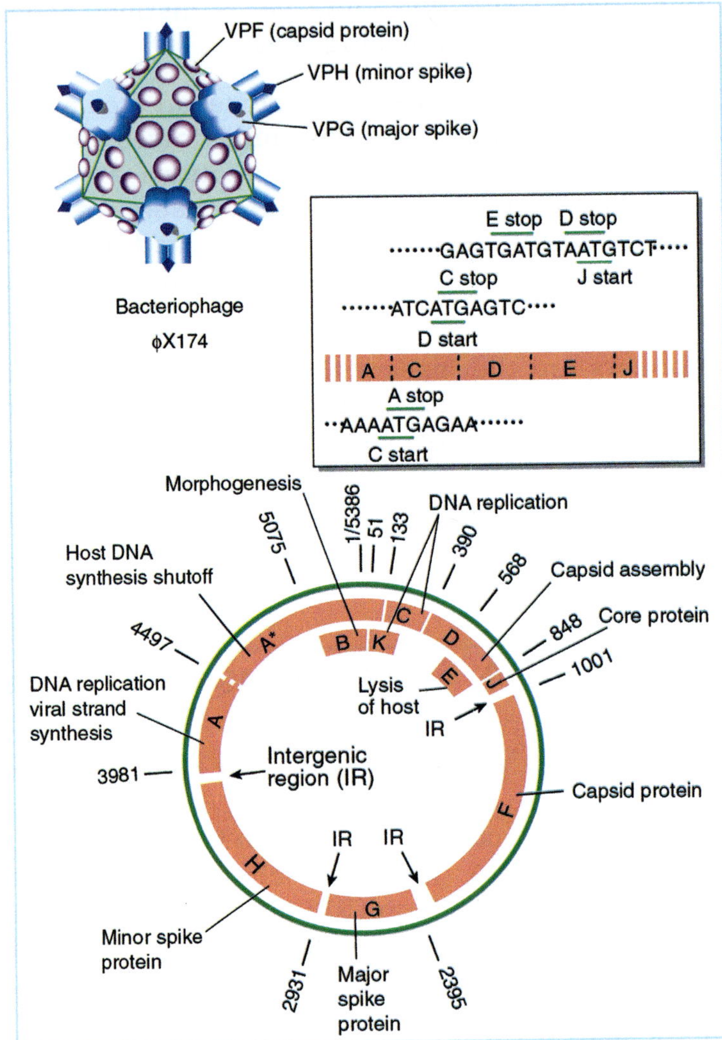

Figure 10.10. *The capsid structure and compressed genome of bacteriophage ΦX174. The capsid is made up of three proteins: major capsid, major spike, and minor spike. In all, a total of 10 genes are compressed into 3.4 kb of ssDNA! This is accomplished by very short intergenic regions and two completely overlapping genes. The functions of the proteins encoded by these genes are listed.*

Source: https://www.blackwellpublishing.com/content/bpl_images/Content_Store/Middle_Sized/Wagner9781405103466/286to310.pdf.

There is frequent overlap between the start and stop signals for translating distinct nearby ORFs. Additionally, the translational reading frames of two genes completely enclose those of other, larger genes. The translational frames are in distinct phases to achieve this configuration. Although many viruses, including the largest ones like herpesviruses and poxviruses, have overlapping genes, ΦX174 has gone beyond this tendency. This bacteriophage has some beneficial advantages due to its compactness, but like all dynamic systems, there are costs. One base change in a region of overlapping genes in a viral genome with such overlaps can impact the functions of two genes rather than just one. Because of this, more mutations would be expected to be fatal than are typically found in viral genomes. This is true for ΦX174, a DNA virus whose sequence is more strongly conserved during replication than most others. It is a laborious task to generate mutations in this virus for genetic analysis (Howley and Livingston, 2009).

The virus is likely less able to adapt to changes in its host and other environmental factors as a result of overlapping genes. In the prokaryotic world, conservatism may have a negative survival value, but the virus's survival ability provides unmistakable proof that any negative effects are outweighed by the effectiveness of gene packaging (Howley and Livingston, 2009).

CASE STUDY

JC VIRUS: CLINICAL PRESENTATION/CASE HISTORY

A 42-year-old male with AIDS (CD4 count = 128) is taken to the emergency room by his sister due to concerns over dramatic personality and mood changes. Upon admission, he was incoherent and confused. A history provided by his sister indicated that his mental status changed state about 6 weeks prior. He complained that his face hurt, and he demonstrated weakness in his right arm and leg and slurring of speech. The patient had no headache, neck stiffness, or fever. The attending physician ordered blood work, as well as a CT scan (x-ray).

Diagnosis

The attending physician was concerned about the evidence of CNS deficits and suspected either an HIV encephalopathy or an AIDS-related infection of the brain. The CT scan of the head had been ordered to look for evidence of lesions in the brain. The CT showed a left temporoparietal cerebral edema (area of inflammation). This indicated a zone of pathology likely caused by an AIDS-related infection. Serology did not provide useful information for diagnosis. Because a number of viruses (and fungi) can cause lesions in the brains of AIDS patients, a brain biopsy of the lesion was ordered and PCR analysis identified JC virus DNA. This suggested the likely etiology was progressive multifocal leukoencephalopathy (PML), which is caused by a reactivation of latent JC virus that is present in some B-lymphocytes, and these can cross the blood-brain barrier. Once there, they can form a focus of active infection, and result in sometimes severe cognitive deficits.

Treatment

There is no specific treatment for JC virus infection or PML. However, the underlying cause of this infection is the ongoing HIV infection in this patient, as indicated by his low CD4 count. The patient was put on HAART therapy to control his HIV, and in several weeks, his CD4 count showed some recovery, and the PML caused by the JC virus infection resolved.

Disease Notes

JC virus is a polyomavirus that is usually acquired as a subclinical infection in childhood. The virus initially infects the tonsils and then spreads through the blood, infecting the kidneys and becoming latent in the renal epithelium. JC virus has also been shown to be latent in the CNS, bone marrow, and peripheral blood cells. Greater than 80%

of adults have antibodies to JC virus, and this infection normally causes no clinical problems except in the immunocompromised (and especially patients with AIDS).

CLASS ACTIVITY

The drawings in the following table represent possible structures for replicating DNA molecules.

Structure	Possbile for adenovirus?

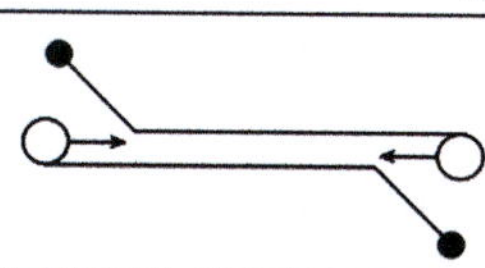	
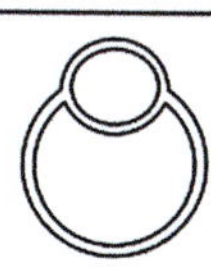	
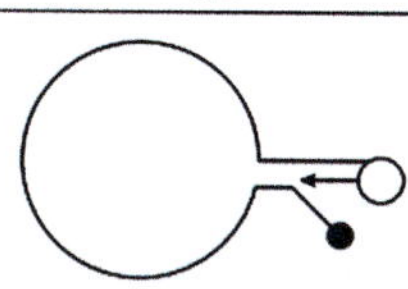	

Indicate which ones might be found if you examined replicating adenovirus DNA isolated from an infected host cell.

Adenovirus DNA replication proceeds in two stages. Suppose that you have an in vitro system that allows you to examine the features of this synthesis. The reaction mixture has all the required viral and host proteins. Predict the effect of the following modifications on the process of the two stages. Use a "+" sign if the stage will occur normally and a "–" sign if the stage will be blocked by the treatment.

SUMMARY

- Viruses with linear genomes face a major problem that also affects the replication of cellular chromosomal DNA. This "end problem" derives from the fact that the primer for DNA replication must be able to recognize short stretches of the viral genome – either through base pairing or through specific DNA–protein interactions.
- Small single-stranded (ss) DNA viruses, like parvoviruses, have solved the problem by encoding a complementary repeat sequence at the end that allows the genome to form a "hairpin loop" at the end; thus, the end of the molecule is not free. A similar solution is seen in the genome structure of poxvirus. Like chromosomal DNA, this linear DNA genome is covalently closed at its ends.
- The term papovavirus stands for "papilloma, polyoma, vacuolating" viruses. Actually, members of the group fall into two distinct families: the papillomaviruses and polyomaviruses. These two groups are similar regarding icosahedral capsids, circular genomes, the ability to remain associated with the host for long periods, and their requirement to specifically alter cell growth in the host cell's response to neighboring cells for virus replication. They differ in genome size and in many details of host cell specificity.
- The polyomaviruses have genomes of approximately 5000 base pairs. Capsids comprise three proteins, usually called VP1, VP2, and VP3. Polyomaviruses can cause tumors in animals and can transform the growth properties of primary cells in culture, especially the cells from animals different from the virus's natural host.
- In rodent (and some other nonprimate) cells, the SV40 virus can infect and stimulate cellular RNA and DNA synthesis by expressing the large T antigen. As noted, this viral protein inactivates at least two cellular tumor suppressor or growth control genes (p53 and Rb).
- Sequence analysis of the bovine papillomavirus genome and the transcripts expressed indicates that early and late transcripts are expressed from a single or limited number of early and late promoters as pre-mRNAs. While the extensive splicing of pre-mRNAs is reminiscent of infections with polyomaviruses, papillomaviruses differ in that early and late promoters are found in several regions within the genome.
- Adenoviruses have complex icosahedral capsids whose proteins are not present in equimolar amounts, with projecting spikes or fibers at the 12 vertices (pentons). The viral genome is encapsidated with a core protein that acts a bit like histone to provide a chromatin-like structure that is condensed in the interior of the nucleocapsid.
- Adenovirus DNA replication begins at either or both ends of the DNA and uses as a primer an 80,000-dalton precursor of the 50,000-dalton viral genome-bound terminal protein. The large priming terminal protein is proteolytically cleaved to the smaller terminal protein found in capsid-associated genomes during

packaging. This is the only known instance where DNA replication initiates without a short RNA primer.

- Adeno-associated virus represents one major group of parvoviruses, known as dependoviruses. It is typically found in association with active adenovirus infections and, occasionally, with herpesvirus infections. The human parvovirus, adeno-associated virus (AAV), is a well-characterized example.

REVIEW QUESTIONS

1. What are the functions of T antigen during the SV40 infectious cycle?
2. What are the key stages in the replication cycle of papovaviruses?
3. What role does the large T antigen play in the replication of SV40 virus?
4. What occurs during an abortive infection of cells nonpermissive for SV40 replication?
5. What are the key physical properties of adenoviruses that facilitate their replication?
6. Describe the replication strategy of parvoviruses and how it differs from other DNA viruses.
7. How does the single-stranded DNA bacteriophage ΦX174 manage to package its genes compactly?

REFERENCES

1. Breitburd, F., et al. (1995). Immunization with virus-like particles from cottontail rabbit papillomavirus (CRPV) can protect against experimental CRPV infection. *Journal of Virology*, *69*, 3959–3963.
2. Dupuy, C., Buzoni-Gatel, D., Touze, A., Le Cann, P., Bout, D., & Coursaget, P. (1997). Cell-mediated immunity induced in mice by HPV 16 L1 virus-like particles. *Microbial Pathogenesis*, *22*, 219–225.
3. Howley, P. M., & Livingston, D. M. (2009). Small DNA tumor viruses: Large contributors to biomedical sciences. *Virology*, *384*, 256–259. https://doi.org/10.1016/j.virol.2008.12.006 (accessed on 4th October 2024).
4. Lai, C. J., & Nathans, D. (1974). Mapping temperature-sensitive mutants of simian virus 40: Rescue of mutants by fragments of viral DNA. *Virology*, *60*, 466–475.
5. Lederberg, S., & Meselson, M. (1964). Degradation of non-replicating bacteriophage DNA in non-accepting cells. *Journal of Molecular Biology*, *8*, 623–628.
6. Rous, P., & Beard, J. W. (1935). The progression to carcinoma of virus-induced rabbit papillomas (Shope). *Journal of Experimental Medicine*, *62*, 523–548.
7. Schiffman, M. H., Bauer, H. M., Hoover, R. N., et al. (1993). Epidemiologic evidence showing that human papillomavirus infection causes most cervical intraepithelial neoplasia. *Journal of the National Cancer Institute*, *85*, 958–964.

8. Sclair, M., Edgell, M. H., & Hutchinson, C. A. III. (1973). Mapping of new *Escherichia coli* K and 15 restriction sites on specific fragments of bacteriophage φX174. *Journal of Virology*, *11*, 378–385.

9. Suzich, J. A., Ghim, S. J., Palmer-Hill, F. J., White, W. I., Tamura, J. K., Bell, J. A., Newsome, J. A., Jenson, A. B., & Schlegel, R. (1995). Systemic immunization with papillomavirus L1 protein completely prevents the development of viral mucosal papillomas. *Proceedings of the National Academy of Sciences of the United States of America*, *92*, 11553–11557.

10. Sweet, B. H., & Hilleman, M. R. (1960). The vacuolating virus, S.V. 40. *Proceedings of the Society for Experimental Biology and Medicine*, *105*, 420–427.

11. Waga, S., & Stillman, B. (1994). Anatomy of a DNA replication fork revealed by reconstitution of SV40 DNA replication in vitro. *Nature*, *369*, 207–212.

Index

Printed and bound by CPI Group (UK) Ltd, Croydon, CR0 4YY

18/09/2026

14965063-0003